AF733264

Stockholm and the Rise of Global Environmental Governance

This unique history examines global environmental governance through the lens of Stockholm, which has played an outsized role in shaping its development. Fifty years before Greta Thunberg started her School Strike for Climate, Swedish diplomats initiated the seminal 1972 UN Conference on the Human Environment that propelled Stockholm to the forefront of international environmental affairs. Stockholm has since become a hub for scientific and political approaches to managing the environmental and climate crisis. Utilizing archival materials and oral histories, Sörlin and Paglia recount how, over seventy years, Stockholm-based actors helped construct the architecture of environmental governance through convening decisive meetings, developing scientific concepts, and establishing influential institutions at the intersection of science and politics. Focusing on this specific yet crucial location, the authors provide a broad overview of global events and detailed account of Stockholm's extraordinary impact. This title is also available as Open Access on Cambridge Core.

Sverker Sörlin is a public intellectual, policy advisor, professor at KTH Royal Institute of Technology and a leading name in environmental history. He is a recipient of the August Prize for Non-fiction.

Eric Paglia is an environmental historian at KTH Royal Institute of Technology researching global environmental governance, the 1972 Stockholm Conference, the Anthropocene, and the Arctic.

Studies in Environment and History

Editors

J. R. McNeill, *Georgetown University*
Ling Zhang, *Boston College*

Editors Emeriti

Alfred W. Crosby, *University of Texas at Austin*
Edmund P. Russell, *Carnegie Mellon University*
Donald Worster, *University of Kansas*

Other Books in the Series

Zozan Pehlivan *The Political Ecology of Violence: Peasants and Pastoralists in the Last Ottoman Century*
Matthew P. Johnson *Hydropower in Authoritarian Brazil: An Environmental History of Low-Carbon Energy, 1960s–90s*
Ellen Arnold *Medieval Riverscapes: Environment and Memory in Northwest Europe, c. 300–1100*
Richard C. Hoffmann *The Catch: An Environmental History of Medieval European Fisheries*
Samuel Dolbee *Locusts of Power: Borders, Empire, and Environment in the Modern Middle East*
Andy Bruno *Tunguska: A Siberian Mystery and Its Environmental Legacy*
Lionel Frost et al. *Cities in a Sunburnt Country: Water and the Making of Urban Australia*
Adam Sundberg *Natural Disaster at the Closing of the Dutch Golden Age: Floods, Worms, and Cattle Plague*
Germán Vergara *Fueling Mexico: Energy and Environment, 1850–1950*

A complete list of books in the series is at the back of the book.

Stockholm and the Rise of Global Environmental Governance

The Human Environment

SVERKER SÖRLIN
KTH Royal Institute of Technology

ERIC PAGLIA
KTH Royal Institute of Technology

CAMBRIDGE
UNIVERSITY PRESS

CAMBRIDGE UNIVERSITY PRESS

Shaftesbury Road, Cambridge CB2 8EA, United Kingdom

One Liberty Plaza, 20th Floor, New York, NY 10006, USA

477 Williamstown Road, Port Melbourne, VIC 3207, Australia

314–321, 3rd Floor, Plot 3, Splendor Forum, Jasola District Centre, New Delhi – 110025, India

103 Penang Road, #05–06/07, Visioncrest Commercial, Singapore 238467

Cambridge University Press is part of Cambridge University Press & Assessment, a department of the University of Cambridge.

We share the University's mission to contribute to society through the pursuit of education, learning and research at the highest international levels of excellence.

www.cambridge.org
Information on this title: www.cambridge.org/9781009177801

DOI: 10.1017/9781009177825

© Sverker Sörlin and Eric Paglia 2025

This publication is in copyright. Subject to statutory exception and to the provisions of relevant collective licensing agreements, with the exception of the Creative Commons version the link for which is provided below, no reproduction of any part may take place without the written permission of Cambridge University Press & Assessment.

An online version of this work is published at doi.org/10.1017/9781009177825 under a Creative Commons Open Access license CC-BY 4.0

When citing this work, please include a reference to the DOI 10.1017/9781009177825

First published 2025

A catalogue record for this publication is available from the British Library

A Cataloging-in-Publication data record for this book is available from the Library of Congress

ISBN 978-1-009-17780-1 Hardback

Cambridge University Press & Assessment has no responsibility for the persistence or accuracy of URLs for external or third-party internet websites referred to in this publication and does not guarantee that any content on such websites is, or will remain, accurate or appropriate.

COVER IMAGES:
Top: Indira Gandhi and Olof Palme. © Scanpix Sweden/AFP/Getty Images.
Gordon Goodman. © Lars Kristoferson.
Greta Thunberg. © Wiktor Badkowski/DPA/Alamy.
Hans Palmstierna. © TT bildarkiv.
Middle: Barbara Ward. © UN Photo Library.
Maurice Strong. © Scanpix Sweden/AFP/Getty Images.
Bert Bolin. © Curt Dahlgren.
Birgitta Dahl. © TT bildarkiv.
Bottom: Johan Rockström. © TT bildarkiv.
Inga Thorsson. © TT bildarkiv.
Sverker Åström. © TT bildarkiv.
Will Steffen. © TedXCanberra.

Contents

List of Figures — *page* vii
Acknowledgments — ix

Prologue: The *Human* Environment – The Stockholm Idea — 1
1 The Stockholm Story: A Progressive Counternarrative — 8
2 Sweden and Nature: The Model Country Paradox — 45
3 Stockholm: A Climate Science Node — 84
4 The Swedish UN Initiative — 119
5 Twelve Days in Stockholm, June 1972 — 143
6 Enter the Earth System — 172
7 Planetary Boundaries and Big Tent Science — 217
8 "Listen to the Scientists": Yes, But What Is the Message? — 265
9 Conclusion: An Environmentalism of the Rich? — 292

Selected Bibliography — 327
Index — 333

Figures

1.1	Stockholm in the early summer of 1972	*page* 18
1.2	The "Elm Battle" in Stockholm, May 1971	26
1.3	Anna Lindh and Kofi Annan in 2002	29
2.1	Selma Lagerlöf at her desk while writing *Nils Holgersson* in the early 1900s	46
2.2	Elin Wägner at Swedish National Radio in 1944	69
2.3	Anna Lindhagen, circa 1920, and cover of an allotment garden advocacy brochure from 1916	73
2.4	The stylized map of the northern section of the Stockholm Royal National City Park	79
3.1	Paul Crutzen at the Max Planck Institute	90
3.2	Carl-Gustaf Rossby with a Jaumotte Aerograph at MIT, circa 1933	92
3.3	Bert Bolin at Stockholm University in 1965	99
4.1	Svante Odén at Uppsala University in 1965	127
4.2	Hans Palmstierna at a Swedish government bacteriological lab in 1965	130
4.3	Inga Thorsson in 1971	132
4.4	Tage Erlander, Olof Palme, and Sverker Åström in 1965	134
4.5	Swedish UN delegation and Crown Prince Carl Gustaf in September 1970 at the twenty-fifth anniversary commemoration of the founding of the United Nations	142
5.1	Maurice Strong at the UN Conference on the Human Environment in Stockholm, June 1972	144
5.2	Barbara Ward (Lady Jackson) and Margaret Mead at the UN Conference on the Human Environment in Stockholm, June 1972	159

5.3 Indira Gandhi addressing the plenary session of the UN Conference on the Human Environment in Stockholm on June 14, 1972 161
5.4 The front cover of the February 1972 inaugural issue of *Ambio – A Journal of the Human Environment* 164
6.1 Excerpt from the front matter of the special Rättvik Conference issue of *Ambio* from 1983 177
6.2 Mostafa Tolba in June 1982 181
6.3 Gordon Goodman and Lars Kristoferson outside the Royal Swedish Academy of Sciences 185
6.4 Jill Jaeger at the International Institute for Applied Systems Analysis Conference in 2015 193
6.5 The front cover of the 1987 World Meteorological Organization/UN Environment Programme report *Developing Policies for Responding to Climate Change* 195
6.6 First board meeting of the Stockholm Environment Institute in September 1989 213
7.1 The 350 advertisement that appeared in the *New York Times* on June 23, 2008 218
7.2 Bo Kjellén and Bert Bolin in 1999 219
7.3 The Tällberg Forum in 2008 221
7.4 Panel discussion at the 2008 Tällberg Forum with Diana Liverman, Jim Hansen, Will Steffen, and Tariq Banduri, moderated by Johan Rockström 222
7.5 Planetary Boundaries diagram appearing in *Nature* in 2009 241
7.6 Brian Walker, Simon Levin, Karl-Göran Mäler, and Partha Dasgupta in Askö, Sweden, in 2002 247
7.7 Panel discussion at the Resilience 2008 Conference with Sverker Sörlin, Elinor Ostrom, C. S. "Buzz" Holling, Uno Svedin, Line Gordon, and Carole Crumley 254
8.1 Johan Rockström at the 2008 Tällberg Forum 266
8.2 Will Steffen on stage in the Beijer Hall at the Royal Swedish Academy of Sciences during the tenth anniversary celebration of the Planetary Boundaries framework in 2019 269
8.3 Kate Raworth in 2018 283
8.4 Greta Thunberg in Stockholm in 2018 286
9.1 Stockholm+50 Conference in June 2022 with Sweden's Crown Princess Victoria and King Carl XVI Gustaf, UN Secretary General António Guterres, Kenyan President Uhuru Kenyatta, Swedish Prime Minster Magdalena Andersson, and other officials 294

Acknowledgments

This book is in part the result of many years of immersion in the human environment of Stockholm. This includes the countless conferences, meetings, and seminars on environment and sustainability issues that we have taken part in, as well as our engagement in numerous research projects, the Swedish public debate, and various governmental processes. At events we have attended over the years, visiting scholars or officials from outside Sweden have often made reference to key meetings held in Stockholm that had served as turning points in the development of a particular scientific discipline or some important political process that was directly connected to the Swedish capital.

As environmental historians, we have previously written about a wide range of institutions, individuals, and events centered on Stockholm, for example, Carl-Gustaf Rossby and the establishment of the International Meteorological Institute at Stockholm University; the 1972 UN Conference on the Human Environment (UNCHE), including the Swedish science diplomacy of its crucial preparatory period; and the history of the international journal *Ambio*, published by the Royal Swedish Academy of Sciences. Moreover, in our academic as well as journalistic work producing environmental radio programs and podcasts, we have had the opportunity to interview and document the oral histories of scores of scientists, activists, diplomats, and other public officials, from Sweden and elsewhere. Despite their disparate backgrounds, many told a similar story on the outsized role Stockholm has played in advancing understanding of the planetary environment and in promoting international action on sustainability issues.

At some point, perhaps a decade ago, it became clear to us that Stockholm has, over the course of some seventy years, indeed punched far above its weight in the evolution of what is today referred to as global environmental governance (GEG). That realization inspired us to write this book. It aims to bring together the histories of an array of parallel processes of institution building, scientific innovation, and political engagement, both before and after the landmark 1972 Stockholm Conference, into a single comprehensive narrative. Given its central place in an otherwise complex international history, it is our hope that the "Stockholm story" told here can serve as a converging lens. Aided by it, we believe, readers will be able to perceive and perhaps appreciate well over a half century of history – perhaps even progress – in a realm of global governance that was critical for the future of humankind in 1972 and is even more so today.

People are central – as they should be in a book about "the *human* environment" – and we have had the privilege to interview many of the main protagonists that animate the pages of this book. Sadly, some of those have since passed away. The American–Australian Earth system scientist and world-class science communicator Will Steffen, who served as Executive Director of the Stockholm-based International Geosphere-Biosphere Programme from 1998 to 2004, was one of them. Will was a colleague and friend, an endless source of insight and inspiration, a tireless Anthropocene entrepreneur, and a diplomatic voice seeking to combine a sound human–Earth relationship with fair sharing and justice among societies and between humans. He was also one of our primary sources of insight into the institutionalization of Earth system science and other processes connected to Stockholm, including the emergence of the Anthropocene concept. For all of this, and much more, we have chosen to dedicate this book to the memory of Will.

The late Swedish diplomat Lars Björkbom shared with us his experiences in helping to establish the Convention on Long-Range Transboundary Air Pollution, and the American physicist Jack Hollander, who has also since passed away, provided valuable details on the founding of the Beijer Institute, where he served as a chairman of the international board during the entire duration of the institute in its original form from 1977 to 1989.

Lars Kristoferson, who served as Deputy Director of the original Beijer Institute as well as the Stockholm Environment Institute during its first few years, has been an important source of information on and insight into institutional development in Stockholm. Phone interviews were conducted with two of the key actors, climate scientists Jill Jaeger and Michael Oppenheimer, who worked at or collaborated with the Beijer Institute in arranging and documenting the seminal Villach and Bellagio climate meetings in the latter half of the 1980s. Over the course of many years and many interviews, Ambassador Bo Kjellén, Sweden's former chief climate negotiator who took part in creating the UN Framework Convention on Climate Change and the Kyoto Protocol, has been an invaluable resource for our comprehension of climate diplomacy. Ambassador Lars-Göran Engfeldt has more than anyone been an essential source due to his first-hand experiences and deep understanding of environmental diplomacy. He can safely be called the world's first career environmental diplomat due to his instrumental role in the Stockholm Conference preparatory period, a trajectory he continued up through Rio 1992 and Johannesburg 2002. The former Swedish foreign ministry official Göran Bäckstrand, closely involved with preparations for Stockholm 1972, also made important contributions to our understanding of that event. Policy expert Bob Munro provided insight into events and international networks connected to Stockholm, including the 1982 Rättvik Conference and its influence on environmental governance well beyond Sweden.

Scientists who have been engaged in the development of disciplines and institutions are a major theme of this book. Thomas Rosswall, Founding Executive Director of the International Geosphere-Biosphere Programme, among many other science leadership positions, and Henning Rodhe, a pioneer in the study of acid rain and climate change across many decades at Stockholm University's Meteorological Institute, greatly enhanced our understanding of a wide range of issues and international processes that they have been deeply involved with. The Lund University ecologist Sören Svensson was another source of insight into Swedish contributions to international scientific networks. Uno Svedin and Måns Lönnroth, who operated at the interface of science, policy, and research funding during their long careers in government-related institutions, provided us with valuable information and perspectives on many of the main themes of this book. Several staff members at the Academy of Sciences in Stockholm were helpful, including Bo Söderström, Editor-in-Chief of *Ambio*, who directed us to the journal's archives, and Carl Folke, Director of the Beijer Institute of Ecological Economics.

In addition, we had previously had the privilege of interviewing many of the individuals appearing in the pages of this book for the radio program and podcast Think Globally Radio, as well as for earlier research projects. These interviews, dating back as far as 2002, have also informed our understanding of the issues and events elaborated on in the chapters that follow. Those that have provided insights, perspectives, and oral histories both inside and outside the parameters of this book project include the late Sverker Åström, Maurice Strong and Malin Falkenmark, Anders Wijkman, Lennart Daléus, Mark Lynas, Johan Rockström, Owen Gaffney, Lars Ingelstam, Kevin Noone, Jan Mårtenson, James Hansen, Anders Turesson, Alexander Crawford, Libby Robin, Leah Aronowsky, and Julia Adeney Thomas. We have presented material from the book in seminars, workshops, and invited talks in multiple institutions, for example: History and Philosophy of Science, and the Center for History and Economics, both at the University of Cambridge; Political Science, Stockholm University; Social Anthropology, University of Oslo; Political Science, Umeå University; History, Roskilde University; Swedish Collegium for Advanced Study (SCAS), Uppsala University; University of Bergen; University of Adelaide (online); Luigi Einaudí Foundation, Turin (online); the Max Planck Institute for the History of Science, Berlin (online); the European Society for Environmental History 2022, Bristol; and the American Society for Environmental History 2023, (Boston) and 2024, (Denver). We are grateful for inspiring meetings and constructive comments.

Valuable comments on earlier versions of the book's chapters were received from Åsa Persson, Henning Rodhe, Jill Jaeger, Lars-Göran Engfeldt, Thomas Rosswall, Uno Svedin, Johan Rockström, Lars Kristoferson, Göran Bäckstrand, Paul Warde, Lars J. Lundgren, Arne Kaijser, and Markus Larsson. Members of the Study of the Planetary Human–Environment Relationship (SPHERE) research project, Sabine Höhler, Jasmin Höglund Hellgren, Erik Isberg, Gloria Samosír, Thomas Harbøll Schrøder, Vanessa Ogle, and Glenda Sluga have provided useful feedback on both the foundational ideas and during the writing of this book. Glenda also spurred our thinking by inviting us to contribute, in 2019, a Stockholm chapter for her volume, coedited with Kate Darian-Smith and Madeleine Herren, *Sites of International Memory* (University of Pennsylvania Press, 2023).

Institutionally, we have benefitted enormously from our fabulous colleagues and PhD students in the Division of the History of Science, Technology and Environment and the KTH Environmental Humanities Laboratory, both at the KTH Royal Institute of Technology in Stockholm. In this vibrant, interdisciplinary setting, SPHERE has been the home of this book project since its start (see: spheregovernance.org). The SPHERE project on the history of GEG has been hosted by the Division from 2018 to 2024 and financed through Sverker Sörlin's Advanced Grant from the European Research Council. In addition, Eric Paglia's work has been supported by a grant from Formas, the Swedish Research Council for Sustainable Development. During the academic year 2022–2023, Sverker worked on the book with a generous grant as a long-term fellow under the Environmental Humanities program at SCAS in Uppsala University. Thanks are also due to the Marcus and Marianne Wallenberg Foundation, which funded a large earlier program (until 2017) on the history of the Swedish Academy of Sciences. In this project, Sverker studied the breakthrough years of "the environment" in the Academy around 1970, research that ignited additional thoughts about the potential of a book on the history of Stockholm as a global environmental node.

Finally, it brings us tremendous pleasure to thank Lucy Rhymer and Rosa Martin at Cambridge University Press and the coeditor of the series that this book is part of, John McNeill, for trust and support in the project from the very start, and to Cambridge University Press's anonymous reviewers for useful comments. Whatever errors remain are exclusively the responsibility of the authors.

Prologue

The Human *Environment – The Stockholm Idea*

Several years into preparations for the 1972 United Nations Conference on the Human Environment (UNCHE), the concept of the "human environment" remained ambiguous. The conference's energetic Secretary General Maurice Strong therefore proposed a clear and concise operational definition: "The term 'human environment' refers to those aspects of man's activities which, by affecting the natural ecological systems of which he is part, affect his own life and well being."[1] Although the term itself today seems anachronistic, at the time "the human environment" articulated a novel conception of the natural world – not simply nature separate from society, but instead entailing the totality of the surroundings that enveloped and sustained human life.

More than fifty years later, it is no longer necessary to attach the qualifier "human" to "environment" to convey the interdependency of the two. The human aspect of "the environment" is largely acknowledged, if not always taken into full account in terms of behavior or policy. Indeed, the conception of the relationship is increasingly becoming that "the environment" *is* "human" and that humans and environment are inseparable. The environment not only surrounds the human, an idea built into the etymology of the word environment and its counter parts in many languages (*environment* in French, *ambiente* in Italian and Spanish, and *Umwelt* in German). Distinct from the notion of "nature," which retains more endogenous autonomy, the environment is also shaped and affected by humans. The environment is moreover deeply

[1] Wade Rowland, *The Plot to Save the World: The Stockholm Conference on the Human Environment* (Toronto: Clarke, Irwin & Company, 1973).

attached to human responsibility and accountability, both from past agency and in relation to future action.[2]

We have chosen to take that essential word "human" as a key point of departure for this examination of the emergence and evolution of what has for the past few decades been called global environmental governance (GEG). It is a book in which the 1972 UN Conference in Stockholm occupies a central position and historical presence, primarily because it has had a lasting impression on modern environmental politics around the world, but also because the conference was held in Stockholm, which in and of itself requires some reflection. What made Stockholm assume this role as the convening venue for what turned out to be the first truly global event in addressing concerns over the incipient environmental crisis that had only entered public consciousness a few years earlier? What were the implications of its taking place in that particular location on the northern periphery of the European continent?

A central presence in this book is Stockholm itself, a city that not only hosted the 1972 environmental conference but has also provided the convening power, institutional ingenuity, and material as well as intellectual resources and leadership for multiple policy and scientific interventions over a much longer period. Stockholm, we argue, connected the post-WWII prehistory of UNCHE 1972 – when "environment" rose to a significant position on the international agenda and became a household word – with the long and winding story of assembling knowledge and politics into what has become the current global architecture for environmental governance. It is very much a story of the "human" element of the environment, of human networks and conventions, as well as of the human impacts on nature. And perhaps above all, an environment that gradually, over a period of several decades, enters the realm of the polity and hence also exits the exclusive domain of the natural where it was once firmly located.

The book examines the distinct human dimension of the history of environmental governance: a social sphere of individuals, the networks they operated within, and the institutions they helped build. Focusing on

[2] "The environment" is historicized in Paul Warde, Libby Robin & Sverker Sörlin, *The Environment: A History of the Idea* (Baltimore, MD: Johns Hopkins University Press, 2018). On the concept of "nature," see Kate Soper, *What Is Nature?* (Oxford: Wiley-Blackwell, 1995); Noel Castree, *Nature* (London: Routledge, 2005); and the eternal classic, Clarence J. Glacken, *Traces on the Rhodian Shore: Nature and Culture in Western Thought from Ancient Times to the End of the Eighteenth Century* (Berkeley & Los Angeles, CA: University of California Press, 1967). On the "surrounding" properties of environment, see also Etienne Benson, *Surroundings: A History of Environments and Environmentalisms* (Chicago, IL & London: The University of Chicago Press, 2020).

people – primarily scientists and experts, diplomats and government officials, environmental activists, and engaged parts of the private sector – we elaborate *a* "human environment," in and related to Stockholm, the capital of Sweden. *This* human environment, of interacting individuals, has made critical contributions to rendering *the* human environment what it has now become, a *governable object*, which is in and of itself a prerequisite for GEG. This Stockholm story, more than an account of a singular event such as the 1972 UN Conference or of advances made in a particular Northern European location, represents a multidecade environmental history that is deeply intertwined with global-scale scientific and political processes such as the rise of the biosphere, biodiversity, and climate change on the international agenda. With individuals and their networks at the center of the narrative, "human environment" in this book carries multifold connotations. It encompasses *humanity*'s physical operating space, the social and political context from which a *humane* sustainability knowledge and initiatives emerged over the course of some seventy years. It also encompasses the *human* sphere that a city and a culture in a particular historical period could provide in its work to make the environment the responsibility of *humans*. Stockholm brought these qualities to bear, and we will try to demonstrate that this was no coincidence. It was part of what Swedish society and politics was in the second half of the twentieth century, a quite human place for a human environment to take shape.

Over the course of our analysis and narrative, we will touch upon several strands of activities that have been undertaken in Stockholm and through Stockholm-centric networks. We will chiefly look at work in or related to science, the systematic search for knowledge. Stockholm has been a major hub of knowledge production and dissemination, and the city's key role at the core of the GEG enterprise simply cannot be understood without taking this into account. Throughout this book, we will argue that Stockholm's function as an important place for environment, climate, and sustainability was to a very large degree derived from Sweden's attributes as a nation with an ambitious policy for science and scholarship, and an ability to creatively move with the times and change focus to stay relevant. Toward the end, we will pay a great deal of attention to how this role can continue to develop and transform as the cultivation of the human–Earth relationship enters new phases in the age of the Anthropocene. We can see how new concepts and ideas are entering a discourse that is constantly widening and also how darker clouds have arisen on the Stockholm horizon.

This book is intended as a contribution to these discussions, on how GEG can further evolve and expand and become more effective. It must also be acknowledged that change is, as always, ongoing. As we have witnessed during the crisis and pandemic period at the start of the 2020s, otherwise announced as the Agenda 2030 decade, unexpected new turns in global affairs are almost a given in the modern world. Swedish domestic politics and social trends are no exception. In a remarkably short period of time, surprising shifts in environment and climate policies have put Sweden's green leadership in peril, a theme we will develop towards the end of the book. Whether this is a lasting trend or a short-term concession to geopolitical and domestic circumstances remains to be seen.

Another consequential type of activity, often linked to the development of knowledge, that this book pays particular attention to is meetings, which in the evolution of GEG have typically engaged multiple groups and categories of actors, including scientists, scholars, diplomats, politicians, intellectuals, artists, activists, civil society, and business leaders. The mix has shifted somewhat over time, as have the scientific disciplines that have been most prominent in environmental debates, and the scope has widened. Much of this work, too, can be seen as reactions to or applications of science, for example, science diplomacy, science policy, science organization, science communication, and practices, as will be shown throughout this book, where Sweden has excelled as part of its GEG-related activities, although a great deal of such efforts have also been more loosely organized. We have striven to identify the meeting places, examine the actors and ideas, and determine how these factors have, over a long period since the immediate postwar years, shaped and conditioned the continually unfolding future of people and planet. The "human environment" is now part of the very large and essential, and also "human," enterprise that we call GEG.

THE SIGNIFICANCE OF STOCKHOLM

First of all, a justification question: Why Stockholm? More than a half century after the seminal 1972 UN Conference, the status of Stockholm as a premier site of international "green memory" has had decades to grow and mature into a de facto brand.[3] Rather than simply providing

[3] Eric Paglia & Sverker Sörlin, "Greening Our Common Fate: Stockholm as a Node of Global Environmental Memory," In: Glenda Sluga, Kate Darian-Smith & Madeleine Herren, eds., *Sites of International Memory* (Philadelphia, PA: University of Pennsylvania Press, 2023), 237–265.

a geographic identity or chronological starting point to commemorate a process that has substantially influenced international relations over the past fifty years, the association of Stockholm and 1972 with the emergence and ongoing evolution of GEG has been gradually strengthened over time. A growing international network of institutions, initiatives, ideas, and individuals have centered on the Swedish capital. Stockholm, as a major node in these networks, has been a stronghold of some of the key findings that became building blocks of a modern understanding of the *human–Earth relationship*. Sweden also continued to serve as a meeting point for scientific workshops with a policy angle and for diplomatic activities with bearing on the environment and climate.

An array of scientific concepts, centers of excellence, and environment-oriented international organizations have enduring links to Stockholm: acid rain, the Anthropocene, the Intergovernmental Panel on Climate Change, the Stockholm Environment Institute, Earth system science and the International Geosphere-Biosphere Programme, the Stockholm Resilience Centre and the Planetary Boundaries framework, and Hans Rosling's Gapminder Foundation and Trendalyzer tool are just a few examples. As host of World Water Week and home to the Stockholm International Water Institute, as well as with the late internationally-renowned hydrologist Malin Falkenmark's innovations such as the Water Stress Indicator, water has been a prominent part of Stockholm's sustainability branding. The living legacy of Stockholm as a location of global environmental leadership in the early 2020s is perhaps most powerfully personified by Greta Thunberg, the teenage activist who in 2018 initiated what has become a worldwide movement – School Strike for the Climate, or, as it was later termed, Fridays for Future – on the steps of the Swedish Parliament. Greta follows in a long tradition in Sweden of promoting public understanding, building innovative institutions, and acting upon scientific knowledge on issues related to the environment and its inextricable relevance for humankind, drawing considerable attention from abroad since at least the American journalist Marquis Childs' book *Sweden: The Middle Way* (1936). Childs, who later won a Pulitzer Prize, launched what became a long tradition of books that celebrated Sweden as a realized utopia of science and fairness, mixed, indeed, with an equally lively series of publications that took the complete opposite view, of Sweden as a state run by a band of "new totalitarians."[4]

[4] Marquis Childs, *Sweden: The Middle Way* (New Haven, CT: Yale University Press, 1936). Roland Huntford, *The New Totalitarians* (New York: Stein & Day, 1972). It may be noted that the left-leaning Childs during his period in Sweden produced no less

A number of important works on international environmental politics devote considerable attention to the 1972 Stockholm Conference. It immediately became a landmark event and historical reference point, and when histories of environmentalism began to appear, Stockholm was rarely omitted. In the late 1980s, John McCormick, who wrote what was arguably the first comprehensive analysis of the global environmental movement, included a discussion on the significance of the Stockholm Conference.[5] His reading was preceded by Lynton Caldwell's book *International Environmental Policy* (1984), in which Stockholm 1972 was portrayed as playing a seminal role. These works were followed by Ramachandra Guha's global environmental history that more decisively included the Global South.[6] By then, in the 1990s and early 2000s, it had become an almost standard feature of any narrative of international environmental politics to include the 1972 Conference. However, very little of this coverage was based on deep engagement with the historical sources, nor was it intended to explain in any more profound way how Stockholm 1972 could have happened, why it took place precisely there, and whether there were any politics of place involved. If the prehistory of the Stockholm Conference was not deeply studied, even less energy was devoted to writing a "posthistory" encompassing the conference's effects and legacies, not to mention Stockholm's more recent links to the rise of GEG.

Rather, the approach in referencing the Stockholm moment was repetitive and ritualistic, reinforcing the significance of the conference and putting forward what was for a long time a linear, evolutionary narrative

than three books that all signaled that there was something to the Swedish model that had to do with balancing power and looking at the collaborative capacity of society. The two others were *Sweden: Where Capitalism Is Controlled* (New York: John Day, 1934) and *This Is Democracy: Collective Bargaining in Scandinavia* (New Haven, CT: Yale University Press, 1938).

5 John McCormick, *Reclaiming Paradise: The Global Environmental Movement* (Bloomington & Indianapolis: Indiana University Press, 1989).

6 Lynton Keith Caldwell, *International Environmental Policy: Emergence and Dimensions* (Durham, NC: Duke University Press, 1984); 2nd ed., 1996. Ramachandra Guha, *Environmentalism: A Global History* (New York: Pearson, 2000). Stockholm also reached out to the Global South, not least through its official development agency, Sida (Swedish International Development Cooperation Agency), and its research collaboration unit SAREC (Sida's Department of Research Cooperation), starting already in the 1960s. David Nilsson & Sverker Sörlin, *Research Aid Revisited: A Historically Grounded Analysis of Future Prospects and Policy Options*, The Expert Group for Aid Studies (Stockholm), Report 2017:7, online: https://eba.se/wp-content/uploads/2017/05/2017-07_Forskningsbist%C3%A5nd_webb_Tillganp.pdf.

of growing and solidifying institutions of environmental diplomacy and the ever-growing sequence of agreements, without offering much detailed insight. Only in recent years has more comprehensive coverage of the history and legacy of the Stockholm Conference appeared, for example, in work by Iris Borowy, Ken Conca, Stephen Macekura, Perrin Selcer, and Simone Schleper.[7] However, there has to date been no comprehensive analysis of the central place of Stockholm in the evolution of GEG. "Place" here connotes both the city's position in the historical narrative of GEG and, in the geographical sense, Stockholm as a pivotal location where environmental knowledge and political initiatives have been cultivated, not only in the early 1970s but also for the better part of a century. This book is about that place.

[7] Iris Borowy, *Defining Sustainable Development for Our Common Future: A History of the World Commission on Environment and Development (Brundtland Commission)* (London & New York: Routledge, 2013). Ken Conca, *An Unfinished Foundation: The United Nations and Global Environmental Governance* (Oxford: Oxford University Press, 2015). Stephen Macekura, *Of Limits and Growth: The Rise of Global Sustainable Development in the Twentieth Century* (Cambridge: Cambridge University Press, 2015). Perrin Selcer, *The Postwar Origins of the Global Environment* (New York: Columbia University Press, 2018). Simone Schleper, *Planning for the Planet: Environmental Expertise and the International Union for Conservation of Nature and Natural Resources, 1960–1980* (New York & Oxford: Berghahn Books, 2019).

1

The Stockholm Story

A Progressive Counternarrative

The United Nations Conference on the Human Environment (UNCHE) is indeed a milestone, even the seminal moment, in the evolution of global environmental governance (GEG). Held in Stockholm in June 1972, and thus commonly known as the Stockholm Conference, it was, however, by no means an isolated event in Sweden's long-term contribution to the politics and science of environment and sustainability. The conference, in fact, followed from a longer-term process that started several decades before 1972, continues to this day, and unfolds into the future.

In this book, we attempt to demonstrate what material, intellectual, civic, political, and other assets Stockholm – as a global hub of leading institutions, innovative concepts, and deeply engaged individuals – has been able to bring to bear in enhancing awareness and understanding of environmental issues and in the development of international governance regimes. Progress on these fronts has, to a large extent, been based on advances in scientific knowledge, something that Sweden has excelled at in disciplines ranging from atmospheric chemistry to ecological economics, as well as in developing and popularizing paradigms and concepts such as resilience, sustainability science, and the Anthropocene. Moreover, a significant part of Stockholm's outsized influence has been the result of combining capabilities from multiple realms, as demonstrated in its contributions to science diplomacy, science organization, and science communication. The attributes we analyze have in effect allowed Stockholm to cultivate and sustain, consciously but more often unconsciously, its status as an indispensable player in global environmental governance over an extended period of time.

Not even knowing for certain what it actually means to be "green" – a complex, hard-to-define concept – it may be even more difficult to conceive of global "green capitals" as a meaningful category. There seems to be established hierarchies of global centers or hubs for other sectors or aspects of society. The way, for example, we think of Frankfurt, London, New York, Singapore, and Hong Kong as global hubs of finance; the Boston area for higher education; or Tokyo, Milan, Paris, and Los Angeles as metropoles of art, fashion, or luxury,[1] Geneva clearly comes to mind as a center and symbol of internationalism, as does Vienna as a venue of environment and security. Vienna, the neutral Austrian capital, is associated with milestones in environmental history, such as the Vienna Convention for the Protection of the Ozone Layer, and became the location of institutions of environmental knowledge like the International Institute for Applied Systems Analysis (IIASA), established in 1972 in nearby Laxenburg. Hence, Vienna is arguably Stockholm's closest analogue, particularly if one were to also include the Austrian town of Villach where, as discussed at length in Chapter 6, a series of important climate science meetings took place during the 1980s.

The literature on such urban hierarchies, their evolution, and explanations for their success grew significantly in the period of intensified globalization in the late decades of the twentieth century. Saskia Sassen famously coined the concept "the global city" in a book with that title in 1991 and went on, like many others, to study financial centers. During this phase of economic globalization, when cities aspired to enhance their competitive position and prestige relative to other locations, scale, centrality, and attractiveness for investment and tourism became important features. Urban competition and world cities also became much theorized by Manuel Castells, Neil Brenner, Doreen Massey, and others in books that attracted a great amount of attention, partly because some of them could be read as manuals on how to succeed in the global economy and achieve status and significance, whether in finance, fashion, art, science, or innovation.[2]

[1] Mario Paris, ed., *Making Prestigious Places: How Luxury Influences the Transformation of Cities* (Abingdon & New York: Routledge, 2018).

[2] On global urban hierarchies and their drivers and processes, see Saskia Sassen, *The Global City: New York, London, Tokyo* (Princeton, NJ: Princeton University Press, 1991). Saskia Sassen, "Global Financial Centers," *Foreign Affairs* 71(1999):1, 75–87. Mark Yeandle & Chiara von Gunten, *The Global Financial Centers Index* (London: City of London, 2013). Neil Brenner & Roger Keil, *The Global Cities Reader* (London: Routledge, 2006). Youssef Cassis, *Capitals of Capital: A History of International Financial Centres,*

There is no index for environmental significance, no established or quantified hierarchy, and no "global cities" of nature conservation or sustainable development. Nor is it our intention to propose any such hierarchy, let alone to crown Stockholm as the champion of such values. The reason we have conducted extensive research and written a book on this topic has much more to do with our general, and nagging, impression that importance, influence, and leadership are significant for success, and in questions on environment and climate, there has clearly been too little success so far. The world is struggling and not doing well while engaged in a dangerous game. Perhaps even worse, positive, progressive forces too often fail in generating the recognition that the issues deserve and, above all, in sparking the organized action that is needed for humanity to change course and thereby lower the burdens put on the planet and many of its inhabitants. It may be that some cities, established sites of authority, are also linked to a perception of the status of certain voices, the trust and credibility of particular institutions that have gained a right to speak credibly on specific issues, and are known for making a difference through their ability to lead and instigate change.

This is not how environmental communication is typically conducted, nor is it how "the environment" is commonly conceived. When we speak of athletics (as in well-known teams in major cities, or especially successful nations in certain sports), science, or finance, we can often agree upon the location of leadership and the criteria of prowess or success. Almost the opposite is true about the environment. The word is, to a very large extent, associated with problems or challenges, and these moreover tend to be "wicked" and entail a long list of dilemmas and goal conflicts. Much of what is attractive, and indeed ingrained, within finance, fashion, and innovation has "environment" situated at the opposite end of the scale, sequestered and out of sight rather than displayed on Fifth Avenue or aboard the private jets and luxury yachts where Wall Street titans and

1780–2005 (Cambridge: Cambridge University Press, 2006). Doreen B. Massey, *World City* (Cambridge, UK: Polity Press, 2013). Neil Smith, *Uneven Development: Nature, Capital, and the Production of Space* (Athens: University of Georgia Press, 2008). From Manuel Castells' vast output, may it suffice to reference two of his visionary review articles on these topics, "European Cities, the Informational Society, and the Global Economy," *Tijdschrift voor Economische en Sociale Geografie* 84(1993):2, 247–257 and "Local and Global: Cities in the Network Society," *Tijdschrift voor Economische en Sociale Geografie* 93(2002):5, 548–558, and his coauthored book with Peter Hall, *Technopoles of the World: The Making of Twenty-First Century Industrial Complexes* (London: Routledge, 1994).

Russian oligarchs park their fortunes and pursue high-consumption lifestyles, a consumerist ideal that in turn trickles down to influencers and shapes the aspirations of millions of would-be emulators.

REMEMBERING DECLINE

Global human–environment history has, on the contrary, so far been mostly associated with sites of disaster, decline, and decay. The Yucatán Maya in Mexico and the Norse Settlement in Greenland are known for premodern cultural demise, while Easter Island and New Zealand in the Pacific are known for ancient extinctions.[3] Emblems of more recent environmental disaster abound: Bhopal, Chernobyl, Three Mile Island, Fukushima, New Orleans, Love Canal, and Minamata Bay, to name a few locations of crisis, contamination, and human tragedy. Entire regions are associated with profound anthropogenic change or upheaval: the Arctic (disappearing sea ice and melting permafrost), the Amazon (rainforest destruction and species loss), Antarctica (Ozone hole and ice reduction), Australia (invasive species), Newfoundland (cod stock collapse), the American West (drought and forest fire), the Sahel (desertification), and the Aral Sea (desiccation).[4] These are some

[3] Alfred Crosby, *Ecological Imperialism: The Biological Expansion of Europe, 900–1900* (Cambridge: Cambridge University Press, 1986). Jared Diamond, *Collapse: How Societies Choose to Fail or Succeed* (New York: Penguin, 2005). On the Maya, Scott Heckbert, Christian Isendahl, Simon Brewer, Vernon Scarborough, Arlen Chase, Diane Chase, Nicholas Dunning, Robert Costanza, Timothy Beach, Sheryl Luzzadder-Beach, David Lentz & Paul Sinclair, "Growing the Ancient Maya Social-Ecological System from the Bottom Up," In: Christian Isendahl & Daryl Stump, eds., *The Oxford Handbook of Historical Ecology and Applied Archaeology* (Oxford: Oxford University Press, 2019), 302–322. On the demise of the Greenland Norse as a climatic event, the classic study is Gustaf Utterström, "Climatic Fluctuations and Population Problems in Early Modern History," *Scandinavian Economic History Review*, 3(1955):1, 3–47.

[4] The literature on these and other similar key, indeed global, sites of environmental disaster is comprehensive and enormous. There are numerous studies on, among others, Love Canal, Bhopal, Chernobyl, and Three Mile Island. Here is a sample of significant works on some of the other main sites. On Fukushima, Sara B. Pritchard, "An Envirotechnical Disaster: Nature, Technology, and Politics at Fukushima," *Environmental History* 17 (2012):2, 219–243. On Hurricane Katrina and previous disasters in New Orleans, Craig Colten, *An Unnatural Metropolis: Wresting New Orleans from Nature* (Baton Rouge: Louisiana State University Press, 2006); Ashley Carse & Joshua A. Lewis, "Toward a Political Ecology of Infrastructure Standards: Or, How to Think about Ships, Waterways, Sediment, and Communities Together," *Environment and Planning A*, 49 (2017):1, 9–28; Joshua A. Lewis & Henrik Ernstson, "Contesting the Coast: Ecosystems as Infrastructure in the Mississippi River Delta," *Progress in Planning*, 129 (April 2019),

of the names at the top of the global hierarchy of "the environment," a greatest hits album of abuse and folly, glorious mistakes, and utmost failure. Very little of what is known about the environment, or for that matter climate, is associated with anything prestigious or progressive, with the possible exception of certain treaties such as the Montreal Protocol, an oft-cited triumph of green diplomacy that has served to repair the hole in the ozone layer that was discovered by British scientists in the 1980s and spurred rapid international action. As a rule, environmental sites have not been associated with accomplishment, quality, or capacity for progress. Rather they are focal points of declensionist, if not "destructionist," or catastrophist narratives. For obvious reasons, although there have been important improvements to local environments as well as periods of optimism in environmental governance, if we look at global trends and consider the planetary scale, most indicators clearly point in the wrong direction. Thus,

1–30. On Minamata Bay, Brett L. Walker, *Toxic Archipelago: A History of Industrial Disease in Japan* (Seattle, WA: University of Washington Press, 2010). On the Arctic, Paul Josephson, "Technology and the Conquest of the Soviet Arctic," *Russian Review*, 70(2011):3, 419–439; Paul Josephson, *The Conquest of the Russian Arctic* (Cambridge, MA: Harvard University Press, 2014); Pey-Yi Chu, *The Life of Permafrost: A History of Frozen Earth in Russian and Soviet Science* (Toronto: University of Toronto Press, 2021); Miyase Christensen, Annika Nilsson & Nina Wormbs, eds., *Media and Arctic Climate Politics: Breaking the Ice* (New York: Palgrave Macmillan, 2013); Batsheba Demuth, *Floating Coast: An Environmental History of the Bering Strait* (New York: W. W. Norton & Company, 2019); Sverker Sörlin, ed., *Resource Extraction and Arctic Communities: The New Extractivist Paradigm* (Cambridge: Cambridge University Press, 2023). The Pacific Northwest (Exxon Valdez) is covered in multiple special studies but also finds a mention in Adrian Howkins, *The Polar Regions: An Environmental History* (Cambridge: Polity Press, 2016). On the Amazon, José Augusto Pádua, "Tropical Forests in Brazilian Political Culture: From Economic Hindrance to Endangered Treasure," In: Fernando Vidal & Nélia Dias, eds., *Endangerment, Biodiversity and Culture* (London: Routledge, 2015), 148–171; José Augusto Pádua, "Brazil in the History of the Anthropocene," In: Liz-Rejane Issberner & Philippe Léna, eds., *Brazil in the Anthropocene* (London: Routledge, 2017), 19–40. On the Antarctic Ozone hole, Sebastian Grevsmühl, "A Visual History of the Ozone Hole: A Journey to the Heart of Science, Technology and the Global Environment," *History and Technology* 33(2017):3, 333–344. On the Newfoundland cod fisheries, Dean Bavington, *Managed Annihilation: An Unnatural History of the Newfoundland Cod Collapse*, with a foreword by Graeme Wynn (Vancouver: UBC Press, 2010). On the Aral Sea desiccation, especially its historical roots, see Maya K. Peterson, *Pipe Dreams: Water and Empire in Central Asia's Aral Sea Basin* (Cambridge: Cambridge University Press, 2019). General background in John R. McNeill, *Something New under the Sun: An Environmental History of the Twentieth Century* (New York: W. W. Norton & Company, 2000). On declensionist narratives and tropes, see Carolyn Merchant, "Declensionist," In: *The Columbia Guide to American Environmental History* (New York: Columbia University Press, 2002), 206.

global environmental governance has yet to deliver on multiple aspects of the mounting planetary crisis, not least on climate and biodiversity.[5]

Such downward trajectories also reflect and underpin a reading of history where human societies repeatedly demonstrate their inability to establish a sound and sustainable relation to the natural world and, just as often, among themselves. They can be located in an environmental narrative of despair and decline that has tended to dominate not only public memory but also professional environmental historiography. In the past, narratives of gloom, or glory, were prominent tropes, often connected with particular features of nature: mountains, deserts, forests, glaciers, marshes, and coral reefs. Whether the trope was that of gloom or glory was determined by the culture and religion of the commentator. Mountains were, for a long while, the locus of foreboding and fear, but more romantic spirits elevated them to places of beauty and majesty.[6] Forests or jungles became cathedrals of worship and unity with nature, as with Henry David Thoreau or John Muir. Just as often, however, they were dreaded and feared in Western culture, from the Brothers Grimm and before, while other cultures held less passionate views. As Roderick Nash recounted in *Wilderness and the American Mind*, when a Malaysian hunter surrounded by a thick tropical forest was asked by Nash's interpreter how to say "I am lost in the jungle," he simply could not understand the question. A Malaysian hunter "did not get lost in the jungle," the interpreter conveyed, in about the same way as a city dweller does not get lost in his apartment. There is nothing to fear.[7]

To speak of "progressivist" narratives on the global scale is very different. In our time, known by many names – the Western-centric "post-war era," or "the Great Acceleration" that accounts for the planetary-scale impact of "humanity" (in reality, mostly its wealthier segments) – and

5 Valérie Masson-Delmotte, Panmao Zhai & Anna Pirani et al., eds., "Summary for Policymakers," In: *Climate Change 2021: The Physical Science Basis: Contribution of Working Group I to the Sixth Assessment Report of the Intergovernmental Panel on Climate Change* (Cambridge: Cambridge University Press, 2021). Partha Dasgupta, *The Economics of Biodiversity: The Dasgupta Review* (London: HM Treasury, 2021). IPCC, 2022: *Climate Change 2022: Impacts, Adaptation, and Vulnerability. Contribution of Working Group II to the Sixth Assessment Report of the Intergovernmental Panel on Climate Change*, [H.-O. Pörtner, D. C. Roberts, M. Tignor & E. S. Poloczanska, et al., eds.] (Cambridge: Cambridge University Press, 2022).

6 Marjorie Hope Nicolson, *Mountain Gloom and Mountain Glory: The Development of the Aesthetics of the Infinite* (Ithaca, NY: Cornell University Press, 1959).

7 Roderick Nash, *Wilderness and the American Mind* (1967), 3rd ed. (New Haven, CT & London: Yale University Press, 1982), xiv.

which might just as well be called "the Age of Environment," environmental decline is not so much a trope in the history of Western ideas about nature as it is now a matter of empirical fact. As such it has become central in the search for new narratives that can capture the full breadth of the story that we need to tell about the human enterprise while giving justice to the enormous knowledge and insight that has been amassed. In essence, environmental declensionism replaced the centuries-long story of progress that has long been the core narrative of modernity.[8] As an undertaking, a progressivist narrative of the environment would immediately run counter to many things we know about the disasters and wrongs that certain individuals and institutions and entire societies have wrought. It would be a *contradictio in adjecto*. Is it possible or at least meaningful? We are convinced that it is important to carry out such a project. Both analytically, to understand what factors can explain environmental progress, and politically, to demonstrate past experiences and suggest pathways forward that can help lead us out of our current quagmire.

Part of the secret is to shift the narrative focus away from the undeniable bad news and instead look at how the instruments of positive change in policy and public understanding became possible. The notion of "progress" as put forward in this book is not meant to evoke some sort of utopian vision of international affairs, and it certainly does not imply that the global environment is at this point firmly entrenched upon a sustainable development pathway. Rather, what we are referring to are the constructive, if incremental and insufficient, steps that have been taken over the past fifty-plus years in establishing the ideas, institutions, and processes that have served as the drivers and catalysts for greater levels of awareness, action, and change in the direction of sustainability and for staying within planetary boundaries. If improvements within the environmental realm thus represent a first-order criterion of progress, the performance of actors and agencies represent second-order criteria, without which advances in the first would not be possible. These two types of narratives are obviously linked. Without overwhelming environmental decline, there would be no impetus for environmental progress. However, we would argue that while the first narrative is well rehearsed, loaded with long-standing tropes and themes, the latter narrative of progressive environmentalism linked to environmental improvement is far

[8] John B. Bury, *The Idea of Progress: An Inquiry into Its Origin and Growth* (London: Macmillan, 1920).

less developed. Indeed, not even within the massive amount of scholarly literature on (global) environmental governance has there been much focus on what has so far been achieved, where progressive steps have been taken, and why precisely in those places. Much of the effort has rather been on discussing how governance instruments and policies can and should be improved, looking mostly toward *the future* rather than examining recent and contemporary *history* as a source of insights, be they on shortcomings and failures, of which there are many, or about progress and the building of institutions and the initiation of processes.

We, of course, concur with the concern over the current state of the planet, and the eagerness to improve environment, climate, and sustainability policies everywhere. But we would also argue that many important achievements have, in fact, been made. Devoting some attention to what has actually worked reasonably well, and explaining why, is in our estimation beneficial, at least as long as we don't fall into the trap of the all too frequent attempts to sell progressivist narratives as a reason to signal that all will go well if we simply continue to seek technological solutions and stay true to the path of urbanization and economic growth. To carefully analyze and make sense of what has favored progress in environmental policy on the global level does not have anything to do with evangelical providence or with the overblown promises of eco-modernism.[9] Sweeping business-as-usual visions of techno-green utopias seen through rose-colored glasses have, if anything, impeded in-depth, sincere studies of the history of environmental governance.

REMEMBERING PROGRESS

Thus, the background of this book is the fact that the scientific history and political geography of advances in environmental governance have not as of yet been very well studied or understood. That is likely due in part to the magnitude of the task. Global environmental governance is a large and complex phenomenon, and few have been tempted to take it on as a scholarly effort in its own right. While the outlook of this present attempt to produce a history of environmental governance is necessarily global, our fundamental approach is to examine it through the

[9] An early example of this line of thinking was Gregg Easterbrook, *A Moment on the Earth* (New York: Penguin, 1996). A more recent one that has circulated widely in public debates is John Asafu-Adjay et al., *An Ecomodernist Manifesto* (2015), www.ecomodernist.org. See also, among the responses, Eileen Crist, "The Reaches of Freedom: A Response to an Ecomodernist Manifesto," *Environmental Humanities*, 7(2016):1, 245–254.

experience of a particular region of the world and from a particular city, which has consistently had a certain, even quite remarkable, centrality in this multi-decade process. As will be demonstrated in subsequent chapters, much of Stockholm's contribution to this history has been to serve as a node and locus of ideas and initiative-taking within an array of international networks and organizations that include, among other constellations, the UN and its specialized agencies. *Hence, this book provides a historical perspective on a trajectory of transnational progress through the lens of a particular geographical location that has played a pivotal role in the evolution of global environmental governance.*

Stockholm, we posit, represents a progressive counterstory to the declensionist narrative endemic to many environmental and policy histories, with the Swedish capital symbolizing the emergence and evolution of international efforts to manage the environment at the global or even planetary level. This process is not exclusively northern, or western, as it has since the preparatory phase of the 1972 Stockholm Conference been deeply intertwined with the development imperatives of the global South. In the years since 1972, an expanding assemblage of national, international, and nongovernmental institutions have been established around the world for addressing the science, policy, and advocacy dimensions of environment and development.

The process has not gone in the direction of strong, centralizing global institutions. Rather, the overarching trend over the decades has been one of "dispersal," a wide multi-layered distribution of institutions and responsibilities in a process that may be better seen as a history of an evolving human–Earth relationship rather than a matter of discrete policy choices.[10] The multitude of conferences, conventions, organizations, and other institutions, ranging from climate accords and science panels to UN-arranged Earth Summits and the Agenda 2030 Sustainable Development Goals, constitute what Frank Biermann has called the "architectures" of Earth system governance.[11] The aim of this book is to shed light on the social origins and history of some of the foundational elements and building blocks of its construction. This includes a range

[10] Sverker Sörlin, Paul Warde, Isobel Akerman, Thomas Harbøll Schrøder, Jasmin Höglund Hellgren, Sabine Höhler, Erik Isberg, Eric Paglia & Gloria Samosír, "The Great Dispersal: The Fall and Rise of Global Environmental Governance," (in review).

[11] Frank Biermann & Rakhyun E. Kim, eds., *Architectures of Earth System Governance: Institutional Complexity and Structural Transformation* (Cambridge: Cambridge University Press, 2020).

of less-heralded events, innovations, and processes facilitated by actors operating through national and transnational networks, bringing both a new narrative arc to the rise of global environmental governance and attention to a number of hitherto less-known case studies, all with a great deal of empirical detail.

Stockholm and the rise of GEG is a real-world history of events, conferences, negotiations, decisions, and activities by national and international organizations, civic movements, scientists, the corporate sector, and many other actors. But it also encapsulates a process of remembering. Stockholm is both a physical space where, for example, institutions are clustered geographically and research is conducted, and a point of reference and locus of memory within the scientific and political story of the environment's ascent on the international agenda. In the parlance of the field of memory studies, Stockholm has become a *noeud de mémoire*, a node of remembrance, in the rising global story of a transforming human–Earth relationship that is now coming of age.[12] We thus also reflect on the problem of *remembering on the global scale*. What does it mean to remember a city and its environmental legacy inscribed within the context of a global story?

Those who share in such remembering are not just those who reside in a global node. To become a site of international memory, the processes of remembering must be spread via media and shared in wider circles. Memory must also last and be passed on over the course of years, decades, and generations; centennial temporalities may even be considered, even if it is perhaps still too early to think of our recent and contemporary history in such timespans. Achieving such status requires at the very least significance, perhaps fame, for the place or process. In the case of Stockholm as a place of environmental memory, possibly more the former than the latter. Although as the environment and, increasingly, climate change gain significance as keywords of the emerging *Weltanschauung*, it can be expected that more remembering in those spheres will take place. After all, as the world is increasingly understood also *qua* environment, *qua* climate, and *qua* vulnerable "planet" – where did this all come from? In the ongoing Anthropocene transformation of global temporalities and emerging, often quantified, rates and trajectories

[12] Michael Rothberg, "Introduction: Between Memory and Memory: From Lieux de mémoire to Noeuds de mémoire," In: Michael Rothberg, Debarati Sanyal & Maxim Silverman, eds., *Noeuds de Mémoire: Multidirectional Memory in Postwar French and Francophone Culture. Yale French Studies* 118/119 (New Haven, CT: Yale University Press, 2010), 3–12.

FIGURE 1.1 Stockholm during UNCHE in the early summer of 1972. Photo: UN Photo/Yutaka Nagata.

of change, we can identify new chronologies – indeed new "spaces of experience" and "horizons of expectation," to use Reinhart Koselleck's concepts.[13] Such fears and hopes of the Anthropocene are closely associated with features of the environment such as climate, biodiversity, the cryosphere, forests, and oceans.

GOVERNANCE IN NEW TIMES OF THE ELEMENTS

These elemental features of the Earth System now present themselves as a set of *environmental times* that increasingly constitute the flow of time and the making of the history of the modern world. This world is not just elemental, material, and natural, it is increasingly interwoven with the human life sphere. More than that, it is, to a perhaps frightening extent, anthropogenic. It is no longer just nature, it is a nature transformed by human action into an environment, and new times that are generated

[13] Reinhart Koselleck, ed. "'Erfahrungsraum' und 'Erwartungshorizont' – zwei historische Kategorien." In *Vergangene Zukunft: Zur Semantik Geschichtlicher Zeiten* (Frankfurt am Main: Suhrkamp Verlag, 1989), 349–372.

in the interference with those embedded in the Earth.[14] We ask: where is this action taking place? In answering that question, we do not look so much at those sites and processes, or even calamities that generate the impacts, that is, the events and anthropogenic environmental change itself. Instead, we look at the work that has been undertaken to understand the change, mitigate its consequences, and redirect it. Originating on the outskirts of world affairs a century ago as ideas circulating in small groups of thinkers and scientists, and not least among those that were affected in nature, on farms, and in their work places, this effort has come to occupy a central position on the global stage and become a common concern and a political issue at the highest levels of international politics.

Although the rise of the environment on the global agenda affects us all, the way we are all part of history, its ascent as a matter of political, scientific, and societal engagement has played out more distinctly in certain contexts. These can be seen as the *arenas* where GEG *practices* take place. Given that this strand of governance is becoming global, it follows that the key arenas are those that allow for national and local issues to converge with international concerns and their associated actors and agencies, which have grown rapidly in number since the middle of the twentieth century. We will focus in particular on meetings, conferences, institutions, and organizations where these kinds of exchanges have taken place. This is not a process where agency, responsibility, contributions, or, for that matter, reticence and friction, are evenly distributed across nations or cities. On the contrary, nor is it an evolution, neatly following a set of principles or laws; and it is certainly not a metaphysical phenomenon whereby "humanity" seeks a new position in relation to "the earth" according to some preordained theological or ideological scheme. It is history and, as such, is always open-ended vis-à-vis the future at any given point in time.

Studying global environmental governance historically means that we need to analyze and understand actions and events, including for example the formation of institutions and the dynamics of meetings, within their wider contemporary context. At the same time, we are writing this history from the vantage point of the 2020s, a moment in time when GEG has advanced dramatically compared to the early days of

[14] Sverker Sörlin, "Environmental Times: Historical and Scientific Temporalities from *Annales* to Anthropocene, 1920s-2020s," In: Anders Ekström & Staffan Bergwik, eds., *Times of History, Times of Nature: Temporalization and the Limits of Modern Knowledge* (New York: Berghahn Books, 2022), 64–101.

"the environment" as a slowly evolving policy concept in the immediate post-World War II years. The United Nations has dubbed it the Agenda 2030 decade, suggesting that it is a critical period for achieving the seventeen global Sustainable Development Goals adopted by the UN in 2015. The SDG on climate, number 13, encompasses ambitious goals for drastic reductions in global greenhouse gas emissions by 2030, agreed upon at the 2015 Paris climate conference and reinforced and sharpened at Conference of the Parties (COP) meetings since then. These facts and these trajectories of hope and vision, alongside fear and despair that the chances of reaching the goals are slim, represent necessary vantage points as we return to the historical events covered in this book. If environment and climate had not become and remained one of the supra-narratives of the modern world, there would not have been any pressing need to write this history.

That need now exists, and it is actually surprising that not more such histories have been written to date. One possible reason is that despite all the current interest in environmental, diplomatic, science, and Anthropocene history, issues related to nature and the environment are still not commonplace for most historians. Much of the GEG history so far has been provided as peripheral information by natural and social scientists primarily studying and writing about other – typically contemporary – topics. This is of course essential information that we have drawn upon extensively. The same goes for biographical and autobiographical literature, and a large number of institutional and thematic reports drafted over decades that are themselves historical sources sometimes offer micro-narratives contextualizing their own relevance.

Writing from the vantage point of Agenda 2030 and approaching the environmental policy past historically, *qua history*, allows us to use all this material simultaneously, assembling it in a way that enhances the possibility of better explaining how we should understand the times we live in at present. How can we understand the agendas and tools of environmental and climate policy and governance? Why are global discourse and negotiations focused and structured the way they are? And how did the issues emerge? Guided by these kinds of questions, we can also improve the way we ask questions about the future. The past does not in any simple way determine or predict what will come in the years and decades ahead. But without a profound and detailed understanding of the rise of GEG, our thinking about the planet's future – and thus the chances of reaching crucial sustainable development goals – will be poorer and more simplistic. Rich, textured, and profound historical understanding is a bit like an insurance policy: you *can* live without it, but only at your own risk.

This book is therefore part of a larger enterprise that no single intervention can cover. Our aim is to connect the present with the early formations of the environment and the human–Earth relationship – a historical phenomenon that, we argue, is one of the most crucial features of the twentieth century, on a par, over the long term, with watershed events like the two world wars and the revolutions in Russia and China. This entails linking the events and agencies of the twentieth century with our current predicament in the first half of the twenty-first – a time when we can read and reflect upon this history of the rise of GEG while we continue to make it part of our contemporary politics.

Like most history, it means offering more than just an array of facts. It will become meaningful only if the dots are connected within a broader historical interpretation, which in turn must be seen from a particular vantage point and offer a particular narrative that gives justice to the past, and at the same time brings meaning and understanding to the present. Setting out to produce such a narrative of the environment, focused not on the environment itself – including the various catastrophes that have befallen it – but rather on the three-quarter century quest to understand it and mitigate the ongoing disruption of planetary systems, requires its own anchoring points, obligatory passage points, and "truth spots" on the map of the world.[15] The arenas of environmental governance – the knots in the warp of time as it were – where broader change can be detected or important steps were taken are useful tools in crafting such a narrative and memory of past events and processes. Some of the important moments are indeed remembered, even celebrated, like the 1972 UN Conference on the Human Environment in Stockholm. Multi-level repercussions – local, national, and global – are required for large-scale memories to form and last; however, as the reader of this book will discover, many if not most of the key incremental steps as well as decisive turning points are *not* very well remembered. They have faded from our minds, which is why it is necessary to return to them as we craft a meaningful GEG narrative.

History and memory stand in a productive, but also not unproblematic, relation to one another. To remember is a "use of the past." As French sociologist Maurice Halbwachs suggested a century ago, to

[15] Thomas F. Gieryn, *Truth-Spots: How Places Make People Believe* (Chicago, IL & London: The University of Chicago Press, 2018).

remember is to form community and to expose yourself to the process of being formed by it.[16] A community is a social entity that remembers together and thus worships, mourns, and rejoices together. It requires agreement on what is worth remembering. We find pronounced versions of these processes among nations, professions, and religions, but they can be observed in some form in most communities, including families, firms, and social movements. As the idea of a planetary human–Earth relationship continues to take shape, communities are now forming around this overarching narrative of the human species. This is an ongoing process for which no established narrative yet exists. Proactive groups and actors have achieved a level of institutional density and acknowledged the importance of the environment and climate. These topics are merging in multiple ways with geopolitics, security, diplomacy, progress, risk management, and ideas of the future. Environmental and climate themes are making their way into the writing of general and world history; and when history changes, the present and outlooks on the future will change, too.[17]

WHY CITIES MATTER

This history plays out on multiple levels. One is obviously global, and certain aspects of history, as well as remembering, are now indeed global if not planetary in scale. As Dipesh Chakrabarty has pointed out, the planetary has become a "humanist category." This idea was percolating for a while but has now, in a relatively short amount of time, become firmly established.[18] Many of the most important institutions in the

[16] Maurice Halbwachs, *Les Cadres sociaux de la mémoire* (Paris: Albin Michel, 1925). Maurice Halbwachs and Lewis Coser, ed. and transl., *On Collective Memory* (Chicago: The University of Chicago Press, 1992).

[17] As is increasingly the case, for example, in ocean history, such as David Armitage, Alison O. Bashford & Sujit Sivasundaram, eds., *Oceanic Histories* (Cambridge: Cambridge University Press, 2017), and in global environmental histories that started to appear around 2000, for example, McNeill, *Something New* (2000), J. R. McNeill & William McNeill, *The Human Web: A Bird's-Eye View of World History* (New York: W. W. Norton & Company, 2003), Guha, *Environmentalism* (2000), Joachim Radkau, *Nature and Power: A Global History of the Environment* (German orig. 2002), English transl. (Cambridge: Polity, 2008).

[18] Dipesh Chakrabarty, "The Planet: An Emerging Humanist Category," *Critical Inquiry* 46(2019):1, 1–31. Sverker Sörlin, "Scaling the Planetary Humanities: Environmental Globalization and the Arctic," In Ursula K. Heise, Jon Christensen & Michelle Niemann, eds., *The Routledge Companion to the Environmental Humanities* (London: Routledge, 2017), 433–442.

environmental arena are global. Some, such as the Nairobi-based United Nations Environment Programme (UNEP), are part of the UN system and have a global reach that transcends the city where the agencies' staff sits. Another example is the Intergovernmental Panel on Climate Change (IPCC) headquartered in Geneva, close to its much-older parent institution, the World Meteorological Organization (WMO).[19] The latter became part of the UN system between 1947 and 1951 with, at the time, a very different rationale and *raison d'être* than monitoring and understanding "the environment" as such. The IPCC reports are almost invariably global, and they make a special effort not to speak from any particular place. It is, and strives to be, a voice from nowhere, reflecting the neutral tone of science, taking all parts of the world into consideration. But as work in the history and social study of science has taught us, there is in reality no voice that can speak from nowhere. Science and knowledge have their own geographies, and knowledge is in a number of respects situated.[20] It is not only data that reflects local conditions and their differences. The stories that are told with data change depending on who collected it, where, and when. Which facts become available depends upon what is being looked for and what research is getting funded and pursued. But perhaps even more important are the conditions for assembling the narratives that the data can sustain.

That is why cities matter. Or, perhaps more specifically, the geographic nodes where facts are assembled and turned into knowledge. In a now-famous book about the social organization of knowledge, Bruno Latour spoke of "centres of calculation."[21] He drew upon colonial science, which for centuries was organized around a protracted collection of data – specimens, samples, and measurements – that were typically brought back from the far reaches of empires to collections in museums, academies, or universities in European capitals and centers of learning.

[19] The IPCC was established in 1988 jointly by WMO and UNEP.

[20] Donna Haraway, "Situated Knowledges. The Science Question in Feminism and the Privilege of Partial Perspective," *Feminist Studies* 14(1988):3, 575–599. David N. Livingstone, "The Spaces of Knowledge: Contributions towards a Historical Geography of Science" *Environment and Planning D: Society and Space* 13(1995):1, 5–34; David N. Livingstone, *Putting Science in Its Place: Geographies of Scientific Knowledge* (Chicago: University of Chicago Press, 2003). Richard C. Powell, "Geographies of Science: Histories, Localities, Practices, Futures," *Progress in Human Geography* 31(2007):3, 309–329. Martin Mahony, "Geographies of Science and Technology I: Boundaries and Crossings," *Progress in Human Geography* 45(2021):3, 586–595.

[21] Bruno Latour, *Science in Action: How to Follow Scientists and Engineers through Society* (Cambridge, MA: Harvard University Press, 1987).

The storytelling of modern science, the sense-making narratives of the world we lived in, was told from these centers, which framed – and tainted – such stories significantly. Science works much better today, but certain elements of this coloring of knowledge by the institutions and political contexts of their production cannot be completely avoided. That is even more the case if we look at what kinds of policy narratives can be told based on scientific understanding. These are even more conditioned, indeed constrained, by the politics of the institutions and the people that are engaged in writing the reports and turning them into policy.

It is in such a perspective that cities matter even more. If there is a "Washington consensus," there should also be a Brussels perspective, a Beijing outlook, a Moscow, Nairobi, and Brasilia standpoint – and, we argue, most certainly a Stockholm view on the environment as a global concern. We are in this book aiming to illuminate that. What Stockholm and Sweden have contributed is of course part of a much larger entirety of GEG. Yet perhaps in contrast to a wave rising and falling with the flow of an entire ocean, an individual city – including its institutions, traditions, norms, and national political culture – can be more independent and relate more proactively to the wider growth of knowledge and general changes in the global environment. In that sense, the environment as such has a geography, not just in the obvious sense that environments and climates differ with latitude and altitude, but because what becomes environmental governance is related to place and, needless to say, time. In this respect, too, the environment is political.

Our argument is thus that Stockholm is interesting because of what it has specifically brought to bear in facilitating the rise and practice of GEG. It represents history in its own right, with a unique combination of parts and elements, infused with human agency at every step. This is also how we think it can be meaningfully told, as an in-depth presentation of a continuous process where the many actions of a multiplicity of actors operating in special circumstances are needed to demonstrate the thrust of the argument. Similar histories can surely be written about other cities or geographies, and some day such stories may emerge on a broader scale. So far, they are few. One example is Richard Walker's *The Country in the City* (2008) on the rich and progressive environmental history of the San Francisco Bay Area, spanning approximately 100 years from the late-nineteenth to late-twentieth centuries. It is a fabulous book, in its approach not without similarities to this one, but it is, above all, a local or regional history. The aspects of governance it encompasses are on the level of the urban region, beautifully traced over generations and across special places in the greater Bay Area – creeks and groves, parks and harbors, wine lands and wetlands – and across ethnic groups, gendered communities, social classes, and a wide diversity of knowledge, traditions, and experiences.[22] Another such book of similar stature is Jens Lachmund's narrative of the modern, green Berlin that emerged, very literally, with *Ruderal* species and all, from the ruins of World War II.[23]

Stockholm has an environmental history of this kind as well. Some of it is captured in prominent works, although none quite as eminent and with such broad spatial and temporal scope as Walker's book on the Bay Area.[24] In our examination of the central role of Stockholm in the rise of GEG, however, that strand of Stockholm's history is of limited interest, although there will be some mention of castles and conference centers where meetings were organized, and of parks and public squares where protesters gathered or talks were held. We would not deny that popular sentiment and a passion for nature and local ecologies have deep roots and may have been important in both the Bay Area and in Stockholm, just as they have proven to be important for environmental engagement elsewhere.

But what interests us is something perhaps even harder to get to the bottom of and write a history about; namely the relationship between a city and something as vast as the "global" and its concomitant environment, and how the governance of these two combined things came into existence over a period of a few generations following World War II. Despite this process being global, it will have certain regional and local

[22] Richard Walker, *The Country in the City: The Greening of the San Francisco Bay Area* (Seattle, WA: University of Washington Press, 2008). A compact, updated version is Richard Walker, "Nature's Popular Metropolis: The Greening of the San Francisco Bay Area," In: Henrik Ernstson & Sverker Sörlin, eds., *Grounding Urban Natures: Histories and Futures of Urban Ecologies* (Cambridge, MA: MIT Press, 2019), 169–200.

[23] Jens Lachmund, *Greening Berlin: The Co-Production of Science, Politics, and Urban Nature* (Cambridge, MA: The MIT Press, 2013). Jens Lachmund, "Regimes of Urban Nature: Organic Urbanism, Biotope Protection, and Civic Gardening in Berlin," In: *Grounding Urban Natures* (Cambridge, MA: MIT Press, 2019), 247–276. Bettina Yvonne Stoetzer, "Ruderal Ecologies: Rethinking Nature, Migration, and the Urban Landscape in Berlin," *Cultural Anthropology* 33(2018):2, 295–323.

[24] From a rich, so far largely journalistic and narrative nonfiction literature on Stockholm urban environmental history in Swedish, for example, Nils-Erik Landell, *Den gröna staden* [The Green City] (Stockholm: Bonnier Alba, 1979); Nils-Erik Landell, *Vattenstaden* [The Water City] (Stockholm: Carlsson, 2006); Gunnar Brusewitz & Henrik Ekman, *Ekoparken: Djurgården Haga Ulrikdsdal* (Stockholm: Wahlström & Widstrand, 1996). Book-length environmental histories of Stockholm available to international readers are basically absent.

sites of particular significance. Stockholm stands as one of the leading candidates to become a preeminent location of both agency and collective memory for this truly global history. Not another site of environmental disaster, which has its own worthy acts of commemoration, but as a key place of discovery for the narrative itself and the ongoing effort to reorient the course of human progress.

ROADS TO THE CONFERENCE

Stockholm 1972 looks very different from the location of the Conference than it does outside Sweden. From a national perspective, many would consider the most significant site of environmental politics connected to early 1970s Stockholm to be a patch of thirteen Scots elm trees outside the Opera House in the public garden of Kungsträdgården.[25] After over a decade of sweeping redevelopment of central Stockholm, including the demolition of hundreds of centuries-old structures, plans to build a subway entrance where the trees stood sparked, in May 1971, a massive protest and clash with local authorities (Figure 1.2). "The Elm Conflict," organized by an action group that would later become part of the local Stockholm chapter of Friends of the Earth, represented a turning point in modern Swedish politics, as it was framed as a matter of democracy

FIGURE 1.2 The "Elm Battle," Stockholm 1971. Activists and engaged citizens opposed the planned destruction of elm trees that stood in the way of a new subway station in the heart of the capital. Photo: Lennert af Petersens. Stockholm City Museum Archives.

[25] Literally, the King's Garden produced herbs and vegetables for the Royal household in the Early Modern period.

and remains a point of reference for subsequent environmental and city planning conflicts in Sweden.[26]

In terms of establishing Stockholm's status as an *international* site of environmental history, less dramatic processes in preparation for UNCHE were at the same time underway in Sweden's capital and elsewhere, the products of which would have substantial influence on scientific, social, and political engagement with environmental issues at the global level. An array of textual artifacts surrounding the Conference represent objects of significance associated with Stockholm and the origin of GEG. In May 1971, while protesters were scurrying up the elm trees in downtown Stockholm, the UNCHE Secretariat in New York was commissioning what would become the unofficial report and "conceptual framework" for the Conference. The volume, a scientific and philosophical work considered the "conference bible," was written for lay readers but based on expertise from diverse realms of knowledge and all global regions.[27] Prominent environmental and internationalist thinkers, Barbara Ward and René Dubos, were enlisted to coordinate input from 152 experts from fifty-eight countries – carefully chosen to represent first, second, and third world perspectives and provide "a view from everywhere" – and ultimately author the report that carried the Stockholm Conference motto, *Only One Earth: The Care and Maintenance of a Small Planet.*[28] As the first global state-of-the-environment report, and an early articulation of the concept of sustainable development, *Only One Earth* had a substantial impact on public opinion and the thinking of political leaders in the industrial world as well as, to a lesser extent, in developing countries.[29] A half century ago, it was among the first official documents at

[26] Details on this conflict are in Daniel Helldén, *Demokratin utmanas: Almstriden och det politiska etablissemanget* [Democracy Challenged: The Elm Conflict and the Political Establishment], PhD diss. (Stockholm University: Department of Political Science, 2005). See also Anders Gullberg, *City: Drömmen om ett nytt hjärta* (Stockholm: Stockholmia förlag, 2001).

[27] McCormick, *Reclaiming Paradise.* Selcer, *Postwar Origins.* Schleper, *Planning for the Planet.*

[28] Barbara Ward & René Dubos, *Only One Earth: The Care and Maintenance of a Small Planet* (New York: W. W. Norton & Company, 1972).

[29] Lars-Göran Engfeldt, *From Stockholm to Johannesburg and Beyond: The Evolution of the International System for Sustainable Development Governance and Its Implications* (Stockholm: Ministry of Foreign Affairs, 2009). Eric Paglia, "The Swedish Initiative and the 1972 Stockholm Conference: The Decisive Role of Science Diplomacy in the Emergence of Global Environmental Governance," *Humanities and Social Sciences Communications* 8(2021):2.

the international scale that laid out the emerging understanding of the Earth as a fragile object that required human stewardship.[30]

The seeds of sustainable development were planted in the principles of the Declaration on the Human Environment, one of the most significant Conference outcomes. Several months before Ward and Dubos were commissioned to prepare *Only One Earth*, the second session of the UNCHE preparatory committee created, in February 1971, an intergovernmental working group to begin drafting a declaration for the Conference. Hans Blix, legal advisor to the Swedish foreign ministry and later head of the International Atomic Energy Agency, was instrumental in negotiating and writing the declaration over the course of several meetings in 1971 and 1972, with the final text – consisting of a preamble and twenty-six principles – being adopted on the Conference's last day, June 16, 1972.[31] Although not a legally binding instrument, the Stockholm Declaration is considered a landmark in the establishment and evolution of international environmental law, with its normative contributions in the realm of environment and development reaffirmed twenty years later by the similarly structured 1992 Rio Declaration.[32] The legacy of Stockholm included the environmental summits in Rio de Janeiro in 1992 and 2012 and Johannesburg in 2002, shaping an arc of diplomatic mega-events that can be traced directly back to the Stockholm Conference. A more modest commemoration was held in the Swedish parliament in June 1997, a mini Stockholm+25 convened by Anna Lindh, then Minister of the Environment, with Maurice Strong among the speakers, and a similar Stockholm+40 event was arranged in 2012.[33] In June 2022, the semi-centennial of UNCHE was marked by the Stockholm+50 conference, a major if much less ambitious UN event that lasted only two days compared to the twelve-day gathering in 1972.

[30] Schleper, *Planning for the Planet*.

[31] Hans Blix, "History of the Stockholm Declaration," In: Myron H. Nordquist, John Norton Moore, & Said Mahmoudi, eds., *The Stockholm Declaration and Law of the Marine Environment* (Leiden: Brill Nijhoff, 2003), 15–24. Louis B. Sohn, "The Stockholm Declaration on the Human Environment," *The Harvard International Law Journal* 14(1973):3, 423–515. Engfeldt, *From Stockholm to Johannesburg*.

[32] Sheila Jasanoff & Marybeth Long Martello, *Earthly Politics: Local and Global in Environmental Governance* (Cambridge, MA: MIT Press, 2004). Biermann, "Global Environmental Governance." K. O'Neill, E. Weinthal, K. Marion Suiseeya, S. Bernstein, A. Cohn, M. Stone & B. Cashore, "Methods and Global Environmental Governance," *Annual Review of Environment and Resources* 38(2013), 441–471.

[33] It may, for transparency, be noted that one of us (Sörlin) was also on the list of speakers at the Stockholm+25 event.

FIGURE 1.3 Anna Lindh and Kofi Annan, photographed in 2002 when Lindh was Foreign Minister of Sweden. Lindh, in a long Swedish tradition, cherished the idea of Sweden as a nonaligned country and also the idea of a strong United Nations. Photo: UN Photo/Eskinder Debebe.

Other international conferences, agreements, and science-policy processes associated with Stockholm and other Swedish locations have followed in the years since 1972. The Stockholm Convention on Persistent Organic Pollutants was signed in May 2001, and Sweden's second-largest city gave name to the 1999 Gothenburg Protocol limiting emissions of multiple chemical pollutants that cause acidification and eutrophication. Adopted within the Convention on Long-Range Transboundary Air Pollution, which Swedish government officials and scientists were instrumental in developing during the 1970s and 1980s, the Gothenburg Protocol also gave rise to a series of international meetings on acid rain science and policy that, despite being held in Gothenburg over the past twenty years, are known as the "Saltsjöbaden Workshops," named after the community on the outskirts of Stockholm where the first meeting took place in 2000.[34] A wide variety of locations across Sweden, ranging

[34] Camilla Andersson et al., *Achievements and Experiences from Science – Policy Interaction in the Field of Air Pollution: Synthesising 20 Years of Research and Outreach, Thinking about Future Needs* (Stockholm: IVL, 2021).

from conference centers and manor houses in and around Stockholm to marine laboratories and field stations as far north as Abisko in the Swedish Arctic – where for two decades the Swedish Research Council, as part of a productive long-term relationship with Vienna-based IIASA, invited twenty to thirty experts to annual retreats for reflecting on issues related to sustainability and systems analysis[35] – have in fact represented important sites for various processes underpinning the evolution of global environmental governance.

Part of the Stockholm narrative is tied to properties that had for many years been associated with Sweden as a democratic template. Sweden represented "the Middle Way," as Marquis Childs had called it in his 1936 portrait of the model modern nation, a society working according to consensus norms and in a spirit of diplomatic give and take.[36] The notion of consensus also extended to aspects of Sweden's foreign policy, which was strongly invested in the United Nations. Reflecting Sweden's postwar commitment to internationalism, Sverker Åström – a renowned Swedish diplomat who played a decisive role in convening a UN environmental conference – even saw national sovereignty as a hindrance to collective global action on the emerging environment issue.[37] In this small northern country, the environment as a societal problem had by the mid-1960s begun to loom large, with Sweden establishing the world's first environmental protection agency in 1967. Compared to other countries, Sweden was early and comprehensive in its attempts to pass legislation on issues such as vehicle pollution, work environments, acidification, and many other questions surrounding the environment and public health.[38] A ban on ozone-depleting chlorofluorocarbons (CFCs) became

[35] Uno Svedin, personal communication, September 23, 2022. Sweden became a member of IIASA in 1974; responsibility for managing the relationship was eventually passed from the Swedish Research Council to Formas in 2001. Torsten Hägerstrand, Anders Karlquist, Sture Öberg, and Uno Svedin were the main Swedish actors involved in arranging the series of Abisko workshops, which ran from 1983 to 2003.

[36] Childs, *Sweden – The Middle Way*.

[37] Sverker Åström, "Global Consensus or Global Catastrophe?," *Ambio* 1(1972):1, 2–5.

[38] Svante Odén, "The Acidification of Air Precipitation and Its Consequences in the Natural Environment," *Ecological Bulletins* vol. 1 (Stockholm: Swedish Natural Science Research Council, 1968); this publication was preceded by his highly influential op-ed. article in the Dagens Nyheter in October 1967 (see Chapter 4). Hans Palmstierna, *Plundring, svält, förgiftning* (Stockholm: Rabén & Sjögren, 1967). On Swedish environmental chemistry and public health in this period, Michael Egan, "Communicating Knowledge: The Swedish Mercury Group and Vernacular Science, 1965–1972," In: Dolly Jørgensen, Finn Arne Jørgensen & Sara B. Pritchard, eds., *New Natures: Joining Environmental History with Science and Technology Studies* (Pittsburgh, PA: University

law directly after the 1987 Montreal Protocol – in which Swedish scientists played a prominent role – and a carbon tax, still anathema in most nations, was introduced already in 1991.[39]

This is a far from comprehensive, even impressionistic, review of some of the factors that make up Sweden's domestic environmental governance performance. There is a well-established tradition of comparative study of environmental governance where such performance is measured and quantified across a range of indicators. Such studies have for the last couple of decades confirmed the impression of Sweden as an environmental pioneer and a high performer. "[T]he most progressive country across all observations is Sweden," a study of environmental policy "adoption" in twenty-four Organisation for Economic Co-operation and Development (OECD) countries tells us. A study focusing on environmental policy "change" repeats that Sweden overall does well, but that the early decades after 1970 were its most pioneering phase while other EU member countries have in many cases advanced more quickly in the 2000s. In a study of "sustainable citizenship," Sweden again ranks high, in line with findings about Sweden as a world leader in "political consumerism."[40]

This is essential information in the sense that Sweden's level of performance on a range of *national* environmental policy indicators cannot be disregarded in a study of its role in GEG. This performance builds what we may call progressivist legitimacy, a reputation among other nations, which helps explain Sweden's remarkable career as a

of Pittsburgh Press, 2013), 103–117; Ellen Griffith Spears, *Baptized in PCBs: Race, Pollution, and Justice in an All-American Town* (Chapel Hill, NC: University of North Carolina Press, 2014).

39 Mats Engström, *Miljöframgångar: Från freonförbud till klimatlag* [Environmental Successes: From CFC Ban to Climate Law] (Stockholm: Fri tanke, 2020).

40 Christoph Knill, Susumo Shikano & Jale Tosun, "Explaining Environmental Policy Adoption: A Comparative Analysis of Policy Developments in Twenty-four OECD Countries," In: Andreas Duit, ed., *State and Environment: The Comparative Study of Environmental Governance* (Cambridge, MA: The MIT Press, 2014), 53–79, on 69; Thomas Sommerer, "Early Bird or Copycat, Leader or Laggard?: A Comparison of Cross-National Patterns of Environmental Policy Change," In: *Duit, State and Environment*, 149–177; Michele Micheletti, Dietlind Stolle & Daniel Berlin, "Sustainable Citizenship: The Role of Citizens and Consumers as Agents of the Environmental State," In: *Duit, State and Environment*, 203–236. Dietlind Stolle & Michele Micheletti, *Political Consumerism: Global Responsibility in Action* (Cambridge: Cambridge University Press, 2013). The perception of Sweden as a global leader of sustainable development, a main strand of environmental policy since the 1980s, was already well established, see Katarina Eckerberg, "Sweden: Progression Despite Recession," In: William M. Lafferty & James Meadowcroft, eds., *Implementing Sustainable Development: Strategies and Initiatives in High Consumption Societies* (Oxford: Oxford University Press, 2000), 209–244.

broker and facilitator on the global arena. In fact, we will argue in favor of such a conclusion throughout the volume. However, we also think it is useful to see the differences in analytical approach. The formalized studies of comparative national governance, with large amounts of performance data, do not in and of themselves tell us much about the complex scientific, diplomatic, and political work carried out by relatively small groups of individuals and institutions over long periods of time that have underpinned Sweden's outsized performance as a global governance agent. This is also a reason why our study is mainly historical, tracing what we see as performance trajectories where the various national properties have explanatory value, including those of a city. Very likely, the reverse is also true. If a country performs less well nationally, its legitimacy as a leader in GEG will weaken. We return to this issue at the end of the book.

SCIENCE AND ENVIRONMENTALISM

A significant aspect of Sweden's early adoption of green policies was the relatively peaceful co-existence of the state with a growing environmental movement, parts of which had strong public support and even attracted the attention of the government. From its infancy, Swedish environmentalism cross-fertilized with protest movements against the Vietnam War and the civilian use of nuclear energy, with young activists such as microbiologist Björn Gillberg, and Lennart Daléus – who would later become leader of Sweden's Center Party and eventually the executive director of Greenpeace Nordic – playing prominent roles.[41] Furthermore, Sweden was one of the earliest European countries to establish, in 1982, a Green party, which entered Parliament in 1988 and has remained there for over three decades.[42] This evolution of what has been described as a Nordic "Green State"[43] does not mean that environmental issues did not also

[41] Daléus was founding chairman of the Swedish branch of Friends of the Earth from 1971 to 1976, during which time he was also employed as information secretary at the Royal Swedish Academy of Sciences. His work in government and environmental advocacy has also included serving as head of department at the Environmental Advisory Council (*Miljövårdsberedningen*) and as international secretary at the Swedish Society for Nature Conservation.

[42] Except for a three-year period in the 1990s when they failed to receive the required 4 percent of the national vote.

[43] Robyn Eckersley, *The Green State: Rethinking Democracy and Sovereignty* (Cambridge, MA: The MIT Press, 2004). Karin Bäckstrand & Annica Kronsell, eds., *Rethinking the Green State: Environmental Governance towards Climate and*

encounter a certain amount of skepticism that has played out in the politics of energy, forestry, and mining, including in relation to the Sami minority of the north.[44] Swedish environmental sins have certainly been committed in all these sectors, not to mention other industries as well as major infrastructure projects that have done significant harm to landscapes and natural systems. And, although environmental goals were introduced comparatively early and remain advanced by international standards, after over twenty years since their inception in 1999, only one of the sixteen politically instituted goals has actually been achieved.[45]

The role of science and scientists has been conspicuous, especially in the early years of establishing the environment as a policy issue, as we will elaborate in some detail in the chapters that follow. Here in the introduction, we will simply provide a basic overview of Swedish scientific performance in the immediate post-World War II decades, a time when Sweden made massive expenditures in science, technology, and medicine. As a formally neutral country during the Cold War, the Swedish state invested heavily in creating an oversized national defense force to underpin the simple strategy of making it too costly for any would-be aggressor to attack Sweden. This in turn justified the massive investments in research and development (R&D) that made Sweden one of the top countries in the world in scientific performance, in absolute terms (not just relative to size!) on par with Italy and Spain, and not far behind much larger countries such as France. Like the other Scandinavian nations, Sweden had a strong international orientation – in society in general, and academia in particular. Denmark and Norway were members of the North Atlantic Treaty Organization (NATO), and Sweden, even as a nonmember, had a well-established *de facto* collaboration with the Western defense alliance throughout the Cold War. All three countries adopted an Americanized popular culture with strong Anglo-Saxon influences in areas such as music, film, media, and design. This made Scandinavian scholars prone to develop a general Western and US-centered orientation. Although this

Sustainability Transitions (Abingdon, Oxon: Routledge, 2015). Andreas Duit, Peter Feindt & James Meadowcroft, "Greening Leviathan: The Rise of the Environmental State?," *Environmental Politics* 25(2016):1, 1–23.

44 Kristoffer Ekberg & Johan Hultman "A Question of Utter Importance: The Early History of Climate Change and Energy Policy in Sweden, 1974–1983," *Environment and History* (2021), online, 1–23, https://doi.org/10.3197/096734021X16245313030028 (retrieved July 24, 2022).

45 Swedish EPA (2021). The new environmental legislation of 1999 was prepared during Anna Lindh's term as Environment Minister 1994 to 1998. See, Engström, *Miljöframgångar*, ch 4.

was most pronounced in the sciences and medicine, it was distinct in the human sciences as well.[46] Some of this science capacity was turned toward the environment, and further nurtured by significant transatlantic links that had started to form in the interwar years and were rapidly strengthened during the Cold War. Some such expertise was also employed in advising the Swedish government and its public agencies.

Sweden had in the 1960s crafted its distinct international persona as a small, scientifically advanced, independent, and neutral – although decidedly Western-leaning – nation. A pronounced version of this image of Sweden as a responsible and "progressive small state" was articulated by Olof Palme when he became Prime Minister in 1969, and throughout his first period in power lasting until autumn 1976. This image in many ways endured under the center-right governments that followed and continued after Palme and the Social Democrats returned to power in 1982. Even after Palme's assassination in February 1986, the image of Sweden's exceptional place in the world to some extent lingered on. It was on this late-1960s foundation, including a major national breakthrough of environmental issues in autumn 1967 (see Chapter 4), that this small country with considerable self-confidence could have the audacity to propose a major intervention on behalf of "the human environment." It was an era of major steps forward, of giant leaps for mankind, of moonrises, great responsibilities, and great expectations.[47] Proposing a conference of this magnitude and scope was an act of ingenuity, boldness, creativity, and

[46] Sverker Sörlin, "Introduction: The International Contexts of Swedish Science: A Network Approach to the Internationalization of Science," *Science Studies*, 5(1992): 1, 5–12. Thomas Schott, "Scientific Research in Sweden: Orientation Toward the American Centre and Embeddedness in Nordic and European Environments," *Science Studies* 5(1992): 1, 13–25. On Norwegian social science research after World War II, see Fredrik W. Thue, *Empirisme og demokrati: Norsk samfunnsforskning som etterkrigsprosjekt* (Oslo: Universitetsforlaget, 1997); Fredrik W. Thue, "Empiricism, Pragmatism, Behaviorism: Arne Næss and the Growth of American-styled Social Research in Norway after World War II," In: Juha Manninen & Friedrich Stadler, eds., *The Vienna Circle in the Nordic Countries: Networks and Transformations of Logical Empiricism* (Dordrecht: Springer, 2010), 219–229.

[47] Robert Poole, *Earthrise: How Man First Saw the Earth* (New Haven, CT: Yale University Press, 2008). Sheila Jasanoff, "Image and Imagination: The Formation of Global Environmental Consciousness," In: Clark A. Miller & Paul N. Edwards, eds., *Changing the Atmosphere: Expert Knowledge and Environmental Governance* (Cambridge, MA: MIT Press, 2001), 309–337. Denis Cosgrove, *Apollo's Eye: A Cartographic Genealogy of the Earth in the Western Imagination* (Baltimore, MD: Johns Hopkins University Press, 2001); Denis Cosgrove, "Contested Global Visions: One-World, Whole-Earth, and the Apollo Space Photographs," *Annals of the Association of American Geographers* 84(1994):2, 270–294.

also hope and confidence in politics and diplomacy – a *human* effort – as a means to tackle complex, international challenges.

The Declaration on the Human Environment, along with the 109-point Action Plan, was adopted on December 15, 1972, by the United Nations General Assembly (UNGA) as part of a series of resolutions on matters related to UNCHE. The adoption of the Stockholm Declaration provided a bookend for the process that was officially launched four years earlier on December 3, 1968, with UNGA resolution 2398. Largely drafted by Swedish diplomat Lars-Göran Engfeldt, UNGA 2398 endorsed "the Swedish initiative" – in the works for a year at that point – to convene the first ever global environmental conference under the auspices of the United Nations. The speech by Sweden's UN ambassador Sverker Åström before the December 1968 vote on the resolution in the General Assembly moreover marked the first instance that concern over climate change was raised in an official UN context.[48] In 1988, twenty years after UNGA 2398, the United Nations Environment Programme – a direct result and lasting legacy of the Stockholm Conference – together with the World Meteorological Organization, established the Intergovernmental Panel on Climate Change, with Bert Bolin as its founding chairman. Professor of meteorology at Stockholm University and a prolific science organizer, Bolin was one of the key Swedish experts who supported the Swedish initiative with scientific insight and played a significant role in bridging the spheres of science and policy during the Conferences preparatory process. As will be demonstrated across most of the chapters in this book, Bolin – perhaps the premier scientific institution builder of his generation – would continue to serve as a catalyst for climate science and governance into the twenty-first century.[49]

FIVE PERIODS OF "ENVIRONMENT"

This book is structured around analyses of what we identify as five especially significant periods in the evolution of GEG. The *first period* is associated with Swedish-born meteorologist Carl-Gustaf Rossby's return to Sweden in the late-1940s after spending the better part of

[48] Engfeldt, *From Stockholm to Johannesburg*. Sverker Åström, *Ögonblick: Från ett halvsekel i UD-tjänst* [Moments: From Half a Century of Service in the Foreign Office] (Stockholm: Bonnier Alba, 1992).

[49] Henning Rodhe, "Bert Bolin and His Scientific Career," *Tellus B: Chemical and Physical Meteorology* 43(1991):4, 3–7. Henning Rodhe, "Bert Bolin (1925–2007) – A World Leading Climate Scientist and Science Organizer," *Tellus B* 65(2013):1: 20583.

several decades at elite academic institutions in the United States (Chapter 3). Once back in Stockholm, Rossby built up an institute – later under the leadership of Bert Bolin – for cutting-edge atmospheric and climate-oriented research that made Stockholm University a major node in the evolution and organization of climate science internationally. The *second period* is centered around 1972 – including several years before and after the Stockholm Conference – when the concept of "the environment," after a quarter century of growing significance, entered a phase of rapid expansion that included a broad search for policy applications on all scales, local as well as global (Chapters 4 and 5). The *third period* encompasses the late-1970s and the 1980s, when global environmental policy matured alongside the rise of "sustainable development," leading to an increased level of institutionalization at the international level. This period also marked the emergence of Earth System science as a distinct paradigm and its multiple implications for GEG, including the interconnections between global geophysical processes – climate change among them – and the already well-established catalogue of acknowledged environmental problems such as air and water pollution, resource depletion, poverty, and demography (Chapter 6). The *fourth period* starts around the turn of the millennium with a process of rapid conceptual pluralization that produced, popularized, and institutionalized ideas such as "the Anthropocene," "resilience," "planetary boundaries," "the Great Acceleration," and the concept of "environmental governance" itself (Chapters 7 and 8).

While this fourth period of GEG is in many respects still ongoing, we can discern the emergence during the 2010s of a new, *fifth period* marked by greater integration across an even broader range of themes and issues than "the environment" could offer in the postwar decades. A key concept in this period is "transformation," or "transition," indicating the necessity to conceive of sweeping systemic change to redress gross inconsistencies in the currently unsustainable human–Earth relationship. Perhaps it also signals a more proactivist approach, recognizing the weak de facto response to global challenges despite generations of rising awareness and pledges for widespread change. But it also encompasses new, or perhaps not so new but reinvigorated, sources of friction and inertia as the world has entered yet another phase of geopolitical tension, energy shortages, and a weakening of democracies. This includes the rise of a nationalist and populist right, which is often anti-environmentalist, climate skeptic, and opposed to the idea of transition, that is increasingly influencing politics in many countries. Sweden, which has for almost a century been seen as a beacon of internationalism, environmental stewardship, and progressive political culture (Chapter 2), has not been left unaffected by these populist trends (Chapter 9).

We argue that Stockholm has had a significant presence in all five of these periods of evolutionary work on and for the global environment, although with significant changes between the periods. Each phase of incremental progress related to GEG encompasses significant institutional formation and conceptual development that directly and indirectly connect Stockholm to processes at larger spatial and political scales. Focusing on these five periods and singling out Stockholm implies, perhaps needless to say, that some aspects of GEG will be examined more in depth, while others will be addressed more in passing. We firmly believe, however, that the selected periods and the activities they encompass are not only representative of the historical drafting and development of GEG but also represent load-bearing elements in its very construction. Also explored are key continuities between the five periods – including individual contributions as well as institutional and conceptual development – that underpin a narrative arc of what already encompasses nine decades, from the 1940s to the 2020s, spanning the entirety of the Great Acceleration so far.

FOUR CONCEPTS OF CONNECTION

Through archival research, oral histories, and a review of published literature, we uncover and examine historically the roles played by some of the key GEG "architects" – diplomats, diverse experts, scientists and science organizers, politicians and government officials including representatives of research funding agencies, activists, communicators, and an array of other actors. In explaining how the architects facilitated the forward motion of GEG over the course of multiple decades, we identify *four facilitating traits* that these actors employed in their efforts to shape the institutional and cognitive structures that social and political initiatives on environmental problems have been built upon. These traits we label as contributing, conceptualizing, connecting, and convening.

Each trait is derived from the Latin prefix *con*, meaning "with" or "together," reflecting the *integrative aspect* of the environment concept. "The environment," around the time its modern conception emerged shortly after World War II, was not in fact so much a new "thing" – a place, a space, or a new theory – as it was a new way of perceiving *relations* between a large number of phenomena in both nature and society.

Humans became *together with* nature, entangled with it rather than separated from it.[50] Among the multiple ways that societies and governments chose to approach this new predicament, the Swedish model, we argue, emphasized precisely this set of *con*-approaches. We therefore apply these four traits as a set of analytical tools to make sense of the empirical story elaborated in the pages ahead.

It would be difficult for a government, society, or city to play any major role in environmental governance without access to a wide and deep pool of intellectual and institutional resources for science and scholarship on topics of environmental significance. Drawing upon the concept of "contributory expertise,"[51] *contributing* in our analysis entails the scientific knowledge that individuals and institutions have contributed in illuminating emerging environmental problems such as acid rain, climate change, ozone depletion, and many other issues. The contribution of financial and in-kind resources, such as providing facilities for international organizations based in Stockholm, is another salient factor in the building of key institutions that have underpinned the evolution of GEG.

Related to the scientific contributions, but still quite distinct, are the innovative ways in which knowledge and understanding of global environmental change have been framed or popularized. *Conceptualizing*, that is to say, crafting new ideas, terminologies, or cognitive lenses to apply in practice by experts, policymakers, and the wider public, has been a long-standing feature of the Stockholm story. Examples of emblematic GEG concepts with strong Stockholm associations include the Earth System, the Anthropocene, the Planetary Boundaries framework (including the "safe operating space for humanity" formulation), and resilience, among others. We posit that compelling, portable concepts are essential to lubricate social interaction and thus serve as heuristic tools that facilitate increased communication and consensus around governance.

Many of the GEG architects featured in this book have been embedded within and adept at establishing international networks and durable institutions that fostered and fortified structures of global environmental governance. This GEG groundwork largely involved a determined effort of *connecting* within and across national, sectoral, and disciplinary boundaries. The environment was in and of itself a boundless

[50] Warde, Robin & Sörlin, *The Environment*.

[51] Harry Collins & Robert Evans, *Rethinking Expertise* (Chicago, IL & London: The University of Chicago Press, 2008).

concept. It could be extremely local, like the life-sustaining microcosm of a miniscule organism, but it could also encompass the entire Earth. On a cognitive level, although the environment tended in the early years to only engage a small subset of natural sciences, it has grown into a societal imperative that transcends disciplinary boundaries, making transdisciplinarity an essential prerequisite for scholarly understanding and political engagement, the capacity of which varies between nations and institutions.

Operating through professional networks, as well as utilizing established institutions such as the International Council for Science, the United Nations Environment Programme, the World Climate Programme, and, in Sweden, the Royal Swedish Academy of Sciences, an array of entrepreneurial GEG architects were able to convene a wide range of gatherings – from small but important expert workshops to major international conferences. These fostered and solidified scientific and political agendas on environmental and sustainability issues. *Convening*, and convening power, is therefore an essential aspect in explaining Stockholm's successful facilitation of a remarkable number of GEG ideas and initiatives. Bringing expertise to bear at various meetings – which served as sites of negotiation and circulation, not least of new concepts – could compensate for a lack of critical mass in a small country with a limited amount of hard power. Convening also served as a way of engaging and alerting the array of new, often pronouncedly internationally oriented, institutions that took root in Stockholm, from the International Meteorological Institute at Stockholm University (1947, expanding in 1955) to the Beijer Institute (1977), Stockholm Environment Institute (1989), and the Stockholm Resilience Centre (2007), as well as secretariats of international organizations such as the International Geosphere Biosphere Programme (1987) and one of the five global hubs of the current Future Earth research program (which replaced the International Geosphere-Biosphere Programme [IGBP] and its sister organizations in 2015). Moreover, Stockholm-based scholarly journals such as *Ambio* and *Tellus*, with more or less articulated GEG agendas, have provided platforms for epistemic and policy communities to take shape and to communicate research to wider audiences. At the interface of development and environment, the Uppsala-based Dag Hammarskjöld Foundation has since the 1960s brought together diplomats and leading development experts – many from the Global South – for high-level conferences and influential reports and in the pages of its in-house journal *Development Dialogue*.

Our aim is not to provide a comprehensive overview of the dispersed GEG edifice.[52] It is rather to gain a degree of granular insight into some of the pivotal aspects of its drafting and development by delimiting our investigation to the particular location of Stockholm within a wider geography of knowledge and political activism. We identify the Swedish capital as a primary node and a knowledge cluster that has played a formidable role, well beyond its relatively small population, in the historical evolution of a polycentric system and a transnational network of experts, officials, and other actors. Our key proposition is that these have fostered expertise and institutions that helped render the entire Earth a governable object – an outcome of the increasing "entangling" of humanity with the interdependent "planetary" features of the human environment.[53] This was preceded by a fairly long period of conceptualizing and configuring the Earth as an *object of knowledge*, which in turn required scientific work in many fields, reaching further back into the early decades of the twentieth century and accelerating in geophysical sciences and cognate fields during the Cold War.[54] Ultimately, through the creative use of environmental numbers and temporalities and synchronizing work performed in multiple fields of science – made possible by rapid digitalization – the planet itself has been turned into a mediated and governable object with the rise of Earth System science since the 1980s.[55] Through these processes, the

[52] But see Sörlin et al., "The Great Dispersal." Other strands of the literature already provide further information, for example, Peter Dauvergne, ed., *Handbook of Global Environmental Politics* (2005), 2nd ed. (London: Elgar, 2013); John S. Dryzek, Richard B. Norgaard & David Schlosberg, eds., *The Oxford Handbook of Climate Change and Society* (Oxford: Oxford University Press, 2011); Amandine Orsini & Jean-Frederic Morin, eds., *Essential Concepts of Global Environmental Governance* (London: Earthscan, 2020); Helge Jörgens, Christoph Knill, and Yves Steinebach, eds., *Routledge Handbook of Environmental Policy* (Abingdon, Oxon: Routledge, 2023).

[53] William E. Connolly, *Facing the Planetary: Entangled Humanism and the Politics of Swarming* (Raleigh, NC: Duke University Press, 2017).

[54] Leah V. Aronowsky, *The Planet as Self-regulating System: Configuring the Biosphere as an Object of Knowledge, 1940–1990*. PhD diss. (Harvard University, Graduate School of Arts & Sciences, 2018). Benjamin W. Goossen, "A Benchmark for the Environment: Big Science and 'Artificial' Geophysics in the Global 1950s," *Journal of Global History* 15(2020): 1, 149–168.

[55] Sverker Sörlin & Erik Isberg, "Synchronizing Earthly Timescales: Ice, Pollen, and the Making of Proto-Anthropocene Knowledge in the North Atlantic Region," *Annals of the American Association of Geographers*, 111(2021):3, 717–728. Erik Isberg & Eric Paglia, "On Record: Political Temperature and the Temporalities of Climate Change," In: Anders Ekström & Hampus Östh Gustafsson, eds., *The Humanities and the Modern Politics of Knowledge: The Impact and Organization of the Humanities in Sweden, 1850–2020* (Amsterdam: Amsterdam University Press, 2022), 259–282, on 260, 266. Johan Gärdebo & Adam Wickberg, *Environing Media* (Abingdon, Oxon: Routledge, 2022).

Earth System emerged from its previous identity as an object seemingly too massive to destroy and too large to dislodge from its natural state of "self-governance," and gradually entered the realm of what John Dryzek once called the "politics of the Earth."[56]

Although Stockholm as a center of GEG activities and initiatives is our primary object of research, the scope of our investigation extends well beyond that geographic location. Moreover, our analysis is not limited to Swedes operating there, as the cast of Stockholm-associated GEG architects has by no means been composed solely of Swedish nationals. Likewise, some Swedes have made major contributions while not physically based in their home country. These two types of "expatriates" represent a significant part of the Stockholm story. The enrollment of cosmopolitan expertise to Stockholm and the knowledge and networks cultivated by Swedes working outside Sweden are indeed crucial factors in explaining Stockholm's impact over the course of multiple decades. With its focus on the individual GEG architects and their transnational networks, this book is largely a study of an *inter-scalar "human environment."* This environment is *simultaneously local and global*: an extended *social sphere* centered on Stockholm that has been deeply engaged in developing and transmitting important elements of the *knowledge*, *ideas*, and *institutions* that represent *structural components of GEG*. In this book, we will encounter all of these.

OUR GUIDING LIGHTS – AND BLACK HOLES

The literature on the history of global environmental governance is still rather limited, and seen from the viewpoint of a particular city, it is almost nonexistent. There is nonetheless a great deal of research to draw upon. The intellectual terrain we have operated within while writing this book has had its own set of guiding lights, but also vast areas of darkness, and some strange black holes with a great deal of gravitational force that has allowed little enlightenment to escape so far. Book-length

[56] John Dryzek, *The Politics of the Earth: Environmental Discourses* (Oxford: Oxford University Press, 1997). There is also an argument to be made for an evolution from "world" to "global" and onward to "planet"; the latter concept being closer to the language of Earth System science. Hence, we can also see a movement from "globality" – or, even, "globalities" in the plural, suggesting a historical sequence of global understandings – to "planetarity." Rens van Munster & Casper Sylvest, eds., *The Politics of Globality Since 1945: Assembling the Planet* (London & New York: Routledge, 2016), see especially the editors' "Introduction," on 3–5.

studies on the topic of GEG, widely construed, have included overviews of the environment as a concept, and of environmental diplomacy in the Anthropocene after 1945.[57] Collections addressing GEG issues, to some extent historically, have started to surface, albeit with long gaps in between publications.[58] The somewhat nebulous idea of governance (including its probably unavoidable ring of neoliberal reform from the 1980s until around the financial crisis of 2008–2009) has in recent years been brought together with the literatures on the Anthropocene and on Earth System science, a perspective introduced only in the previous decade.[59] However, none of this cross-disciplinary work has engaged with the geography of knowledge and institutions, and, moreover, there does not yet exist much systematic research on the origins and historical aspects of GEG.

On the conceptual history of "the environment," some of which examine its relation to the idea of "the Anthropocene," several new studies have appeared in recent years.[60] There is also a sizeable literature on environmental movements and organizations; many titles are historical, and most are regional or national. Few such works are global in coverage, although early examples of international histories include the above-mentioned books by John McCormick (1989) and Ramachandra Guha (2000). A fine example of how a national perspective, even from a small country, can be brought to bear on global developments is Peder Anker's recent book on Norway's environmental engagement (2020).[61]

57 John R. McNeill & Peter Engelke, *The Great Acceleration: An Environmental History of the Anthropocene since 1945* (Cambridge, MA: The Belknap Press of the Harvard University Press, 2016). Carolyn Merchant, *The Anthropocene and the Humanities: From Climate Change to a New Age of Sustainability* (New Haven, CT & London: Yale University Press, 2020).

58 James Gustav Speth & Peter Haas, eds., *Global Environmental Governance: Foundations of Contemporary Environmental Studies* (Washington, DC: Island Press, 2006). Wolfgang Kaiser & Jan-Henrik Meyer, eds., *International Organizations and Environmental Protection: Conservation and Globalization in the Twentieth Century* (New York & Oxford: Berghahn Books, 2017).

59 Frank Biermann, "Global Environmental Governance: Conceptualization and Examples," *Global Governance Working Paper* No 12. Amsterdam, Berlin, Oldenburg, Potsdam: The Global Governance Project (2004). www.glogov.org. Frank Biermann, *Earth System Governance: World Politics in the Anthropocene* (Cambridge, MA: MIT Press, 2014). Frank Biermann & Eva Lövbrand, eds., *Anthropocene Encounters: New Directions in Green Political Thinking* (Cambridge: Cambridge University Press, 2019). Biermann & Kim, *Architectures of Earth System Governance.*

60 Benson, *Surroundings*. Warde, Robin & Sörlin, *The Environment.*

61 Peder Anker, *The Power of the Periphery: How Norway Became an Environmental Pioneer for the World* (Cambridge: Cambridge University Press, 2020).

Methodological nationalism is the norm, which often means looking inward while using comparisons and influences from the outside to broaden the analysis. Using nations, let alone cities, to look outward and "globally" mirror the environment is uncommon, although Julia Adeney Thomas has done so using Japan as her prism.[62] There is a great deal of literature, some of it historical, on the environment, sustainability, and the developing world, including justice and legal aspects.[63] But again, these books tell other stories than we do, and links to Stockholm 1972 are typically sparse and tend to repeat what is already known.

A rapidly growing field is the humanities and social sciences working on the Anthropocene, with literally dozens of book-length studies in several languages being published in less than a decade. A few of the Anthropocene volumes clearly go deeper into some elements that we also examine, for example, the history of certain aspects of Earth System science. But typically, these do not make very much of global governance issues, let alone historicize them. The same goes for several works published on the history of the environmental, climate, and geophysical sciences.[64] We share with them their deep curiosity about sprawling new areas of knowledge and the interface between science and policy. However, our interest in science leans more toward understanding the rise of governance rather than the development of scientific knowledge itself. Besides, this body of work pays little attention to GEG and does not attempt to construct a narrative arc of its historical evolution. Another field with tangential relevance is that of the modern

[62] Julia Adeney Thomas, "Using Japan to Think Globally: The Natural Subject of History and Its Hopes," In: Ian Jared Miller, Julia Adeney Thomas & Brett L. Walker, eds., *Japan at Nature's Edge: The Environmental Context of a Global Power* (Honolulu: University of Hawai'i Press, 2013), 293–310.

[63] For example, Borowy, *Defining Sustainable Development*. Eve Croeser, *Eco-Socialism and Climate Justice: A Neo-Gramscian Analysis* (Abingdon, Oxon: Routledge, 2021).

[64] Spencer Weart, *The Discovery of Global Warming* (2003), new ed. (Cambridge: Harvard University Press, 2008). James Fleming, *Fixing the Sky: The Checkered History of Weather and Climate Control* (New York: Columbia UP, 2010). James Fleming, *Inventing Atmospheric Science: Bjerknes, Rossby, Wexler, and the Foundations of Modern Meteorology* (Cambridge, MA: The MIT Press, 2016). Jacob D. Hamblin, *Arming Mother Nature: The Birth of Catastrophic Environmentalism* (New York: Oxford University Press, 2013). Joshua P. Howe, *Behind the Curve: Science and the Politics of Global Warming* (Seattle, WA: University of Washington Press, 2014). Mike Hulme, *Why We Disagree about Climate Change: Understanding Controversy, Inaction and Opportunity* (Cambridge: Cambridge University Press, 2009). Mike Hulme, *Climate Change* (Abingdon: Routledge, 2021). Naomi Oreskes, *Science on a Mission: How Military Funding Shaped What We Do and Don't Know about the Ocean* (Chicago: University of Chicago Press, 2021).

history of environmental expertise and advice, including a recent volume that tries to understand the relative tardiness of climate science and other policy-relevant science fields to influence decision-making. In the same category falls Susan Owens' important book on the Royal Commission on Pollution in the United Kingdom.[65] In conclusion, we can say that there is now a fairly sizeable literature on science and cities that has appeared in the last thirty years. Less has been written on the geography of policy influence and very little on the combination of these two fields of study: the geography of environmental science and policy in modern history.

[65] Michael Oppenheimer, Naomi Oreskes, Dale Jamieson, Keynyn Brysse, Jessica O'Reilly, Matthew Shindell & Milena Wazeck, *Discerning Experts: The Practices of Scientific Assessment for Environmental Policy* (Chicago: The University of Chicago Press, 2019). Susan Owens, *Knowledge, Policy, and Expertise: The UK Royal Commission on Environmental Pollution 1970–2011* (Oxford: Oxford University Press, 2015).

2

Sweden and Nature

The Model Country Paradox

In 1909, Swedish author Selma Lagerlöf received the Nobel Prize in literature (Figure 2.1). She had written widely in many genres. Her most recent major work was a two-volume reader for Swedish primary schools, *The Wonderful Adventures of Nils* (1906–1907). The book became an instant success. Not only was it used in schools around the country for the better part of the twentieth century, but it was also read by the general public as a narrative geography of Sweden.[1] Soon translated into dozens of languages, the book served as a saga version of a real country, a strange northern kingdom straddling the Arctic Circle, with deep forests and shining glaciers, close to the North Pole. It mixed historical, cultural, and economic aspects of the old nation, now rising to a new industrial prowess based on rich natural resources, engineering ingenuity, social cohesion provided by strong social movements, and old moral and civic standards sometimes summarized with the single word "Lutheran," but perhaps better captured as "trust."

It was profound change, deep modernization, presented as fable and fantasy. From the back of a goose, high above the long north-south stretch of the Swedish landscape, five-year-old Nils, the son of ordinary farmers in the south, could look down on the wonderland that was his home. There were fields, there were factories, and there was – as Nils gradually discovers as the goose carries him northward towards the newly-established iron-mining town of Kiruna – anything that a boy, or a king or queen for that matter, could dream of. In fact, not a single

[1] Erland Lagerroth, *Landskap och natur i Gösta Berlings saga och Nils Holgersson* (Stockholm: Bonniers, 1958).

FIGURE 2.1 Selma Lagerlöf, Nobel Prize-winning author and early conservationist, at her desk while writing *Nils Holgersson* in the early 1900s. Photo: Swedish Labour Movement's Archives and Library, Stockholm, and the Landskrona Museum.

member of the entire population of Sweden could possibly find anything lacking in such a richly endowed country that was surely a gift from some higher authority. Could any nation be better?

Nonetheless, in the same year as she received her Nobel Prize, Selma Lagerlöf also became a member of the Swedish Society for Nature Conservation (SSNC). Today, an environmental NGO with over 200,000 members, the SSNC started as a small elite organization with academic naturalists, legal specialists, intellectuals, and industrialists among its ranks.[2] It frequently provided expert advice to the government. Several of those among the top strata of the SSNC were members of the Royal Swedish Academy of Sciences, which in the absence of a public agency for the national parks of Sweden – the first ten were inaugurated the very same year, 1909 – served as the semi-official authority on issues to do with nature. This meant that nature was a strong interest of the state.[3] Sweden was the first country in Europe to establish national parks for the protection of nature. Public opinion in favor of national parks had emerged in Sweden back in the 1880s with reference to Yellowstone and

[2] Désirée Haraldsson, *Rädda vår natur!: Svenska Naturskyddsföreningens framväxt och tidiga utveckling*, Bibliotheca Historica Lundensis 63 (Lund: Lund University 1987).

[3] Lars J. Lundgren, *Staten och naturen: Naturskyddspolitik i Sverige 1869–1935*, Part 1: 1869–1919, Part 2: 1919–1935 (Brottby: Kassandra, 2009–2010).

other such protected areas in the United States. The immediate inspiration for the final push came from the debate in Germany, although similar parks were not established there until much later.[4]

The involvement of the state should not be seen as a sign that all was well in Swedish nature. In fact, the opposite was the case, which was the reason for Selma Lagerlöf to join the SSNC. Like many others, across the socially and economically grossly unequal Swedish population,[5] she was upset with the state of nature in the country that she had researched so carefully and knew so well from travels across all of its twenty-five provinces. Sweden's forests were cut down at alarming rates with weak schemes to replant them, until new legislation was enacted in 1903 to rescue and secure the long-term use of crucial timber resources.[6] Rivers and lakes were polluted to such an extent that fish populations were impacted, as was the health of people when swimming or doing laundry – usually outside for most of the year in rural areas where most people still lived – in affected water bodies.[7] Even in the cities, evidence mounted that increasing portions of the population lived under unhealthy conditions, especially children. A debate soon raged on the dangers of urban life, largely driven by the political far right, which asserted that the blood and soil of the land

[4] Björn-Ola Linnér & Ulrik Lohm, "Administering Nature Conservation in Sweden during a Century: From Conwentz and Back," *Jahrbuch für Europäische Verwaltungsgeschichte* 11(1999): 307–334. Bo Sundin, "Environmental Protection and the National Parks," In: Tore Frängsmyr, ed., *Science in Sweden: The Royal Swedish Academy of Sciences 1739–1918* (Canton, MA: Science History Publications, 1989). Samuel P. Hays, "From Conservation to Environment: Environmental Politics in the United States since World War II," In: Kendall E. Bailes, ed., *Environmental History* (Lanham: University Press of America, 1985). On early international agreements, Sherman Strong Hayden, *The International Protection of Wild Life* (New York: Columbia University Press, 1942).

[5] Erik Bengtsson, Anna Missiaia, Mats Olsson & Patrick Svensson, "Wealth Inequality in Sweden, 1750–1900," *European Economic History Review* 71(2018):3, 772–794.

[6] Pär Eliasson & Sven G. Nilsson, "'You Should Hate Young Oaks and Young Noblemen': The Environmental History of Oaks in Eighteenth- and Nineteenth-Century Sweden," *Environmental History* 7(2002):4, 659–677. Pär Eliasson & Erik Törnlund, "Swedish State Forestry, 1790–2000," In: K. Jan Oosthoek & Richard Hölzl, eds., *Managing Northern Europe's Forests: Histories from the Age of Improvement to the Age of Ecology* (New York: Berghahn Books, 2018), 248–287. Svante Lindqvist, "Naturresurser och teknik: Energiteknisk debatt i Sverige under 1700-talet," In: Tore Frängsmyr, ed., *Paradiset och vildmarken: Studier kring synen på naturen och naturresurserna* (Stockholm: Liber, 1984).

[7] Lars J. Lundgren, *Vattenförorening: Debatten i Sverige 1890–1921* (Lund: Gleerups, 1974). Idem, *Miljöpolitik på längden och tvären: Några synpunkter på svensk miljövård under 1900-talet* (Solna, 1989).

were the true roots of a sound and healthy nation. It was in the smoky cafés of the city that dangerous radical ideas emanated, threatening to undermine the stamina and resistance of the innately loyal and patriotic rural Swedes.[8]

NATURE – A NATIONAL ISSUE OF THE HIGHEST ORDER

Names like that of Selma Lagerlöf, along with August Strindberg, the foremost in Swedish literature, drew attention to the issue of nature conservation, which became one of the most pressing issues in Swedish political debate in the decades around 1900. This was also the period of Sweden's rapid industrialization. There was of course a connection between these two major developments in Swedish society. Neither agriculture nor fisheries lacked environmental effects, but it was the industrial transformation and the expanding technologies of energy, including heating cities with coal, gas, and oil, that affected nature in an array of tangible ways.[9] In addition, this period also saw the privatization of state forests and the liberalization of forest and resource legislation that, alongside the mechanization of agriculture, gave industry advantages and favored large-scale farming over small-scale farming.

Propelled by a series of bad harvests in the 1860s, the structural changes in the economy caused mass unemployment and the emigration of more than a quarter of the population of around five million up until the outbreak of World War I in 1914. Although wealth also grew considerably in certain periods, there was a widespread perception that Sweden was a country in deep crisis. The visual image of the nation was not just Nils' bird's-eye view of its beauty and natural resources. If you opened the daily newspapers, especially the satirical press, and if you came across pamphlets and novels on the state of Swedish forestry, you would find cartoons of a nature stripped to the bones and starving people in communities left in disarray, easy prey for the infamous peddlers of single tickets to the United States.

[8] Herman Lundborg, *Rasbiologi och rashygien: Nutida kultur- och rasfrågor i etisk belysning* (Stockholm: Norstedts, 1922). On biological racism in general and Lundborg's version in particular, see Maja Hagerman, *Käraste Herman: Rasbiologen Herman Lundborgs gåta* [Dear Herman: The Enigma of the Race Biologist Herman Lundborg] (Stockholm: Norstedts, 2015).

[9] Torsten Hägerstrand & Ulrik Lohm, "Sweden," In: B. L. Turner II, W. Clark, R. W. Kates, J. F. Richards, J. T. Mathews & W. R. Meyer, eds., *The Earth as Transformed by Human Action: Global and Regional Changes in the Biosphere over the Past 300 Years* (Cambridge: Cambridge University Press, 1980), 605–622.

Nature was becoming a national issue of the highest order. It was also an issue linked to many other social and economic problems of the day. Self-subsistence was still common in Sweden as part of a household's income. Forests and wetlands were regarded as common lands. In the sparsely populated country – with less than a tenth of the population density of Denmark or the Netherlands and a fifth of that of Germany and the United Kingdom – private property was in general not a hindrance to pick berries, fish in lakes, and keep cattle grazing in the forest. Nature had a commonality to it that gave the issue resonance when it was addressed by elites. In Swedish nationalism, nature already played a part and in the period of pronounced and vitalized nationalism sparked by the industrial transformations, nature became a unifying force, an element that many Swedes had a relationship with, and were now actively encouraged to cherish. They learned the value of nature from schoolbooks, such as Lagerlöf's, but also from their own experience of it, as an ultimate safeguard against starvation, although for many, North America remained a more attractive alternative.

Nature was protected at least in part because it was loved and respected. It was also given a language, both written and visual. Not unlike many other countries in Europe and the Western world, nature was nationalized through a series of intervening and constructing institutions: schools, museums, visual art, music, and literature, also intended to impose Swedishness on minorities such as the Sami, through a special kind of "Nomad schools" system, and the Torne River Valley Finns in the north.[10] Swedish nature turned out to become an iconic centerpiece of an imagined community, with an intensity that was exceptionally strong and profound. Swedes came across as an avid, almost genetically predisposed "nature-loving people," *naturälskande folk*, as Sweden's second Nobel Prize-winning author Verner von Heidenstam called it.[11]

[10] Benedict Anderson, *Imagined Communities: Reflections on the Origin and Spread of Nationalism* (London: Verso, 1983). Ernest Gellner, *Nations and Nationalism* (London: Blackwell, 1983). Otso Kortekangas, "Controlling the 'Wilderness' through a Rationalised Reindeer Husbandry: The Establishment of the Sámi Nomad School in Sweden, 1906–1917," *Nordic Journal of Educational History* 10(2023):1, 27–39.

[11] Verner von Heidenstam, "furor och djurgårdsekar susa över svensk lyrik," In: Kate Bang & Fredrik Böök, eds., *Samlade skrifter*, vol. 23 *Uppsatser och tal* (Stockholm: Bonniers, 1943), on 208. Also: "If anything should be defined as particularly Swedish, it is the sense of unity with nature..." (*ibid.*). Quoted in Sverker Sörlin, "Natur och kultur: Om skogen och fosterlandet i det industriella genombrottets Sverige," *Lychnos* (1981–82), 87–144, on 134. Also quoted in David Thurfjell, *Granskogsfolk: Hur*

A massive body of literature in multiple genres was published in the decades around 1900 that articulated and spread the new nature-based nationalism. One, odd but telling, example was statistician Gustav Sundbärg's remarkably nonstatistical and much-debated book, "The Swedish People's Character" (*Det svenska folklynnet*), published as part of the massive parliamentary inquiry into emigration that had become a de facto national trauma. Like so much else of the knowledge production surrounding the awkward, emigrating, and despairing Swedes, it focused on the disproportionate passion for nature in this country, which, Sundbärg argued – coining what would become a long-lived stereotype – was also a nation of "engineers," thinking rationally, if not terribly deeply. Swedes were loners, lacking psychological interest, but they could get things organized, and nature was their central interest, he maintained, and few disagreed.[12]

This line of thought brought out sympathies and passions among Swedes, and the sheer mass of this pseudoscientific propaganda helped build a deepening sense of true Swedish identity for the new Swedish nation that emerged with industrialism. Some of it was, at least in part, designed to instill in Swedes such love for their beautiful country with poetry of "pine and oak" that they might hold back their impulse to emigrate. Nature was also a source of the raw materials that underpinned Sweden's new industrial wealth, from minerals and timber to turf and hydropower. Although there were deep controversies over the best way to utilize these resources – to prioritize industry or build self-sustaining agrarian societies, the classical conflict of *Industriestaat gegen Agrarstaat* – it could not be denied that the wealth of nature provided the option to choose. Whatever path the nation chose to embark upon – turning nature's bounty into a commodity or preserving nature's beauty as an ecological asset – nature was the core of the national project. With nature, Swedes would become rich, materially or spiritually, or even both.

naturen blev svenskarnas religion [Fir Forest People: How Nature Became the Religion of Swedes] (Stockholm: Norstedts, 2020). For a still eminently readable general background, see Staffan Björck, *Heidenstam och sekelskiftets Sverige: Studier i hans nationella och sociala författarskap* (Stockholm: Natur och Kultur, 1946). More on the literary heritage of Swedish forests and forestry in Johannes Ekman, *Skogen i vårt inre: Utmark och frihetsdröm* [The Forest Inside Us: Outfield and Dream of Freedom] (Stockholm: Carlsson, 2008).

12 Gustav Sundbärg, *Det svenska folklynnet: Aforismer* [The Swedish National Character: Aphorisms] (Stockholm: Norstedts, 1911).

NATURE AS NATIONAL(IST) IDEOLOGY

The fervor with which this natural nationalism was practiced in the decades around 1900 never fully returned. The war poured cold water over some of the sentiments. Emigration stopped almost completely during the war years and never returned to pre-war levels despite the considerable social tensions and economic hard times under which the Swedish population suffered toward the end of the war. Instead, democracy and full voting rights for women and men, decided in 1919 and realized in the 1921 elections, signaled a turn toward social and economic affairs. Nonetheless, nature as a theme stayed on, although it turned more social. During the war, travel was difficult and in the interwar years, "Social nature protection" became a catchphrase, especially from the political left, where there had always been some skepticism toward the most solemnly patriotic versions of the turn of century nature worship. Instead, periurban nature was presented as the vernacular alternative. Housewives of all classes were mobilized to take children outdoors.[13] Schools were required to apply "outdoor days" to make sure that healthy traditions were passed on to the coming generations. The SSNC advocated arranging accessible walking paths and recreation facilities near towns and cities for the large majority of families who couldn't afford adventurous trips to the high mountains or accommodation in resort hotels. Social and urban planners soon followed suit. Swedes were expected to go to the forest to pick berries and mushrooms, and they did. Vernacular subsistence habits such as fishing and hunting were turned into widespread pastimes, and they established their own, soon sizable organizations with memberships that by mid-century reached more than a million, out of a total population of little more than six million.

Friluftsliv became the new term for the out of doors. By the early 1930s, with the rise of the soon famous Swedish welfare state, which drew such interest abroad, nature practices were becoming integrated into welfare thinking and politics. The name for that was "socialt friluftsliv," or "social out of doors." It was advocated particularly by the political left. Physicians, educators, and social reformers argued that social cohesion and public health, as well as general education (in Swedish *bildning*, after the German word *Bildung*), would improve

[13] C. Tornbjer, *Den nationella modern: Moderskap i konstruktioner av svensk nationell gemenskap under 1900-talets första hälft* [The National Mother], PhD diss. (Lund University: Dept of History, 2002).

if people experienced nature firsthand and not just through books. Social movements and their rising civic, but partly publicly supported, study organizations, under the name *folkbildning* (general education for the people), were among the pioneers. The practice of friluftsliv, however, went beyond organizations and became common also on the household level. It was featured in the Stockholm exhibition in 1930, a showcase of Swedish modernity, and in 1936, friluftsliv received its own special exhibition in Ystad, a city on Sweden's southern coast. It demonstrated how nature practices had already become a civic, commercial, and political issue of some status.[14]

In this way, the first half of the twentieth century was a long period of "education in nature" that reached the better part of the Swedish nation. Already close to nature thanks to a vast territory with a thinly spread population and small cities – the largest, Stockholm, had a mere 300,000 inhabitants in 1900; London had five million, while Paris and Berlin had three and two million, respectively – Swedes were accustomed to nature and culturally prepared to appreciate it. It was a precondition for both health and wealth, and for welfare, in a period that also saw a soon-to-be world-famous welfare state take form, for which the nation took great pride. The two went together in progressive legislation enacted in the early 1940s on rights of access to lakes and rivers, which followed an already widespread "everyman's right" to waterways and almost all land, private and public.[15] A comprehensive infrastructure of paths and trails emerged as new practices of pastime nature use expanded in the twentieth century, with Sweden as a pronounced version of an international trend.[16] The growing cities also enjoyed an

[14] Lena Eskilsson, "Fritid som idé, struktur och praktik: Rätten till lättja eller friluftsliv i folkhemmet," *Historisk tidskrift* 120(2000), 29–53. Klas Sandell & Sverker Sörlin, eds., *Friluftshistoria – Från "härdande friluftslif" till ekoturism och miljöpedagogik* (2000), new ed. (Stockholm: Carlsson, 2008). T. H. Beery, "Nordic in Nature: Friluftsliv and Environmental Connectedness," *Environmental Education Research* 19(2013):1, 94–117. Ebba Lisberg Jensen & Pernilla Ouis, *Det gröna finrummet: Etnicitet, friluftsliv och naturumgängets urbanisering* [The Green Living Room: Ethnicity, *friluftsliv* and the Urbanization of Out of Doors] (Stockholm: Carlsson, 2014).

[15] Klas Sandell & Peter Fredman, "The Right of Public Access: Opportunity or Obstacle for Nature Tourism in Sweden?," *Scandinavian Journal of Hospitality and Tourism* 10(2010):3, 291–309. Claes Bernes & Lars J. Lundgren, *Use and Misuse of Nature's Resources: An Environmental History of Sweden*, English transl. Martin Naylor (Stockholm: Swedish Environmental Protection Agency, 2009).

[16] Daniel Svensson, Katarina Saltzman & Sverker Sörlin, eds., *Pathways: Exploring the Routes of a Movement Heritage* (Winwick, Cambridgeshire: The White Horse Press, 2022).

active planning movement with green areas and parks as a fundamental feature.[17] Part of this national nature project was to bridge social divides and foster *Gemeinschaft* and cohesion. Ethnic and racial divides were less emphasized, not because Sweden lacked immigrants but because their numbers were modest and integration was largely successful. Racial debates around who had access to nature, well known from colonial and postcolonial contexts,[18] were not at the forefront until the very end of the century. Nature was a unifying, nation-building project, widely embraced in Swedish society. It enjoyed official support and rested upon a growing number of civil society institutions such as popular movements, out-of-doors associations, and sports clubs.

This modest, vernacular cult of nature was part of a Swedish exceptionalism that only expanded in the coming years. Sweden remained neutral during World War II – a position widely supported then, but critically discussed later[19] – with unusually close and congenial relationships between industry and trade unions and between both these interests and the state.[20] It was in fact this corporatist model of close ties and collaboration between different camps and classes of Swedish society, and between its geographically and socially quite diverse provinces, that admirers from abroad found so unlikely, advanced, and "modern," to quote what is likely the most recurring term in representations of Sweden. The iconic Stockholm Exhibition of 1930 became the epitome of this image. Enormously popular, with more than four million visits in a country of six million inhabitants, the exhibition employed a nationalist narrative, stressing history, to define the "Nordic race" and evoke a glorious, industrial future built on nature's resources.[21] Nature

[17] Mattias Qviström, "Shadows of Planning: On Landscape/Planning History and Inherited Landscape Ambiguities at the Urban Fringe," *Geografiska Annaler: Series B, Human Geography* 92(2010):3, 219–235.

[18] Carolyn Finney, *Black Faces, White Spaces: Reimagining the Relationship of African Americans to the Great Outdoors* (Chapel Hill, NC: The University of North Carolina Press, 2014).

[19] Klas Åmark, *Att bo granne med ondskan: Sveriges förhållande till nazismen, Nazityskland och Förintelsen* [Neighbor to Evil: Sweden's Relation to Nazism, Nazi Germany, and the Holocaust] (Stockholm: Bonnier, 2011); Åmark, "Sweden: Negotiated Neutrality," In: Richard J. B. Bosworth & Joseph A. Maiolo, eds., *The Cambridge History of the Second World War*, vol. 2 Politics and Ideology (Cambridge: Cambridge University Press, 2015), 350–374.

[20] Bo Rothstein, "Social Capital in the Social Democratic Welfare State," *Politics & Society* 29(2001):2, 207–241.

[21] Allan Pred, *Recognizing European Modernities: A Montage of the Present* (London: Routledge, 1995), on 99.

as scenery, beauty, replenishing source, or as ecology and preservation was not at the forefront. Seen from a later vantage point, the interest among foreign commentators was not directed toward the exceptional relations between people, state, and nature that were nurtured as a vital part of this somewhat corporatist version of a cohesive nation and its identity. They focused more on the social and economic dimensions, perhaps because in other countries, nature was not always considered part of the welfare state agenda.

This means that significant early contributions in Sweden to what was later termed "environmentalism" or ecological thought did not become very well known internationally. This was also true of another strand of nature interest that served as a countercurrent to mainstream modernism, but flourished in certain environments, especially in the 1930s and 1940s. While the expanding connections between nature and welfare were developed largely from the reform-oriented parts of the political spectrum – social democrats and liberals – a somewhat more skeptical approach arose from within the much smaller communist left, which had more urban tastes and, embracing class confrontation, disliked the unifying nationalist overtones ascribed to nature.

The stronger and more persistent countercurrents were conservative and informed by hierarchical and socio-biologist figures of thought. These nuances in ideologies around nature started to appear in the Swedish discussion around the time of World War I. A more profound refraction between distinctly opposite positions only became manifest when democracy was fully established in Sweden, and after the Revolution in Russia in 1917, the rise of Communism, and the harsher ideological climate of the 1920s. In this conservative strand, nature was mixed with ideas of race and biology. The first tendencies in this direction had no particular politics. It has been observed that ecology and "Darwinist" camouflage were present in Swedish landscape painting from around 1900, for example, in the work of wildlife painter Bruno Liljefors. At the time, however, no one could have imagined that his work would become much appreciated in the nature-loving Third Reich, which in and of itself could not have been imagined in 1900. Nor could anyone have guessed that Liljefors only a few decades later would win a prestigious prize named after a leader who would orchestrate and symbolize some of the most atrocious evils ever committed, "der Grosse Preis Adolf Hitlers." The Prize was presented to an aging Liljefors, who was himself in attendance, in 1937 at the Internationale Jagdausstellung in Berlin, a major event in the Nazi calendar that year. A pet project of Reichsforstmeister and Reichsjägermeister Hermann Göring, the exhibition had animal painting and photography as one of its major themes and brought together the entire Nazi elite, including Hitler, Himmler, and Richard Walther Darré, the German Minister of Agriculture.[22]

Conservationist care for nature and animals was present in many European countries and in North America as well. Germany is a case in point. Mountain landscapes had a particular prominence. They were the subject of an aesthetic cult, and *Bergfilme* was a popular genre. In Nazi ideology, mountains and their summits had a special aura. The alpinist was a cherished social character, with his athletic prowess embodying both the longing for purification and the physical strength that the Nazis worshipped. Germany dispatched well-equipped expeditions to the Himalayas to search for the original Aryan *Heimat* and to demonstrate their obsession with the highest summits of the world. Nature was a hierarchy and at the top, closest to heaven, was the *Übermennsch*.[23] According to the same logic, aristocratic animals such as eagle, bear, and moose ought to be preserved, and landscapes and waterways serve as the nourishing soil and blood of the nation.[24]

Such sentiments, fundamental for the Nazi regime, were distinctly less common in Swedish conservation circles, although they did exist in less pronounced forms and with ideas of a "Nordic" race and lifeworld as a common denominator.[25] Conservative nature lovers argued that Lapland in the north should be reserved as the ultimate wilderness where local people, including the reindeer herding Sami, should not be allowed to hunt charismatic birds and megafauna – the kind of "policing nature" policy that was practiced overseas among colonial elites. These outliers within the Swedish debate, with prominence on the political right and in certain academic circles, including museums,[26] were predominantly

[22] Karen Wonders, "Wildlife Art and the Nazis: The Case of Sweden's Bruno Liljefors (1860–1939)," *Lychnos* (2006), 70–86.

[23] Christopher Hale, *Himmler's Crusade: The Nazi Expedition to Find the Origins of the Aryan Race* (Hoboken, NJ: John Wiley & Sons, 2003). Franz Josef Brüggemeier, Mark Cioc & Thomas Zeller, eds., *How Green Were the Nazis?: Nature, Environment, and Nation in the Third Reich* (Columbus, OH: Ohio University Press, 2005). Frank Uekoetter, *The Green and the Brown: A History of Conservation in Nazi Germany* (New York: Cambridge University Press, 2006).

[24] Ludwig Heck & Lutz Heck, *Tiere in der Natur und Kunst* (Dresden: Zwinger Verlag, 1942). David Blackbourn, *The Conquest of Nature: Water, Landscape, and the Making of Modern Germany* (New York & London: W. W. Norton & Company, 2006).

[25] Tora Byström, *Nordens Frihet: Samfundet, tidningen, kretsen* (Lund: Sekel Bokförlag, 2009).

[26] Sverker Sörlin, "Ideologi i kamouflage: Politiska tendenser i Stig Wessléns dokumentära konstnärskap," In: Erland Mårald & Christer Nordlund, Skytteanska Samfundet,

male and came across as puritan cousins of the aristocratic "Teddy Bear Patriarchy," which Donna Haraway has identified as central to fauna conservation.[27] The male white upper class, who saw it as their prerogative to visit Africa to shoot elephants and rhinos, did not consider their hunt as a mere pastime but sublimated their pursuits as primarily moral and scientific.[28] The Godfather of this particular element of Western imperialism was President Theodore ("Teddy Bear") Roosevelt, who helped establish national parks and monuments such as Crater Lake and the Grand Canyon.[29] His hunting adventures were documented in *Roosevelt in Africa* (1910), an early prototype of the genre of safari films.[30]

REALIST SUSTAINABILITY

Post-1945 developments gradually marginalized most of what remained of the conservative, patriotic version of nature. Its ideological foundations had become discredited and impossible due to the war and the complete disaster, not only for aggressive nationalism but also for ideologies of race.[31] The tradition still played some role in the popularization of nature and in connecting some of the Swedish self-understanding with strands of science, travel, art, film, and photography. The conservative appreciation of nature, ideologically cautious and diffident, that remained in the postwar decades pragmatically sided with mainstream welfare modernism, although it was less prone to take a vanguard role in legislation and activism. Nature, later "environment," thus became one

eds., *Kamerajägaren: Stig Wessléns skildringar av naturen och det samiska*, Skrifter 65 (Umeå: Skytteanska Samfundet, 2010), 203–244. Eva Silvén, *Friction: Ernst Manker, Nordiska Museet, and Sami Cultural Heritage* (Lund: Nordic Academic Press, 2021), on 225–238 (in Swedish, with a summary in English, on 273–274).

27 Donna Haraway, *Primate Visions: Gender, Race and Nature in the World of Science* (London & New York: Routledge, 1989).

28 R. P. Neumann, "Dukes, Earls and Ersatz Edens: Aristocratic Nature Preservationists in Colonial Africa," *Environment and Planning D: Society and Space* 14(1996):1, 79–98. William M. Adams, *Against Extinction: The Story of Conservation* (London & Sterling, VA: Routledge, 2004); idem, "Nature and the Colonial Mind," In: William M. Adams & Martin Mulligan, eds., *Decolonizing Nature: Strategies for Conservation in a Postcolonial Era* (London & Sterling, VA: Routledge, 2003), 16–78.

29 Stephen Pyne, *How the Canyon became Grand: A Short History* (New York: Penguin, 1998).

30 Gregg Mitman, *Reel Nature: America's Romance with Wildlife on Film* (Seattle, WA & London: University of Washington Press, 2009).

31 Johan Östling, *Nazismens sensmoral: Svenska erfarenheter i andra världskrigets efterdyning* (Stockholm: Atlantis, 2008).

of several policy areas – such as defense, security, science, technology – where consensus reigned. Pragmatic solutions were prioritized, and as a gift in return, the Social Democratic Party, in government, save summer 1936, from 1932 to 1976, although in coalition during the war, made sure nature protection would not seriously oppose the interests of industry and resource extraction.[32] The party was always a broad church and although a more radical, environmental wing grew in the 1960s, it inevitably had to compromise with more traditional versions of modernity, more palatable to the trade unions and the low-income strata of society that remained the party's chief voter base.[33]

A pragmatic version of "eco-modernism" gained hegemony in Sweden, long before such a concept was even coined in the 1980s. A seminal work in that tradition was *The Politics of Environmental Discourse* (1995) by the Dutch political scientist, Maarten Hajer. It was a study of acid rain in Western Europe, but what made it influential was a simple and attractive policy idea: the environment is no reason to hold modernization back, but rather to speed it up. It came at a moment when such solutions were on the rise. Better incentives, distinct goals, certification of methods, and innovation in policy and technology would contribute to the efficient use of natural resources. Above all, economic growth did not stand in opposition to a healthy environment, an idea that had been talked about before, but in Hajer's version was presented more coherently.[34] Although not precisely congruent with Social Democratic orthodoxy, these were ideas that they could just as well employ, and they did.[35] Domestically, the party had been constantly navigating issues of nature and the environment. When environmentalism grew in the postwar decades, the Social Democrats certainly did not lead the way as a political force on the issue, although many individuals within or close to the party were important voices. Momentum was brought to politics from the outside by popular movements, scientists, intellectuals, and artists.

[32] Erland Mårald & Christer Nordlund, "Modern Nature for a Modern Nation: An Intellectual History of Environmental Dissonances in the Swedish Welfare State," *Environment & History* 26(2020):4, 495–520.

[33] Jonas Anshelm, *Socialdemokraterna och miljöfrågan: En studie av framstegstankens paradoxer* [The Social Democrats and the Environment: A Study of the Paradoxes of Progress] (Stehag: Brutus Östlings bokförlag, 1995).

[34] Maarten A. Hajer, *The Politics of Environmental Discourse: Ecological Modernization and the Policy Process* (Oxford: Oxford University Press, 1995).

[35] Brendan Hailey, "Social Democracy and Ecological Modernization: Swedish Lessons for a Green Industrial Policy," *Major Essay in Ecological Economics* (Toronto: York University, 2005).

Only in the 1960s did "the environment" begin to gain a significant position in the party. Olof Palme, who was a member of the government since 1963 and became party leader and prime minister in 1969, included the environment in his talks, articles, and books as the environment increasingly grew in importance. Still, he was always careful to connect environmental issues with more traditional Social Democratic concerns such as democracy, human rights, peace, and welfare. Whether addressing an audience at a May Day manifestation or speaking as head of the Swedish delegation at the 1972 Stockholm Conference, he found the nonenvironmental topics relevant to each audience that made the environment seem crucial to them. In and of itself, the environment was of little significance.[36] Other features of Swedish social and political life also go a long way toward explaining the politics of what we might call realist sustainability. The instrumentalist paradigm that sustainability, launched by the UN report *Our Common Future* (1987), sought to redress, perhaps even to challenge, had had a long time to grow and become established.

A tradition of political arithmetic had already gained ground in the late seventeenth and early eighteenth centuries, well before the tradition of sustained yield that arrived in Sweden from the German states in the nineteenth century.[37] Political arithmetic portrayed nature in this powerful Baltic nation as well-endowed and a divinely privileged repository of people (labor power), plants, and animals. Such ideas were present in an unusually articulated form in the patriotic economic botany of Linnaeus, who was an early voice of "ecological" and circular ideas avant la lettre. One example is his short thesis, *Oeconomia Naturae* (Husbandry of Nature, 1749), although he just as well fits a more extractivist-oriented "cameralist" approach to natural resources that also runs deep in Swedish nature practices.[38] In that respect,

[36] Olof Palme, www.olofpalme.org/1972/05/01/forsta-maj-tal-pa-hasslo/; www.olofpalme.org/1972/06/06/anforande-infor-fns-miljokonferens/. Palme archive, electronic collections. The Labour Movement Archives and Library in Stockholm (ARAB).

[37] Karin Johannisson, *Det mätbara samhället: Statistik och samhällsdröm i 1700-talets Europa* [With a Summary in English: Society in Numbers: Statistics and Utopias in the Eighteenth Century Europe] (Stockholm: Norstedts, 1988). Tore Frängsmyr, John L. Heilbron, & Robyn E. Rider, eds., *The Quantifying Spirit in the Eighteenth Century* (Berkeley, CA: University of California Press, 1990).

[38] On Linnaeus's proto-ecological thought and its economic "mechanism," Tore Frängsmyr, "Den gudomliga ekonomin: Religion och hushållning i 1700-talets Sverige" [The Divine Economy], *Lychnos* (1971–1972), 217–244, and Donald Worster, *Nature's Economy: A History of Ecological Ideas* (1977), 2nd ed. (Cambridge & New York: Cambridge

Linnaeus was a forerunner of both ecological modernization and care for nature as a divine gift. Industrialism in the nineteenth and twentieth centuries reinforced particularly the former tradition, placing pride in the fact that Sweden was rich in natural resources on which to build a new and glorious future.[39]

The oft-proclaimed sensitivity for Nordic nature and its mythological dimensions displayed in the arts was growing alongside, or rather on top of, industrialized nature. It was the rise of the industrial economy, professional elites, and an urban middle class with deepening nationalist aesthetics and values, which formed the social basis of esteem for nature and, thus, for preservationist and conservationist policies in the first decades of the twentieth century. However, these groups were also quite sensitive to the demands of industry, and responsibility for the productive capacity of the nation was sometimes used as an argument to sacrifice precious forests, waterfalls, or marshes. Nature was the national common ground and the essential backbone of wealth, in particular on Sweden's northern periphery, where the aboriginal Sami population was affected by continuous territorial and judicial setbacks as resource extraction rolled north to exploit its forests, iron ore, peat bogs, and waterfalls for energy, and land for agriculture.[40]

The environmental issue has loomed large since the mid-1960s and onward. Compared to other countries, Sweden was early and comprehensive in its attempts to deal legally with issues such as vehicle pollution, work environments, acidification, and many other issues around the environment and public health. A full-time cabinet minister for the environment was appointed in 1986. The most significant characteristic of Sweden in this respect may be the relatively peaceful co-existence of the state with a growing environmental movement, which had considerable public support and attracted the attention of center-left parties and often the government. The green movements also enjoyed an

University Press, 1994), 31–44. On cameralism, Lisbet Koerner, *Linnaeus: Nature and Nation* (Cambridge, MA: Harvard University Press, 1999).

39 Sverker Sörlin, "State and Resources in the North: From Territorial Assertion to the 'Smorgasbord State'," In: E. Carina H. Keskitalo, ed., *The Politics of Arctic Resources: Change and Continuity in the "Old North" of Northern Europe* (Abingdon, Oxon: Routledge, 2019), 38–61.

40 Patrik Lantto and Ulf Mörkenstam, "Sami Rights and Sami Challenges: The Modernization Process and the Swedish Sami Movement, 1886–2006," *Scandinavian Journal of History* 33(2008):1, 26–51. Bo Sundin, "Environmental Protection and the National Parks," In: Tore Frängsmyr, ed., *Science in Sweden: The Royal Swedish Academy of Sciences 1739–1918* (Canton, MA: Science History Publications, 1989).

expert role, dating back to 1909 with the SSNC as well as with the Samfundet för hembygdsvård (SfH), Society for Heimat, founded in 1916. If the interest of the SSNC was "nature," SfH was engaged with "landscape," especially landscapes where structures and infrastructure were to be built, or when older buildings or locations of cultural heritage were threatened by new development. Architects, archaeologists, planners, engineers, and other "spatial" and "temporal" experts were at the forefront, rather than natural scientists. Working closely with the widely spread Heimat movement, their structural role was the same as that of the SSNC: not opposing development, but brokering interests, finding a way forward.[41]

This does not mean that the wave of radical environmentalism in the 1960s and 1970s did not represent something novel, in scope, intensity, and argument. It simply operated within an already rich institutional context with the capacity to co-opt and mainstream new initiatives and bend them to fit under the pragmatic umbrella of state authority. Furthermore, Sweden's Green Party was formed. Although the Scandinavian countries' role as early environmentalists and vanguard ecofeminists has faded somewhat into the background, it is an important feature to keep in mind if we wish to understand the rise of Swedish environmentalism. Before the 1980s, some of these green energies were obviously present in other political parties, most of which had to cede sizable numbers of voters and activists to the new Greens, with some exception for parties on the right, the Christian Democrats and the market liberal Moderate Party.

THE GREEN *FOLKHEM*

As we shall see in the following chapters, Sweden after 1972 kept working at the forefront of the international environmental agenda. At home, the pace forward was, albeit solid, not quite as impressive. The hegemonic Social Democrats, back in 1982 after six years of rare center-right government, continued to stay in power for another quarter century, except for three turbulent years under the Moderate Party leader Carl Bildt from 1991 to 1994.

[41] Gunilla Lindberg & Mats Rolén, eds., *Återblickar: Samfundet för hembygdsvård 75 år.* Special issue of *Bygd och natur* (1991). Maria Björkroth, *Hembygd i samtid och framtid 1890–1930: En museologisk studie av att bevara och förnya* [Past, Present and Future: A Museological Study of Museums and the Preservation of Environment in Sweden c. 1890–1930], PhD diss. (Umeå University: Department of Culture and Media, 2000).

Environmental issues kept rising on the agenda. In 1995, the newly installed Prime Minister Göran Persson launched a political idea that he called "the Green *folkhem.*" It alluded to a 1928 speech in Parliament by then party leader Per Albin Hansson. In his speech, later to become what is likely the most cited speech in Swedish history, Hansson had drawn upon a concept advocated by Swedish conservative thinker Rudolf Kjellén about a natural *Heimat*, a state that was at the same time geopolitically pragmatic, corporatist, fair, and caring.[42] *Folkhem* was the formative political idea that stood behind the kind of society described by Marquis Childs, and that the world admired with the Swedish model. It also became a concept used in Sweden, where it became so hegemonic that when historian Bo Stråth in 1993 wrote about the road then embarked upon by Sweden toward membership in the European Community, he could entitle his book *Folkhemmet mot Europa.* The word "mot" means "against." The close *folkhem* community was distinctly local, national, and anti-cosmopolitan. This was in contrast to the more technocratic term "the Swedish model," which, according to lexical data Stråth cites, was rather used retrospectively from the 1970s about the tradition in Sweden to seek consensus rather than confrontation, especially in the labor market and in avoiding strikes. A "model" could travel and be applied elsewhere, a "home" could not. Stråth also notes that the likely first use of the concept was as late as 1967, by French radical politician and journalist (founding editor of *L'Express*) Jean-Jacques Servan-Schreiber in his international bestseller *Le défi américain*: "le modèle Suèdois n'est ni Américan, ni Japonais."[43] The original understanding that Sweden's was a "middle way" had remained intact since the 1930s.

[42] Rudolf Kjellén, *Nationell samling: Politiska och etiska fragment* (Stockholm: Geber, 1906), *Ett program: Nationella samlingslinjer* (Stockholm: Geber, 1908), *Staten som livsform* (Stockholm: Geber, 2016). Peter Davidsen, *The Emancipation of Political Science: Contextualizing the State Theory of Rudolf Kjellén, 1899–1922*, PhD diss. (Helsinki University: Department of Philosophy, History, and Art studies, 2021). Lars Trägårdh, "Varieties of Volkish Ideologies: Sweden and Germany 1848–1933," In: Bo Stråth, ed., *Language and the Construction of Class Identities: Scandinavia and Germany after 1800* (Gothenburg University, 1990). Idem, "Statist Individualism: On the Culturality of the Nordic Welfare State," In: Øystein Sørensen & Bo Stråth, eds., *The Cultural Construction of Norden* (Oslo: Scandinavian University Press, 1997). Seppo Hentilä, "The Origins of the Folkhem Ideology in Swedish Social Democracy," *Scandinavian Journal of History* 3(1978):1–4, 323–345. Hans Dahlqvist, "Folkhemsbegreppet: Rudolf Kjellén vs Per Albin Hansson," *Historisk tidskrift* 122(2002):3, 445–465.

[43] Bo Stråth, *Folkhemmet mot Europa* [The "People's Home" vs. Europe] (Stockholm: Tiden, 1993), 205–206. Jean-Jacques Servan-Schreiber, *Le défi américain* (Paris: Denoël, 1967).

Göran Persson employed the *folkhem* brand – still useful, especially when the Swedes finally dared tie themselves to the mast of the European ship – for a new project, to package and sell the rising norm of "sustainable development." The point was to deliberately hold on to homespun ideals but transform them appropriately. As usual, at no cost to the economy or people's standard of living. The economy would grow, and Sweden's deep economic crisis should be overcome. It was also a time when the environment was no longer, as during the Cold War decades, discussed as only a problem. Was it a thing of the past? Solutions, both technological and market-driven, would be increasingly forthcoming, some argued, most intensely from liberals and conservatives. American journalist Gregg Easterbrook's influential book *A Moment on the Earth* (1995) was built on such eco-modernist optimism, congruent with the generally sanguine Western mood of the post-1989 moment. Easterbrook adapted Maarten Hajer's academic approach and interesting hypothesis to deliver a downright verdict. Just add innovation and market forces, as Easterbrook speculated, and environmental regulation would be superfluous. Consumption was not a burden for nature, it was a driver of innovation that helped fund "best practices" according to the neoliberal handbook that had been rehearsed since the breakthrough years under Thatcher and Reagan.

The Swedish idea of the Green *folkhem* was surely articulated with a healthy distance to reckless neoliberalism. At the same time, it was not the old regulating policy that was presented. Even Scandinavian Social Democracy was changing. More than twenty years had passed since Stockholm 1972, new winds were sweeping through all policy areas, and the environment was no exception. It was a sign of Göran Persson's political instincts that he tried to find a compelling idea, and name, for his basically pragmatic environmental politics. He calculated that sustainable development was the only game in town, the *juste milieu* between right-wing denialism and green radicalism, the path Social Democrats always favored. But it just didn't fly this time. It was ridiculed in the liberal press and it seemed heavy-handed, a Keynesian crisis antidote with heavy investment on environmental sanitation of old mass housing suburbs, although it was essentially a continuation of slowly evolving Swedish policies. It followed very much Palme's prescription: always make sure that green politics work in synergy with other reforms that will win votes.

Such a pragmatic approach had worked in helping to prevent the destruction of the ozone layer. Sweden played a key role in a longer-term process that led up to the 1987 Montreal Protocol, an international agreement meant to phase out ozone-depleting substances such as chlorofluorocarbons (CFCs). Swedish scientists and government officials had at that point been working on the issue for many years. The Academy of Sciences had through its environmental committee taken a keen interest in ozone depletion in the mid-1970s, including organizing a conference in November 1976. What is more, CFCs also contributed to the greenhouse effect; the Academy's interest in addressing such atmospheric pollutants thus demonstrated an early policy presence of the climate issue. On ozone issues, Sweden would work through UNEP, which in 1981 initiated efforts to create a global convention. Chemical meteorologist Henning Rodhe at Stockholm University, who worked closely with Paul Crutzen, later winner of the Nobel Prize for his research on ozone, gave a series of media interviews in 1986 and wrote an op-ed article in the *Dagens Nyheter* in October 1987. From the sidelines, environmental organizations supported policies to phase out ozone-depleting substances. Industry was, perhaps predictably, obstinate. The CEO of the Swedish industrial company Electrolux predicted that the unilateral ban in Sweden would lead to the demise of industry and lost jobs. Within a few months, he had turned 180 degrees on the policy, proclaiming that industries across Sweden were working to implement the ban. Political leadership was necessary. At first, the political right opposed the ban, while the Social Democrats and the Liberal Party wanted to introduce incentives and leave the issue to the market. The Left and the Centre Party, a de facto quite "green" political force at the time, were most decisively in favor of a radical elimination of CFCs. Finally, the government listened to the science and the growing choir of voices and changed its position. A contributing factor was the Green Party, which was by then an established presence in Swedish politics. The young party enjoyed tailwinds as the 1988 elections drew closer, with an enigmatic mass death event of seals on the Swedish west coast and a postponed phasing out of Swedish nuclear plants. It was also the long, hot summer in the United States, where Jim Hansen of NASA gave testimony before Congressional hearings in which he connected the extreme temperatures to an anthropogenic "Greenhouse effect."[44]

[44] Mats Engström, *Miljöframgångar: Från freonförbud till klimatlag* [Environmental Successes: From Prohibiting Freones to Climate Legislation] (Stockholm: Fri tanke 2020), 13–25, gives a comprehensive story of Sweden and the CFC's. More details in Mathias Martinsson, *The Ozone Layer and Risk Society: A Study of the Swedish Political Discussion Concerning the Ozone Layer 1968–1992*, PhD diss. (Linköping University: Faculty of Arts and Sciences, Technology and Social Change, 2001; in

THE IMPORTANCE OF KNOWLEDGE

In the decades that followed, other major political decisions were taken on taxes, chemicals, forestry, biodiversity, and finally on climate, where a comprehensive climate legislation framework was enacted by Parliament in 2017. These were not the only issues that were addressed, but they all sit well with the overarching pattern of Swedish environmental politics in the decades following the UN 1972 Conference, especially the fundamental role of knowledge and consensus culture in combination. Sweden was among the very first countries to introduce a carbon tax in 1991. By then, economic incentives had been debated in Sweden for some time. Just like in the case of ozone a few years earlier, pragmatic collaboration made the difference. Politicians from a broad spectrum of parties acted in liaison with government experts, civil servants, and research institutes, all raising their voices in favor of a carbon tax. Scientists such as Bert Bolin (climate) and Karl-Göran Mäler (ecological economics) offered their advice and had access to politicians through advisory groups. The SSNC, again, was active in providing proposals and documentation. The time was ripe. In the raging economic crisis in Sweden in the early 1990s, a broad tax reform was needed anyway and other energy taxes could be lowered as the carbon taxes were introduced. This stimulated domestic biofuels and propelled the phasing out of oil and gas from the heating sector.

Knowledge was central to progress in all these cases. This speaks in favor of an argument that has been made about Swedish domestic environmental politics in recent years and previously about both Germany and France. The core of this argument has to do with the circulation of knowledge. The modern history of *knowledge* is, in this reading, distinctly different from its counterpart history of *science*. The latter would typically talk about where, how, and why new knowledge *emerges*, especially based in the sciences. The former would rather focus on where, how, and why knowledge *spreads* and how this diffusion and uptake of new knowledge and ideas *affect societies on a larger scale*. Without this wider circulation of knowledge, Sweden's "environmental turn" would not have been conceivable, nor would

Swedish). James E. Hansen, "The Greenhouse Effect: Impacts on Current Global Temperature and Regional Heat Waves." *Statement to the United States Senate, Committee on Energy and Natural Resources*, June 23, 1988. https://web.archive.org/web/20110822055700/http://climatechange.procon.org/sourcefiles/1988_Hansen_Senate_Testimony.pdf.

similar developments have been possible in other countries without similar diffusions of game-changing concepts and ideas.[45]

How could this be possible in Sweden? Fundamentally, because knowledge had been a principal priority. Not only was it identified as a central element in securing a smooth and efficient transition to democracy in the interwar period, but knowledge was also part of Sweden's postwar neutrality politics. It required a strong defense to deter large adversaries from devoting significant resources in considering an attack on Sweden. In turn, this demanded superb science and technology to maintain military capabilities. Swedish research and industrial policies were highly militarized and prioritized at the top political levels.

A complicating factor was that knowledge was not neutral. Isolated facts perhaps were, but the broader conclusions for policy certainly were not; nor were the narratives and metaphors that most people live by.[46] Media controversies stimulated environmental knowledge, especially in a country where newspaper readership was very high. Biochemist and socialist Hans Palmstierna's attack against scientists that worked for industry sparked intense debate and was one of the controversies that politicized the environment in the early 1970s. Another controversy concerned the future, or rather futures studies. How should such studies be carried out? What about prognostications to adapt the economy and society for global competition? Should they be normative, aimed at finding pathways to lasting peace and a healthy planet? Topics like these touched on broad and powerful interests and provoked responses grounded in ideological convictions. In this way, the realm of "the environment" spread even further and could not be bounded by the areas of expertise of ecologists or chemists. It spread into all imaginable parts

[45] David Larsson Heidenblad, *The Environmental Turn in Postwar Sweden: A New History of Knowledge*, transl. Arabella Childs (Lund: Lund University Press, 2021). Anna Kaijser & David Larsson Heidenblad, "Young Activists in Muddy Boots: Fältbiologerna and the Ecological Turn in Sweden, 1959–1974," *Scandinavian Journal of History* 43(2018):3, 301–323. Jens Ivo Engels, *Naturpolitik in der Bundesrepublik: Ideenpolitik und politische Verhaltensstile in Naturschutz und Umweltbewegung 1950–1980* (Paderborn: Schöning, 2006). Frank Zelko, *Make It a Green Peace!: The Rise of Countercultural Environmentalism* (New York: Oxford University Press, 2013). Frank Uekötter, *The Greenest Nation?: A New History of German Environmentalism* (Cambridge, MA: MIT Press, 2014). Michael Bess, *The Light-green Society: Ecology and Technological Modernity in France, 1960–2000* (Chicago, IL & London: The University of Chicago Press, 2003). James Secord, "Knowledge in Transit," *Isis* 95(2005):4, 654–672.

[46] George Lakoff & Mark Johnson, *Metaphors We Live By* (Chicago, IL: The University of Chicago Press, 1980).

of society and impacted ever more interests and social groups, not just those living downstream or downwind from point sources of pollution. A result of these significant societal changes was the emergence of the so-called "alternative movement" – *alternativrörelsen* – for which the environment was a primary common denominator.

On the other end of the spectrum, established interests also mobilized. In his 1972 book *Futurum exaktum*, physicist Tor Ragnar Gerholm critiqued what he perceived as obscurantist "cultural pessimism" in the growing flora of environmental books and pamphlets. Some media, not least the business press, talked about "doomsday prophecies." Others riposted, like the engineer and novelist Sven Fagerberg, who saw Gerholm's book as "a lightly masked partisan contribution, intended to support the established industrial interests."[47] Seen from the retrospective vantage point of today, the positions and fault lines of the debates may seem awkwardly simple. However, it is essential to see them, on the contrary, as working elements in an ongoing process of historical change. They were shaped in a creative phase of intense debate over ideas and political theory in the years around 1970 when "the environment," a distinct topic since the late 1940s, made a quantum leap to become something it had never been before: an issue of societal change. This has remained ever since and has only been reinforced with the rise of the climate issue.

In these breakthrough years, millions of Swedes were part of a massive shift in consciousness as environmental issues in their widest sense were used to shed new light on the world they lived in, including their own country. This is also the way we can understand why knowledge became so important. What knowledge was, and how it was defined and by whom, became controversial. Physicists such as Gerholm and (the meteorologist) Bert Bolin stood on either side of a deep, wide trench. They shared insight into the laws of physics, but it meant totally different things to them. Defining and defending the "right" knowledge became essential.

Having come this far, we can summarize some of the factors that explain the solid, although far from spectacular, progress of Swedish domestic environmental politics. The first prerequisite, just like in many other countries, was the environmental turn itself, and we have identified the central importance of high-quality and widely available knowledge for

[47] Tor Ragnar Gerholm, *Futurum exaktum* (Stockholm: Aldus, 1972). Larsson Heidenblad, *Environmental Turn*, 193–199, Fagerberg's quote on 198.

this turn to take place and gain some hegemony in society. Strong leadership or smart tactical moves would have had no influence had this not been the case. Through the wide circulation of knowledge, Sweden had already passed a mental "tipping point," to apply Malcolm Gladwell's concept.[48] The public had learned just as well, if not better, than the economists that the environment and climate were important enough to protect by raising taxes. Underneath this lay yet another layer of insight, gained in the long-term building of the welfare state and *folkhem*: a relatively high confidence in taxes *above* international average levels. To these underlying factors should be added some general conditions for success, established over many years of research to identify the causes of effective environmental policies. Successful countries such as the Nordics, Germany, the United Kingdom, and the Netherlands are characterized by strong public institutions. Like Sweden, they tend to have functional research policies and robust institutions for education and knowledge dissemination: high enrollment levels in higher education, active popular movements, civic engagement, transparent and independent mass media, and, not least, strong public service media.[49]

In addition, there is a need for bold, committed politicians who are themselves knowledgeable enough to realize the importance of long-term environmental politics that deliver few short-term gains: politicians who can build political trust by linking the unavoidably technocratic articulation of what is good for nature to issues that citizens would care about. Environment and climate are, to begin with, gender and equity issues. The SUV, red meat, and big game hunting are predominantly masculine domains, and the Nordics seem to be no exception.[50] Women all over the world are disproportionately vulnerable to water scarcity and poisonous particles in households. Where girls and women are educated, their autonomy and influence grow, and the environment improves. The take

[48] Malcolm Gladwell, *Tipping Point: How Little Things Can Make a Big Difference* (New York: Little, Brown and Company, 2000).

[49] M. J. Wolf, J. W. Emerson, D. C. Esty, A. de Sherbinin, Z. A. Wendling, et al., *2022 Environmental Performance Index* (New Haven, CT: Yale Center for Environmental Law & Policy, 2022), epi.yale.edu. Andreas Duit, ed., *State and Environment: The Comparative Study of Environmental Governance* (Cambridge, MA: The MIT Press, 2014). See also references earlier, Chapter 1, note 39.

[50] Jonas Anshelm & Martin Hultman, "A Green fatwā?: Climate Change as a Threat to the Masculinity of Industrial Modernity," *NORMA: International Journal for Masculinity Studies* 9(2014):2, 84–96. O. Krange, B. P. Kaltenborn & M. Hultman, "Cool Dudes in Norway: Climate Change Denial among Conservative Norwegian Men," *Environmental Sociology* 5(2019):1, 1–11.

home message seems to be that, first, an environmental turn must take place in a society, but to bring it to fulfillment, environmental policies that converge with broader sustainability concerns are required.

TACIT PROPERTIES – ARTS, CULTURE, GENDER

The widening social space of "the environment" and the gradual expansion of what might be called relevant expertise was an ongoing process and one that has continued into our own time. The process is as visible on the international stage as it is in a single country. Perhaps even more, with the rise of Earth system science, a genuinely interdisciplinary area of expertise that has influenced environmental governance, especially after the turn of the millennium. Expertise has also expanded into the social sciences, humanities, religion, and arts, as causes of environmental challenges are ultimately social and human and hence solutions must be too.[51] How has Sweden responded? In fact, as historical work on the era of the environment is now increasing, it is becoming more and more evident that the arts and the *Geisteswissenschaften* have played bigger roles than was previously assumed.

Perhaps tellingly, the earliest conscious effort to talk about the environment in modern terms in Sweden was arguably two books written not by a male scientist but by a female author, Elin Wägner, in the early 1940s (Figure 2.2). Wägner was a feminist, pacifist, regionalist, novelist, and journalist, and the second woman to take a seat in the Swedish Academy (from 1944). She was also an avid educationist, co-founding the Fogelstad Folk High School for Women in the 1920s, and a passionate gardener who argued for more ecologically friendly agriculture. With her Fogelstad feminist colleague, the large landowner Elisabeth Tamm, she co-authored the agrarian pamphlet *Peace with the Earth* (Fred med Jorden, 1940). From that work, she moved on to write a novel, *Alarm Clock* (Väckarklocka, 1941), in which she combined a critique of the "male" world war with an even more radical

[51] Sverker Sörlin, "Reconfiguring Environmental Expertise," *Environmental Science and Policy* 28(2013):1, 14–24. Paul Warde & Sverker Sörlin. "Expertise for the Future: The Emergence of Environmental Prediction c. 1920–1970," In: Jenny Andersson & Eglė Rindzevičiūtė, ed., *The Struggle for the Long Term in Transnational Science and Politics: Forging the Future* (Abingdon: Routledge, 2015), 38–62. Michael Oppenheimer, Naomi Oreskes, Dale Jamieson, Keynyn Brysse, Jessica O'Reilly, Matthew Shindell & Milena Wazeck, *Discerning Experts: The Practices of Scientific Assessment for Environmental Policy* (Chicago, IL & London: The University of Chicago Press, 2019).

FIGURE 2.2 Elin Wägner, early feminist with ecological ideas. Her anti-war pamphlet *Väckarklocka* [Alarm Clock] (1941) is generally considered Sweden's first articulation of a program for a responsible lifestyle that cared for the qualities of unspoiled nature and home-grown food. Photo: Sören Hoffman/SVT.

critique of the misconstrued economy and norms of modern, militaristic society. A central theme in both books were issues that would be called "environmental": resource waste, pollution, food security, erosion, and decaying soils.

These ideas kept reverberating long after her death in 1949. In the 1940s and 1950s, some of the most pronounced environmental critique concerned the harnessing of rivers with large hydropower dams, an issue also addressed by poets and novelists. In 1970, a dramatic conflict raged over the Vindelälven River. The most beloved of Swedish poets and singer-songwriters, Evert Taube, then in his eighties, composed a song called "Land of Angels," coining that concept (*änglamarken*) and another, "Earth of Heavens" (*himlajorden*). Vindälven was one of only four of Sweden's major rivers that were, ultimately, not harnessed for hydropower, unlike the dozens that were.

Issues of the environment and peace continued to mobilize artists, authors, and public intellectuals. Although it was scientists, doctors, and engineers who were typically asked to serve as experts, the wider public was probably more susceptible to being moved by artistic contributions. Environmental film became almost a sub-genre in its own right in the 1960s and 1970s. An emblematic example is *The Apple War* (Äppelkriget, 1971), set in a scenic apple-growing corner of southeast Sweden. In the film, the German capitalist Jean Volkswagner arrives in the cozy Swedish countryside setting with his PR man, and the two start dreaming of Deutschneyland, a theme park connected by mighty freeways to attract tourists. Local resistance and magic (!) manage to halt the plans, fully in line with the radical green politics of the two directors, Hans Alfredson and Tage Danielsson, leading figures of the Swedish intellectual humor stage in the second half of the twentieth century.

Environment-oriented journalism also reached a wide audience. Barbro Soller wrote in the *Dagens Nyheter* in the middle of the 1960s about fish and eggs contaminated by mercury. She was known in particular through a retrospective DN series entitled "Dirt-Sweden thirty years later" ("Lort-Sverige 30 år efteråt," 1968–69) in collaboration with photographer Stig A. Nilsson. It departed from a highly controversial and extremely influential series of radio shows in 1938 on housing and hygienic conditions in the Swedish provinces. The series was based on a long car journey across the nation produced by author and social visionary Ludvig Nordström, an early proponent of hardcore medical and engineering solutions who was in almost any conceivable way the opposite of his contemporary Elin Wägner. Nordström also turned the radio show transcripts into a best-selling book.[52] Three decades later, Soller's finding was that while homes and hygiene had improved enormously, a new large-scale and more complex set of problems had appeared in "the environment" that were more or less unknown to Nordström (e.g., mercury, dioxins, DDT, acid rain, soil erosion, overfishing, and overgrazing). Moreover, Soller noted that the concept of "the environment" had itself emerged in the intervening decades. She functioned initially as a mediator for engaged scientists, but increasingly came to serve as a more independent environmental voice toward the public – a new role in media at that time.[53]

[52] Ludvig Nordström, *Lort Sverige* (Stockholm: KF:s förlag, 1938). Otto Fagerstedt & Sverker Sörlin, *Framtidsvittnet: Ludvig Nordström och drömmen om Sverige* (Stockholm: Carlsson bokförlag, 1987).

[53] On Soller, see Larsson Heidenblad, *Environmental Turn*, 21–24, 78–92.

Another significant public intellectual on environmental issues was the historian Birgitta Odén at Lund University. Starting out as a quite traditional historian – an expert of Early Modern public finances – in the 1960s and 1970s, she transformed her professional and public profile to take on the role of a visionary entrepreneur and key driver in launching environmental history as a sub-field of the discipline in Sweden and the Nordic countries. She established contacts with funding agencies, served as an expert and government advisor, and engaged in planning, using the high level of trust she enjoyed to spread humanities-based environmental expertise across wide social and political circles. She also had close relations with her younger siblings, both of whom were scientists. Her sister Angelica von Hofsten was a botanist, and her brother Svante was a chemist who became famous for establishing acid rain as a scientific and political issue in 1967. Birgitta and Svante Odén were both together in a meeting with the Defence Research Institute and its General Director Martin Fehrm in the fall of 1967, a crucial moment during the environmental turn in Sweden. It was also an unlikely place to be for a historian at the time, but it was revealing for how expertise was assembled in new ways in response to a changing environment and political climate.[54]

Wägner, Tamm, Taube, Alfredson, Danielsson, Soller, and Odén are in some ways representative of the type of expertise usually referenced in the mediated history of the environment: intellectuals (all of them) and scholars (Odén, indirectly Soller as well as Wägner). However, to an even greater extent, the large majority of these early leading environmentalists were also artists. They used words, music, and images creatively and convincingly. Hence, they reached out. They became known to multiple, in some cases enormous, audiences for whom they translated the insights gained by scientists working and publishing in styles and contexts where ordinary citizens had less access and could not be *moved* in the way that they were moved by Soller's reporting, or by a movie like *The Apple War*.

This speaks to the very concept of environmental expertise. It cannot exclude an artistic and communicative dimension, more so than in other

[54] David Larsson Heidenblad, "Miljöhumaniora på 1960-talet?: Birgitta Odéns miljöhistoriska initiativ och skissernas historiografi," *Scandia* 85(2019):1, 37–64. Idem, "Environmental History in the 1960s?: An Unsuccessful Research Application and the Circulation of Environmental Knowledge," *History of the Humanities* 6(2021):2, 635–647. Idem, *Environmental Turn*, 92–112. Fredrik Bertilsson, "Biopolitisk beredskap: Den beteendevetenskapliga försvarsforskningens betydelse i den svenska krisberedskapen," *Historisk tidskrift* 142(2022):1, 29–50.

fields, because of the almost endless complexity of the environment, in the way that it is seamlessly interwoven with society, culture, and human behavior. The arts and humanities do the kind of work that explains how environmental knowledge can penetrate society and popular culture and become alive in mass media, popular movements, and politics, as well as in language and concepts. It is when knowledge and ideas enter societal institutions infused with agency that they gain the defining power that can in fact influence environmental futures. In these social processes, the voice of the philosopher or folk singer is just as important as that of the ecologist, the historian's narrative is as directional as the chemist's formula, and the conceptual analysis by the linguist is just as effective as the atmospheric calculations by the meteorologist. And above all, they have all needed each other in order to instigate societal change.

The reason that we elaborate upon the artistic and aesthetic presence of the environment in Swedish public life is that we see it as an important feature of the history we wish to tell in what follows of this book. It leads on to a central analytical question: whether these factors were of any real significance in explaining the success and contribution of Stockholm and Sweden in the rise of global environmental governance? We think they were, especially if we combine them with other tacit dimensions of the green transformation. One such aspect is gender. In science, women were always a minority, and still are. The famous Wenner Gren Foundation-funded meeting in Princeton, New Jersey, in 1955, "Man's Role in Changing the Face of the Earth," was attended by seventy-two men and one (1) woman, a biologist from India.[55] In the arts and humanities, it was always much different. Among the, admittedly, randomly composed group of seven Swedes above, four are women. It seems obvious that gender diversity and the presence of multiple forms of art and expertise have been essential for reaching out to a wider public in Sweden. Does the same hold true globally?

While our scoping of Stockholm-centric environmental networks suggests that a majority of the individuals comprising the "human environment" were men, a number of influential women have also been closely connected to the Stockholm story of environmental progress. As the most well-known recent environmental activist in the world, Greta Thunberg is an obvious example.[56] Also, one of the earliest "environmental

[55] William L. Thomas Jr. ed., *Man's Role in Changing the Face of the Earth* (Chicago: University of Chicago Press, 1956).

[56] Greta Thunberg, *No One Is Too Small to Make a Difference* (London: Penguin Books, 2019). "Greta Thunberg: A Year to Change the World." *BBC*, April 12, 2021.

FIGURE 2.3 Anna Lindhagen in a winter fur coat, c. 1920. Lindhagen was born into a family of lawyers, scientists, and politicians engaged in the urban politics of Stockholm. Alongside her regular position as a social worker and as an elected member of the Stockholm city council, she took on issues of urban beautification, social justice, and reform. She was a founder of the movement for allotment gardens in Stockholm in 1906. The booklet was published on the movement's tenth anniversary in 1916. Anna Lindhagen in fur coat and the cover of an allotment garden advocacy brochure 1916. Photo: Stockholm City Museum Archives.

activists" was from Stockholm, Anna Lindhagen, who pioneered the protection of public Stockholm parks and advocated their use for allotment gardens to allow low-income families to grow their own vegetables and get a small recreational space of their own (Figure 2.3).[57]

Other women who made early and significant contributions include Alva Myrdal, scholar, cabinet minister, diplomat, and winner of the Nobel Peace Prize, who advocated disarmament and led the process toward the establishment of the Swedish Institute for Futures Studies with a report published in 1972.[58] Inga Thorsson, a diplomat and stateswoman, initiated discussions that led to the Stockholm Conference.

[57] Sverker Sörlin, "Anna Lindhagen: Politikens fragmentisering och hundra år av ensamhet," In: Håkan A. Bengtsson & Lars Ilshammar, eds., *Demokratins genombrott: Människor som formade 1900-talet* [The Democratic Breakthrough] (Lund: Historiska Media, 2018; paper back ed. 2019), 158–180.

[58] Sissela Bok, *Alva Myrdal: A Daughter's Memoir* (Reading, MA: Addison-Wesley, 1991). Jenny Andersson, "Choosing Futures: Alva Myrdal and the Construction of Swedish Futures Studies 1967–1972," *International Review of Social History* 51(2006):2, 277–295.

Former Social Democratic environment minister Birgitta Dahl held a central role in policy formation for decades.[59] So did the author, journalist, and liberal politician Kerstin Anér – like all those mentioned earlier, a staunch advocate of disarmament – and Birgitta Hambraeus of the Center Party, who was strongly engaged with energy and environmental issues during her long career in politics. The arguably most internationally well-known Swedish woman politician in the twenty-first century, Anna Lindh, was minister of the environment in the 1990s before becoming minister of foreign affairs, a post she held when she was assassinated in a department store in central Stockholm in September 2002.

In many fields, there was a broadening of female leadership, for example, in diplomacy, typically a male bastion. Former environment ambassador Viveka Bohn helped establish and champion the environment in the foreign office, which had little tradition of science and other areas of relevant expertise. The hydrologist Malin Falkenmark has been a leading figure in water and sustainability issues for over fifty years and was a key advisor during preparations for the Stockholm Conference. In the humanities, historian Birgitta Odén used her prestige and networks in science and academia to promote environmental history. Environmental journalism was pioneered in the 1960s and 1970s by Ami Lönnroth in *Svenska Dagbladet* and Åsa Moberg in *Aftonbladet*.[60] Crown Princess Victoria, following in the footsteps of her father, King Carl Gustaf XVI – an outspoken nature lover and avid hunter – has promoted sustainability issues internationally and drawn attention to the climate crisis during her extensive travels in the Arctic.

Women's rights and family policies (e.g., generous parental leave) were politically already more established in Sweden than possibly anywhere else in the world in the 1970s and 1980s. At that time, the Swedish – and indeed Scandinavian – position as an environmental leader remained still largely unquestioned, as was the country's reputation as a cauldron of official feminism. Both environmentalism and feminism were disproportionately represented as ideologies in the academic community, and environmental issues enjoyed strong scholarly links and networks. The greening of the curriculum provided women with new

[59] Martin Hultman, Ann-Sofie Kall & Jonas Anshelm, *Att ställa frågan – att våga omställning: Birgitta Hambraeus och Birgitta Dahl i den svenska energi- och miljöpolitiken 1971–1991* (Lund: Arkiv, 2021).

[60] On Åsa Moberg and her role in environmental debates on forestry 1970–1975, see Karl-Göran Enander, "Framväxten av en skoglig miljöpolitik," *Skogshistoriska Sällskapets Årsskrift* (2002), 98–119.

entry points into the sciences, where women had been relatively scarce before. Amplifying forces worked both ways. The relatively prominent position of women in politics and public life in Sweden turned out to be an asset for the environment. It is well known that women tend to be more pro-environment and that their presence influences decisions and institutions in this regard.[61] When looking for possible explanations for the early and strong environmental interest in Sweden, the role of feminism is an important factor, along with that of popular movements. Since its founding in the early-1980s, the Green Party has had many characteristics of a popular movement, and it also adopted the principle of dual spokespersons, one female and one male, which underscored the perception that green was also feminist.

INFLUENCE THROUGH MATTER – THE SPATIAL DIMENSION

There was an array of hard and soft underlying factors that contributed to Stockholm's ability to play an outsized role in the development of foundational elements and architectures of international environment and sustainability science, politics, processes, and institutions. One was Sweden's conception of itself as a *föregångsland* – a progressive pioneering country. This view was shared by many outside observers of the *Swedish societal model*; the image of Sweden as a forerunner in *environmental policy*.[62] Delegations arrived, from the Fabian Society in the United Kingdom to journalists, planners, and architects.[63] Social, scientific, political, and cultural characteristics played influential roles in supporting this image. Sweden's *geopolitical position* as a nonaligned state, neutral in war (until 2009), and with a

[61] L. Ramstetter & F. Habersack, "Do Women Make a Difference?: Analysing Environmental Attitudes and Actions of Members of the European Parliament," *Environmental Politics* 29(2020):6, 1063–1084.

[62] Lennart J. Lundqvist, *Sweden and Ecological Governance: Straddling the Fence* (Manchester: Manchester University Press, 2004). Henrik Selin & Björn-Ola Linnér, *The Quest for Global Sustainability: International Efforts on Linking Environment and Development* (Cambridge, MA: Center for International Development, Harvard University, 2005).

[63] Marquis Childs, *Sweden – The Middle Way* (London: Faber & Faber, 1936). Marquis Childs, *This Is Democracy: Collective Bargaining in Scandinavia* (New Haven, CT: Yale University Press, 1938). Margaret Cole & Charles Smith, eds. *Democratic Sweden: A Volume of Studies Prepared by Members of the New Fabian Research Bureau* (London: Fabian Bureau of Social Research, 1938).

strong commitment to the United Nations and international development, was crucial for how the country navigated the international system. It also meant that other countries could be confident that Sweden had few obvious national interests at stake that created complications for various initiatives. Signing on to something led by Sweden was relatively low risk.

Sweden, and not least Stockholm, had a strong *association with scientific excellence.* This was due in the first place to remarkable scientific performance, especially from the 1940s, when most of the leading nations in Europe were hampered by the war and then reconstruction, and up to the 1980s, when Cold War research and development investments – including very large defense budgets – supported Sweden's strong position in science and technology. Sweden's image as a country of cutting-edge science has also been greatly enhanced by the glittering global brand of the Nobel Prizes, awarded by the Nobel Foundation and the Royal Swedish Academy of Sciences since 1901. Every year on December 10, Alfred Nobel's birthday, the world's attention is fixed upon the Swedish capital for the award ceremony and Nobel Banquet. The prizes are a significant national asset that is exploited with increasing care by the Nobel Foundation, a mycelium of the establishment in Swedish society. Through their direct link to the Academy of Sciences, a promoter of environmental research since the early 1970s, the Nobel Prizes and the prestige they entail can also be seen as bestowing additional credibility upon Stockholm as a leading international center of environmental knowledge.[64]

As home to, in certain periods, prominent members of the great generation of eighteenth- and nineteenth-century naturalists such as Linnaeus, Swedenborg, and Berzelius, the Swedish capital has in fact for centuries been *a center of knowledge production* in its own right. As we have already noted in Chapter 1 and will further elaborate in coming chapters, this role has only grown, and greater Stockholm, including the neighboring university town of Uppsala, is now among the most prolific science regions in Europe. A growing body of literature in recent years has suggested that knowledge has been a key to the growth and centrality of urban areas since the Renaissance, but especially in the period

[64] Robert Marc Friedman, *The Politics of Excellence: Behind the Nobel Prize in Science* (New York: Times Books, 2001). Sverker Sörlin, "The Heart of the Planet" [On KVA and the Environment], In: Karl Grandin, Johan Kärnfelt & Solveig Jülich, eds., *Knowledge in Motion: The Royal Swedish Academy of Sciences and the Making of Modern Society* (Stockholm & Göteborg: Makadam, 2018), 411–416.

of science-based technologies and globalization.[65] Stockholm is a city of "knowledge clusters" and has benefited from the change away from spaces of assets to "spaces of flows" that dominate in the knowledge economy.[66] This literature has usefully pointed out the advantages cities enjoy in advancing knowledge and catalyzing activity, both nationally and internationally, and how the role of cities has changed in recent decades.[67] Increasingly, cities now take on roles in combating climate change, forming alliances of the willing to become greener and cleaner. In the United States, this trend has often been in stark opposition to the federal government's inaction on climate since the signing of the Kyoto Protocol in 1997.

This literature also allows us to distinguish conditions particular to Stockholm and to make comparisons to other historical or contemporary centers of scientific and political initiative, such as London, Geneva, Berlin, and Copenhagen, the latter hosting the European Environmental Agency, or to places of concentrated specialized expertise, for example, Silicon Valley.[68] It has long been noted in the literature on the history of universities and the urban geography of knowledge that knowledge production and innovation tend to cluster, as do other sectors of society and branches of industry and commerce. The "Latin Quarter" phenomenon of cities dates back to the Middle Ages. Similar to how such areas tended to become established around cathedrals and colleges, today's knowledge clusters congregate around science and industrial parks. What is perhaps less observed is how these urban concentrations can also serve

65 Peter Hall, *Cities in Civilization* (Oxford: Blackwell, 1998).

66 Manuel Castells, *The Informational City: Information Technology, Economic Restructuring, and the Urban-Regional Process* (Oxford: Basil Blackwell, 1989). *Manuel Castells & Peter Hall, Technopoles of the World: The Making of Twenty-first Century Industrial Complexes* (London: Routledge, 1994). H.-D. Evers, "Knowledge Hubs and Knowledge Clusters: Designing a Knowledge Architecture for Development," MPRA Paper 8778 (University Library of Munich, Germany, 2008).

67 David Livingstone, *Putting Science in Its Place: Geographies of Scientific Knowledge* (Chicago: University of Chicago Press, 2003). Robert Mayhew & Charles Withers, eds., *Geographies of Knowledge: Science, Scale, and Spatiality in the Nineteenth Century* (Baltimore, MD: Johns Hopkins University Press, 2020).

68 Ivan Morus, Simon Schaffer & James Secord, "Scientific London," In: Célina Fox, ed., *London: World City, 1800–1840* (New Haven, CT: Yale University Press, 1992). Thomas Söderqvist, Jan Faye, Helge Kragh & Frank Allan Rasmussen, *Videnskabernes København* (Copenhagen: Roskilde Universitetsforlag, 1998). Anna Lee Saxenian, *Regional Advantage: Culture and Competition in Silicon Valley and Route 128* (Cambridge, MA: Harvard University Press, 1994). On the European Environmental Agency, www.eea.europa.eu/en.

as mental and social nodes of transformation. High densities of knowledge tend to foster progressive ideas and a spirit of embracing change. In this respect, it may be useful to note that Stockholm, for very particular reasons, has enjoyed a particularly high concentration of its intellectual and institutional resources for innovation in certain parts of the city.

However essential, this was never planned. The modern growth in research and higher education in Sweden, like in most European countries, has been public, funded, and controlled largely by the state. When Stockholm began its ascent as a city of knowledge in the nineteenth century, earlier universities were in small provincial and ecclesiastic centers such as Uppsala and Lund, or on the outskirts of the Swedish Empire in Greifswald, Turku, Helsinki, and Tartu. It came naturally to locate universities on crown land. Such spaces were in abundance in Stockholm thanks to the existence of *Djurgården* (literally: the Animal Park), which had been royal hunting grounds since the Middle Ages. In this area in the northeastern sector of the city, public institutions have been continuously created: royal castles, large botanical gardens, and from circa 1800 military facilities. In the nineteenth century, hospitals and leisure activities expanded in this area, along with buildings for agricultural experimentation.

In the twentieth century, there was a literal explosion of buildings constructed for the KTH Royal Institute of Technology and Stockholm University. The building boom also included sports facilities such as the 1912 Olympic Stadium, a ski jumping facility and a skating rink, as well as several smaller colleges of arts and music, a veterinary college, a forest institute, the massive Museum of Natural History, the new Royal Swedish Academy of Sciences building from 1914, and Gustaf Arrhenius' Nobel Institute for Physical Chemistry from 1910. Public agencies followed, such as the state-owned Swedish Radio and Swedish Television, concert halls, public archives, and the Sports and Gymnastics College. Later, in the early 1970s, a modernist "Film House" was built devoted to reel media, as were digital archives, a range of public research institutes, and, not least, large facilities and arenas for outdoor and indoor sports, including equestrian sports, slalom, tennis, and electric-lit trails for training and recreation. In the same green corridor, embassy compounds for the Nordic nations and other significant countries were also built (Figure 2.4).[69]

[69] Thomas Hall & Martin Rörby, *Stockholm: The Making of a Metropolis* (London: Routledge, 2009). Thomas Hall & Ola Andersson, *Frescati: Huvudstadsuniversitet och arkitekturpark* (Stockholm: Stockholm University, 1998). Jenny Beckman,

FIGURE 2.4 Map showing parts of the Stockholm Urban National Park. These northwestern areas of the Swedish capital used to be royal land and as state property became home to the large majority of the scientific institutions of Stockholm that we discuss in this volume. Stylized map version for public spaces by Jan Wahlman for The Royal Palaces, Stockholm.

As home to the King and Royal Court, Stockholm was the capital of Sweden, and it turned out to be significant. But not for the King's power. That formal authority had been gradually dismantled with the 1809 constitution and was further diminished with the establishment of full democracy and parliamentarism in 1921. In the early 1970s, all vestiges of Royal authority were completely removed with the so-called Torekov Compromise. In the small hamlet Torekov on Sweden's west coast, representatives of the political parties in the ongoing inquiry into a new Swedish constitution agreed in August 1971 to let Sweden remain a kingdom on the condition that the King would only uphold ceremonial roles and never interfere with (party) politics. The proposal

Naturens palats: Nybyggnad, vetenskap och utställning vid Naturhistoriska riksmuseet, 1866–1925 [Nature's Palace: Buildings, Science and Exhibitions at the Swedish Museum of Natural History, 1866–1925] (Stockholm: Atlantis, 1999). Julia Lindkvist, *Att bygga kunskapsstaden: En studie av högskolornas framväxt i Stockholm 1850–1960* (Stockholm: KTH, 2008). Idem, *Det krönta huset: Kungliga svenska vetenskapsakademiens byggnad i Frescati* (Stockholm: Academy of Sciences, 2002).

was presented in 1972.[70] These events belonged in the *Zeitgeist* of the Stockholm 1972 moment. These were times of upheaval. Symbolically, and emotionally, the final remnants of an *ancien régime* were laid to rest, which was fully congruent with the strong sense that something new was emerging and should be embraced. Even the Parliament and the Royal Court were part of the transformation. Winds of change were sweeping the land, and although the compromise was reached in an old countryside hotel, the shift itself took place in the capital. Sweden was Stockholm, and Stockholm was Sweden.

This massive concentration of public institutions and facilities in one city, in a very small dense fraction of the country's vast territory, only increased during the course of the second half of the twentieth century. It also became Sweden's largest assemblage of diverse and complex innovations, with most of the knowledge-related institutions and enterprises situated on royal property. Moreover, they were all literally located on the other side of the street – a wide Paris-style boulevard – from the largest urban center in the country, the downtown area of Stockholm. Named Vallhallavägen, in the Norse style common to this posh part of the city, it was laid out in the 1880s, with direct inspiration from Haussmann's plan for Paris, as part of a massive deconstruction effort to remake Stockholm into a city of "air and light," in the words of the rising literary star of the country, August Strindberg, who incidentally lived parts of his life in the French capital. The grand boulevard marked the – informal, as there was no problem crossing it – boundary to the existing city with its modern, reformed grid plan. On the city side, the streets were lined with everything a modern metropolis could encompass. This included, importantly, in the core areas near the royal castle, public agencies, political institutions and parties, large and small organizations, and representatives for literally all kinds of special groups and interests in Swedish society. There were also all kinds of media, from newspapers and book publishers to broadcast media and, from around 2000, social media, podcasters, and influencers.

The importance for Stockholm of this concentration of research and higher education, and the diverse urban fabric of the city, has not been formally estimated; it has rather been more assumed than assessed. However, the importance of such concentration for innovation has been analyzed for creative cities elsewhere in the world, and the implications of such research clearly shows that major cities and innovation

70 SOU 1972: 15 *Ny regeringsform, ny riksdagsordning.*

go well together. Not least due to the fact that they serve as import havens for all kinds of new ideas and impulses.[71] This includes the new environmental ideas that emerged in the middle of the twentieth century. The concern over "nature" and "environment" was widespread across Sweden, and often abroad, from the Amazon to the Aral Sea, but the leading voices, pens, keyboards, and cameras in the country were in Stockholm, or at least had to travel to the capital to gain recognition for themselves and their causes.

A SMARTNESS GAME?

In the literature on the history of cities, it is often remarked that urban areas serve as gravitation points. Once one venture becomes established there, it provides an argument to situate another activity in the same place. Path dependencies are strong and typically accelerate with size. Together, they tend to build reasons for people to actively go there. Aristotle is often quoted: "It [the city] comes to be for the sake of life, and exists for the sake of the good life."[72] Just as importantly, people gather in cities to make themselves seen and heard, to gain influence, and to argue their case, whatever that case might be. Cities are stages, increasingly used for mediation. In Sweden, as in for example France and the United Kingdom, but unlike in federal Germany and countries with strong regions such as Spain and Italy, the tendency was for activities of all kinds, not least media, to be disproportionately concentrated in the capital.

This is about understanding one of the more important aspects of international politics, societal mobilization, and scientific progress in the second half of the twentieth century and beyond. Over the course of this period, we argue, Stockholm has attained the status of not only a symbol but also a functional node of environmental science and governance, perhaps as the first among several other possible locations. Contenders would include New York, with the United Nations

[71] Sverker Sörlin & Gunnar Törnqvist, *Kunskap för välstånd: Universiteten och omvandlingen av Sverige* (Stockholm: SNS Förlag, 2000). Morus & Schaffer, "Scientific London." The importance of big cities as import havens for ideas and innovations was explored in governmental reports in the 1980s that paved the way for a positive interpretation of the growth of Stockholm. See the final report from "The Big City Inquiry" (1988–1990): SOU 1990: 36 *Storstadsliv: Rika möjligheter – hårda villkor: Slutbetänkande* (Stockholm: Fritzes, 1990).

[72] Aristotle, *Politics*, translated by H. Rackham. Loeb Classical Library 264 (Cambridge, MA: Harvard University Press, 1932), 1252 b27–30.

headquarters, and the San Francisco Bay Area, with its waves of unique contributions to preservation and conservation, and the planning of green spaces by Frederick Law Olmsted and the rise of a green counter culture.[73] Or Oslo, home to Arne Naess' eco-philosophy and the city of the Nobel Peace Prize, is also recognized as the 2019 Green Capital of Europe for its green city planning.[74] These places, and similarly inspired cities around the world, belong within a wider urban movement to modernize sustainably and use their status as green cities to fast-track carbon-free futures, thereby building a new kind of competitiveness for the twenty-first century. In doing so, they articulate and build new generations of urban ecologies.

They are also, in that regard, competitors, and Stockholm is no exception. In a series of campaigns and statements, Stockholm has presented itself as the ultimate smart and creative city, especially in the context of northern Europe, but sometimes with global ambitions. Its efforts have not been completely farfetched. The growth of dotcom and IT companies and startups in gaming, music (Spotify), apps, and similar tech/entertainment economic sectors has led the city to think highly of itself, and it has also placed near the top of the global rankings of the "new economy." The European Commission chose Stockholm as the Green Capital of Europe in 2010 when they first launched that award. In 2013, the Organisation for Economic Co-operation and Development (OECD) recognized Stockholm as a world-leading city when it comes to water quality, waste handling, and reducing carbon emissions. Stockholm was number one in ICT development and urban sustainability in 2014 and again in 2016, and ranked number four in *The Monocle*'s Quality of Life survey the following year.[75]

[73] Conca 2015, Selcer 2018.

[74] Richard Walker, *The Country in the City: The Greening of the San Francisco Bay Area* (Seattle, WA: University of Washington Press, 2009). Idem, "Nature's Popular Metropolis: The Greening of the San Francisco Bay Area," In: Henrik Ernstson & Sverker Sörlin, eds., *Grounding Urban Natures: Histories and Futures of Urban Ecologies* (Cambridge, MA: MIT Press, 2019), 169–200. Peder Anker, *The Power of the Periphery: How Norway Became an Environmental Pioneer for the World* (New York: Cambridge University Press, 2020). Arne Naess, *Ecology, Community, and Lifestyle*, transl. David Rothenberg (New York: Cambridge University Press, 1989). Mark Luccarelli & P. G. Röe, *Green Oslo: Visions, Planning, Discourse* (London: Elgar, 2012). Jens Lachmund, *Greening Berlin: The Co-Production of Science, Politics, and Urban Nature* (Cambridge, MA: The MIT Press, 2013).

[75] https://monocle.com/film/affairs/quality-of-life-survey-2014/, visited January 22, 2022. Ericsson Networked Society City Index, http://mb.cision.com/Public/15448/2245037/93894148bfbf1118.pdf, visited January 22, 2022, on p. 14. https://ec.europa.eu/environment/europeangreencapital/winning-cities/2010-stockholm/. Five years later the

Being "smart" does support the image of Stockholm as a forward-looking urban region, which dovetails nicely with its green urban persona. The city is obviously an important provider of infrastructure and services, and it is interested in supporting and expanding research and entrepreneurship. The city itself, however, has played a minor role in the story we tell. Although this urban version of a "socio-technical imaginary"[76] holds some sway and is clearly relevant, the green smart city is not the primary version of Stockholm that this book is about. It is also somewhat unclear what the relation of the city itself might be to global environmental governance. In many respects, it is rather the properties of Sweden as a nation, and as a progressive northern state, that have offered the qualities of "con."

Stockholm has been the venue, the meeting place, and the carrier of the larger brand. Partly because it is the capital with the institutional density of government, agencies, institutions, academies, organizations, environmental and other NGOs, popular movements, foreign embassies, universities, research institutes, and think tanks, most within easy access of each other in a truly international atmosphere. To the outside world, "Sweden" has been "Stockholm." In fact, official representatives of the city have been remarkably rare in meetings, reports, or even as meeting hosts, as the reader will become aware when the empirical detail of "environmental Stockholm" is being rolled out in the coming chapters. Stockholm has, on the contrary, been a grateful recipient of the power of the brand. The status of Stockholm on the global stage is not simply due to ABBA, Ingmar Bergman, a handful of hockey players, and a well-known welfare state: it is intimately connected to the emergence of what has become one of the largest and most consequential areas of international politics and governance.

City of Stockholm put together a report on the Green Capital award: https://international.stockholm.se/globalassets/stockholm-first-european-green-capital--2.pdf (2015). In 2020 Stockholm was appointed smartest city in the world according to one ranking which was quickly announced on the website of Smart City Sweden, a multi-partner state funded platform, indicative of the kinds of green smartness competition that is the current form of regional development concept in certain kinds of cities. https://smartcitysweden.com/stockholm-named-worlds-smartest-city/

[76] Sheila Jasanoff & Sang-Hyun Kim, eds., *Dreamscapes of Modernity: Sociotechnical Imaginaries and the Fabrication of Power* (Chicago & London: The University of Chicago Press, 2015).

3

Stockholm

A Climate Science Node

July 2, 1957, was a bright summer day in Kiruna, a remote mining town in northernmost Sweden. Surrounded by low mountains full of rich veins of iron ore, Kiruna sits in the Swedish Arctic and in the heart of *Sápmi*, the homeland of the Sámi, a hunting and reindeer-herding Indigenous people.[1] On this particular day, an unusual set of high-ranking guests had gathered, linked by their connections to the International Geophysical Year 1957/58 (IGY) and to American security interests. Among those present was Lloyd V. Berkner, one of the IGY initiators, then vice president of the National Science Foundation (NSF). He was among a small group of scientists who had been present on April 5, 1950, when physicist James A. van Allen and his wife Abbie threw a dinner party in their suburban Washington, DC, home, where the idea of a third international polar year (IPY) was first voiced by Berkner.[2] The previous Polar Years were held in 1882/83 and 1932/33. The upcoming one was envisioned a quarter century ahead of schedule because the 1930s depression had

[1] The Sámi are a diverse Indigenous ethnic community in Sápmi, a wide transboundary area in northern Scandinavia, Finland, and North-Western Russia. Reindeer herding, emblematic of the image of Sámi, is the main preoccupation of only a minority of the Sámi in Sweden, most of whom live outside Sápmi and belong in families that may not have conducted reindeer herding at all. Lars Ivar Hansen & Bjørnar Olsen, *Hunters in Transition: An Outline of Early Sámi History* (Leiden: Brill, 2014).

[2] Alan A. Needell, "Lloyd Berkner and the International Geophysical Year Proposal in Context: With Some Comments on the Implications for the Comité Spéciale de l'Année Géophysique Internationale, CSAGI, Request for Launching Earth Orbiting Satellites," In: R. D. Launius, J. R. Fleming & D. H. DeVorkin, eds., *Globalizing Polar Science: Palgrave Studies in the History of Science and Technology* (New York: Palgrave Macmillan, 2010), 205.

shrouded so many efforts in the previous IPY, including those by the United States, but also because of the Cold War and the mounting security interests around the poles, especially the Arctic. Also present, both in Kiruna and at the Van Allen dinner, was J. Wallace Joyce, who headed the American IGY office at the NSF.

The reason the Americans had made the long trip was the launch of the Kiruna Geophysical Observatory (KGO). But since the eighteen-month IGY, starting July 1, 1957, and ending on December 31, 1958, to allow summer season fieldwork at both poles, had no official opening, the Kiruna event could be seen as an unofficial version with some of the leading lights present, with little fuss and few headlines. The Americans celebrated the successful start of a new and promising observatory that might contribute to the IGY and certainly was part of Sweden's IGY effort. They were also firmly involved in the new Swedish institute, regardless of the IGY. The NSF provided some of its basic instrumentation, and the US Air Research and Development Command funded much of the research during its first decade of operation. Indeed, for several years, US funding exceeded 50 percent of the funds. The Americans wanted the geographically well-placed Swedes to monitor the activities of Sputnik, as soon as it was launched in October 1957, and they received assistance in monitoring Soviet nuclear tests in a comprehensive research program.[3] On this special opening day, NSF also announced a 25,000-dollar donation to the new institute.[4]

The observatory was headed by Director Bengt Hultqvist, a young aspiring physicist, and news of it was published in top-level science journals.[5] On the first board of the Kiruna Observatory were the highest noblesse of Swedish physics, including plasma physicist Hannes Alfvén, later recipient of the 1970 Nobel Prize in Physics, and Rolf Sievert, a radiation specialist at the Karolinska Institute who had since the war years advocated a radiophysics institute in Lapland.[6] There were good

[3] Sverker Sörlin & Nina Wormbs, "Rockets and Reindeer: A Swedish Development Pair in a Northern Welfare Hinterland," In: Per Lundin, Johan Gribbe & Niklas Stenlås, eds., *Science for Welfare and Warfare: Technology and State Initiative in Cold War Sweden* (Sagamore Beach, MA: Science History Publications, 2010), 131–152, on 140–141.

[4] "Kiruna Geofysiska Observatorium embryo till ett vetenskapligt centrum i norr," *Norrbottens-Kuriren* July 3, 1957.

[5] Bengt Hultqvist, "The Kiruna Geophysical Observatory, Sweden," *Nature* 180(1957): 828–830; idem, "Swedish Geophysical Observatory," *Science*, 126(1957):3276, 691.

[6] Hans Weinberger, *Sievert: Enhet och mångfald: En biografi över den svenska radiofysikens, radiobiologins och strålskyddets grundare Rolf Sievert* (Uppsala: Almqvist & Wiksell, 1990); idem, "Physics in Uniform: The Swedish Institute of Military Physics,

scientific reasons for the Swedes to establish the Observatory. An Arctic location meant less polluted skies and optimal conditions to study solar wind and other solar phenomena, including the northern lights. No less tangible were the geopolitical circumstances prompting the Americans to be of assistance, adding yet another, and particularly well-placed, center in the vicinity of the Soviet Union, to their network of Cold War observation sites in Europe and beyond.[7] The Arctic was a hot zone in the Cold War, potentially a Theatre of War, should tensions between the superpowers spiral beyond control.[8]

Ironically, these were the crass geopolitical circumstances, when science, politics, and local elites met in modernist, peaceful, non-aligned *Folkhem* Sweden on July 2, 1957 – to host the event that can be seen, in retrospect, as the unrecognized opening of the IGY. To further emphasize the role of Kiruna as a temporary hothouse of IGY creativity, we may note that Lloyd Berkner in his Kiruna hotel room drafted a six-page handwritten letter inviting Stockholm glaciologist Hans W. Ahlmann, then President of the International Geographical Union, to take the lead in organizing the post-IGY organization of Antarctic science. The idea that took shape in Berkner's mind under the midnight sun in Kiruna was in essence an early blueprint of what would become the Antarctic Treaty regime of a continent "by and for science." Ahlmann responded with regret, however, that he was too busy to take on the challenge.[9]

A GEOPHYSICAL PLANET IS EMERGING

The IGY was one of the largest single, coordinated actions of science ever to take place. It mobilized sixty-seven nations and an estimated 60,000 scientists and staff in operations around the world that focused on establishing a comprehensive understanding of the entire geophysics of the planet: from the stratosphere and the ionosphere, down through the

1939–1945," In: Svante Lindqvist, ed. *Center on the Periphery* (Canton, MA: Science History Publications, 1993). Hans Weinberger, "Sievert, Rolf Maximilian," *Svenskt biografiskt lexikon*, 156(2003): 150–156.

7 John Krige, *American Hegemony and the Postwar Reconstruction of Science in Europe* (Boston, MA: The MIT Press, 2006).

8 Ronald Doel, "Constituting the Postwar Earth Sciences: The Military's Influence on the Environmental Sciences in the USA after 1945," *Social Studies of Science* 33(2003):5, 635–666.

9 Lloyd V. Berkner to Hans W. Ahlmann, July 4, 1957, Ahlmann Collection, vol. 22, and Ahlmann to Berkner, July 13, 1957, Ahlmann Collection, vol. 20, both in Royal Swedish Academy of Science Archives, Stockholm [henceforth: KVA Archives].

atmosphere, and the oceans down to the sea floor. A major interest was the cryosphere, although that concept, coined already in 1923 by Polish scientist Antoni Dobrowalski, was not much in circulation.[10] Nor was the biosphere, Vladimir Vernadsky's 1925 invention, and comparatively little research was devoted to the life sciences during IGY.[11] Nevertheless, the IGY took the entire "environment," as the US military called it, under its command. Jacob Hamblin has called it "arming Mother Nature," and certain regions, such as the Arctic, mountain regions, and the tropics, became of particular strategic interest in this armament quest.[12] In retrospect, it can also be seen as an early effort to approach the entire planet not as a mere piece of stone in the universe but as an *elemental* property. The Earth consisting of fluids, fields, and forces that were held together as a gigantic system – or rather a system of systems, with a wide set of interacting temporalities of change[13] – that was fundamentally possible to study, understand, predict, and, perhaps, also manage, although that last aspect would turn out to be quite a challenge.

Sweden had already gained a certain reputation as a partner in geophysical research and it was no coincidence that the Americans chose Kiruna and Sweden as a point of departure for the IGY. Establishing the Geophysical Observatory was a logical step on the way of building a Swedish scientific superpower on the northern periphery. Sweden had remained neutral when her neighboring Norway and Denmark joined NATO in 1949, and, with Finland closer to the Soviet sphere, Swedish policy became to maintain the awkward middle space by building a military force that was big and strong enough to make an enemy attack if not impossible at least so costly that it wouldn't be worthwhile. This in turn required massive investment in relevant scientific research and

[10] Antoni Boleslaw Dobrowolski, *Historia naturalna lodu* [The Natural History of Ice] (Warsaw: Kasa Pomocy im. Dr. J. Mianowskiego, 1923), with a summary in French. R. G. Barry, J. Jania & K. Birkenmajer, "A. B. Dobrowolski – the First Cryospheric Scientist – and the Subsequent Development of Cryospheric Science," *History of Geo- and Space Sciences* 2(2011):1, 75–79.

[11] Vladimir Vernadsky, *La biosphere* (Paris: Librairie Félix Alcan, 1929), Russian orig. 1926. Engl. Transl. Vladimir I. Vernadsky, *The Biosphere*, ed. Mark McMenamin (New York: Copernicus, 1998).

[12] Jacob Darwin Hamblin, *Arming Mother Nature: The Birth of Catastrophic Environmentalism* (New York: Oxford University Press, 2013).

[13] Erik Isberg, "Multiple Temporalities in a New Geological Age: Revisiting Reinhart Koselleck's *Zeitschichten.*" *Geschichte und Gesellschaft* 46(2020):4, 729–735. Sverker Sörlin & Erik Isberg, "Synchronizing Earthly Timescales: Ice, Pollen, and the Making of Proto-Anthropocene Knowledge in the North-Atlantic Region," *Annals of the American Association of Geographers* 111(2021):3, 717–728.

defense technologies. Geophysics was one of these relevant research areas, perhaps the largest of all, along with atomic research in the quest to develop Swedish nuclear weapons.[14]

Environment and climate, as terms, were hardly mentioned at all in Swedish defense and security planning in these years and they were not part of the language used in the context of the July 2 ceremony in Kiruna either. Still, the links to Sweden's emerging trajectory as a strong performer in these domains were there, however implicit. In this chapter, we will corroborate this perhaps mysterious statement. It means embarking on a history that will take us back some time in the Swedish geophysical and environmental field sciences, to periods when "environment" and "climate" were marginal and had not even acquired their modern understanding and significance. The year 1957 could be seen as a watershed year. Environment and climate were then already in the air, and not just there. Climate change, or rather the possibility of it, was a topic of concern in the IGY. It was part of the planning of IGY as a truly global enterprise, which spread as a "Big Science" model to other field sciences as well.[15]

President of the Swedish IGY committee was Carl-Gustaf Rossby, a world-leading meteorologist and a pioneer in atmospheric science. A key turning point came in August, the same year that Rossby died of a heart attack at age 58, only weeks after the Kiruna event. Leadership of the International Meteorological Institute (IMI) was passed to his PhD student and protégé Bert Bolin, who only a year after his PhD had to unexpectedly take up the legacy of his supervisor, assuming the chairmanship of the Meteorological Institute Rossby had started at Stockholm Högskola (from 1961 Stockholm University) and also the presidency of the Swedish IGY committee. Bolin went on to become one of the most

[14] Thomas Jonter, *Nuclear Weapons Research in Sweden: The Co-operation Between Civilian and Military Research, 1947–1972*, SKI Report 02:18 (Stockholm: Swedish Radiation Safety Authority, 2002). Jonter, "The Swedish Plans to Acquire Nuclear Weapons, 1945–1968: An Analysis of the Technical Preparations," *Science & Global Security* 18(2010):2, 61–86.

[15] Christy Collis & Klaus Dodds, "Assault on the Unknown: The Historical and Political Geographies of the International Geophysical Year," *Journal of Historical Geography* 34(2008):4, 555–567. Elena Aronova, Karen S. Baker & Naomi Oreskes, "Big Science and Big Data in Biology: From the International Geophysical Year through the International Biological Program to the Long Term Ecological Research (LTER) Network, 1957-Present," *Historical Studies in the Natural Sciences* 40(2010):2, 183–224. Goossen, "A Benchmark for the Environment." On IGY as a case of scientific internationalism, see Frank Greenaway, *Science International: A History of the International Council of Scientific Unions* (Cambridge: Cambridge University Press, 1996), 149–159.

influential climate scientists of the second half of the twentieth century and a respected science organizer and institution builder. A mere decade later, Bolin himself was one of the leading lights in climate change research, paving the way for the 1972 United Nations Conference of the environment in Stockholm. In 1988, he became the founding chairman of the UN Intergovernmental Panel on Climate Change (IPCC). Something had been cooking underneath the crass Cold War surface in Kiruna on July 2, 1957.

The focus of this chapter is on the institutional foundations of international climate science several decades before climate change became a major issue of international environmental politics. It centers on Stockholm University, where, at the turn of the twentieth century, Svante Arrhenius pursued pioneering work on the effects of carbon dioxide on global temperature, and the founding of the International Meteorological Institute by the internationally renowned Rossby in 1955. This followed Rossby's return to Sweden from the United States in 1947, and his initiative also in 1949, to establish the influential climate journal *Tellus*, originally published by the Swedish Geophysical Society and by IMI since 2007. Since its founding, and with dedicated funding from the Swedish government, IMI served to bring leading climate scientists to Stockholm for shorter and longer sabbaticals to share expertise and collaborate with Swedish colleagues. Together with *Tellus* – where seminal articles on the carbon dioxide theory of anthropogenic global warming were published[16] – IMI and its cadre of climate experts made Stockholm University a central node in the evolution of climate science.

This was a particularly important period of international scientific collaboration due to the 1957–1958 International Geophysical Year, during which Charles David Keeling began his tracking of atmospheric carbon dioxide concentrations. He, too, published early results in *Tellus*,[17] including what can be considered an early incarnation of the Keeling Curve, and continued his calculations during a sabbatical year at IMI in

[16] Gilbert N. Plass, "The Carbon Dioxide Theory of Climate Change," *Tellus* 8(1956):2, 140–154. Roger Revelle & Hans E. Suess, "Carbon Dioxide Exchange between Atmosphere and Ocean and the Question of an Increase of Atmospheric CO_2 during the Past Decades," *Tellus* 9(1957): 1, 18–27. Plass also published in more popular fora, such as *Scientific American* and his research hit major news media, for example, Waldemar Kaempffert, "Warmer Climate on the Earth May Be Due to More Carbon Dioxide in the Air," *New York Times*, October 28, 1956. www.nytimes.com/1956/10/28/archives/science-in-review-warmer-climate-on-the-earth-may-be-due-to-more.html.

[17] Charles David Keeling, "The Concentration and Isotopic Abundances of Carbon Dioxide in the Atmosphere," *Tellus* 12(1960):2, 200–203.

FIGURE 3.1 Netherlands-born Paul Crutzen, Nobel Laureate in Chemistry in 1995 for his work on the ozone hole, was essential in the formation of Stockholm into a global hub of environmental and climate science. He was employed by Stockholm University as a computer programmer in 1959 and wrote foundational papers from the year 2000 proposing the Anthropocene as a new geological epoch. Photo: Courtesy of the Archives of the Max Planck Society.

1961–1962, working together with Bolin. The late 1950s also included the incidental hiring of a certain Dutch civil engineer specializing in bridge construction, Paul Crutzen (Figure 3.1), who had been living in the central Swedish city of Gävle with his Finnish wife. Answering a newspaper ad for a computer programmer, Crutzen would become employed by the Meteorological Institute, where he also took courses and subsequently began his PhD training with Bert Bolin as his supervisor. During his time at Stockholm University, Crutzen would also contribute important early articles on stratospheric ozone depletion to *Ambio*, a journal based at the Royal Swedish Academy of Sciences for which he also served on the editorial board.[18] After graduating in 1973, Crutzen fairly rapidly became recognized as a world-leading atmospheric scientist. He was

[18] Ole John Nielsen & Merete Bilde, "Reflection on Two Ambio Papers by P. J. Crutzen on Ozone in the Upper Atmosphere," *Ambio* 50(2021):1, 40–43.

awarded a Nobel Prize in Chemistry in 1995 (co-explaining the thinning of the ozone layer) and launched on the global scene both the idea of "nuclear winter" (that he first termed "twilight at noon" in 1982) and "the Anthropocene" (in 2000).[19]

LINKING THREE CENTURIES OF STOCKHOLM CLIMATE CHANGE

In December 1956, Carl-Gustaf Rossby's portrait graced the cover of *Time Magazine*, and inside the writer did not mince words: "The history of modern meteorology is inescapably paralleled by Rossby's career."[20] This was at the time of the breakthrough for computerized weather forecasts, the beginning of the jet aircraft era, and the height of the Cold War, and Rossby had a role in all of them. To be heralded as the foremost expert in this field was remarkable, and the fact that it was, of course, an American perspective that was expressed in *Time* did nothing to diminish this because the United States at the time had achieved a world-leading position, in no small way thanks to this "likable, high-spirited, round-faced Swede." Most of the leading meteorologists in the United States and the world in the twentieth century were students or colleagues of Rossby (Figure 3.2), and he himself founded some of the world's foremost meteorological institutions at MIT and the University of Chicago. His own name lives on today in the term "Rossby waves" – the most powerful waves in the atmosphere, carrying jet stream energies in a regular circumpolar pattern from west to east – which he identified and co-developed a mathematical formula for. His name and career were also a main, yet little known, lead into the career of Sweden and Stockholm as a hub of environment and climate.

Born and bred in the province of Gotland, an island in the Baltic, Carl-Gustaf Rossby had been a student at Stockholm Högskola in the late 1910s, where he met Bergen meteorologist Vilhelm Bjerknes, former professor at the Högskola, who that year returned to hold a lecture. Rossby was mesmerized and started what was to become a long and meandering international career with Bjerknes at Bergen's vibrant meteorological institute, where the concept of the "Polar front" was invented,

19 Paul J. Crutzen & John W. Birks, "The Atmosphere after a Nuclear War: *Twilight at Noon*," *Ambio* 11(1982):2–3, 114–125. Paul J. Crutzen & Eugene F. Stoermer, "The 'Anthropocene'," *Global Change Newsletter* 41(2000):17–18.

20 "Man's Milieu," *Time* 68 (December 17, 1956): 68–79.

FIGURE 3.2 After a more than twenty-year long career in the United States, Carl-Gustaf Rossby returned to his native Sweden and did foundational work to establish an internationally leading climate science node in Stockholm. A pioneering advocate for the modern orthodoxy of carbon dioxide climate forcing, he was the quintessential science organizer. He also recruited Bert Bolin as his PhD student. Photo: Courtesy of the MIT Museum, Cambridge, Massachusetts.

a term Bjerknes had taken from the World War I trenches and applied to the behavioral patterns of the atmosphere.[21] Rossby threw himself into balloon experiments, storm predictions, and cartographic analysis – all the while losing himself in the kind of frenzy he demonstrated in every task he took on for the rest of his life. His innovations included assigning colors to the two types of fronts: blue for cold fronts and red for warm ones. He subsequently left for Germany where he worked at the Prussian *Aeronautisches Observatorium* in Lindenberg, close to Berlin, and in Leipzig where the Bjerknes school had good connections. Another stay

[21] Friedman, Robert Marc, *Appropriating the Weather: Vilhelm Bjerknes and the Construction of a Modern Meteorology* (Ithaca, NY: Cornell University Press, 1989).

in Bergen followed in summer 1922, when, before incredulous colleagues, Rossby tried to launch aerological balloon ascents of the type he had recently encountered in Germany. Returning to Stockholm, he started working as a meteorologist at the Swedish Meteorological and Hydrological Institute (SMHI). He participated in a series of scientific expeditions to the North Atlantic, where he gained firsthand experience with ice floes, a "witch's dance" around the vessel's hull, to the North Sea, and to Portugal and Madeira.[22] He gained his Licentiate degree in mathematics at Stockholm University in 1925 but, somewhat typical of his high-speed career, never completed his doctorate, although he found the mathematics he learnt very useful.[23]

The next stop was the US Weather Bureau in Washington, DC. When he left for the United States in 1925, he carried in his pocket a letter of recommendation signed by no less than Svante Arrhenius, Nobel Laureate in Chemistry in 1903 and by the 1920s hailed as an outstandingly broad thinking scientist with genius qualities. Arrhenius had been interested in climate change himself, suggesting in a famous article in 1896 that historic changes in climate may have caused the ice ages that had been so important for the Nordic countries. Climate change could, by implication, also explain the geographical features of Sweden, not least the fertile soils caused by postglacial seas covering land that rose as the several thousand-meter thick ice receded.[24]

Through his career that spanned almost forty years, Rossby as a person thus connects Bert Bolin, the IPCC, and the climate Nobel Peace Prize of 2007 – that went to Al Gore and Bolin and his IPCC

[22] Tor Kvinge, "The 'Conrad Holmboe's' Expedition to East Greenland Waters in 1923," *Årbok for Universitetet i Bergen* (1963).

[23] Biographical details in this and following paragraphs, see James Roger Fleming, *Inventing Atmospheric Science: Bjerknes, Rossby, Wexler, and the Foundations of Modern Meteorology* (Cambridge, MA: The MIT Press, 2016); H. R. Byers, "Carl-Gustaf Rossby, the Organizer," In: Bert Bolin, ed., *The Atmosphere and the Sea in Motion: Scientific Contributions to the Rossby Memorial Volume* (New York: Rockefeller Institute Press & Oxford University Press, 1959), 56–59. Sverker Sörlin, *Carl-Gustaf Rossby 1898–1957* (Stockholm: Royal Swedish Academy of Engineering Sciences, 2015).

[24] Svante Arrhenius, "On the Influence of Carbonic Acid in the Air upon the Temperature of the Ground," *The London, Edinburgh, and Dublin Philosophical Magazine and Journal of Science* 41(1896):251, 237–276. Sverker Sörlin, "Rituals and Resources of Natural History: The North and the Arctic in Swedish Scientific Nationalism," In: Michael T. Bravo & Sverker Sörlin, eds., *Narrating the Arctic: A Cultural History of Nordic Scientific Practices* (Canton, MA: Science History Publications, 2002), 73–122. Christer Nordlund, "'On Going up in the World': Nation, Region and the Land Elevation Debate in Sweden," *Annals of Science* 58(2001):1, 17–50.

colleagues – directly with Arrhenius and the formation of the Nobel Prize institution more than a century earlier. The Rossby–Arrhenius–Bolin atmospheric chain of brains also links climate science with the formation of Stockholm as an academic center. Arrhenius was for five years the President of Stockholm Högskola and was passionate about the capital as the center of a new kind of useful knowledge that he was convinced could only be forged in a major modern city.[25] He had been severely critiqued by his former colleagues at the University of Uppsala, then a small college town, where he received his PhD. He quickly left, or fled, Uppsala for his European *Wanderjahre* and never returned. Instead, he took up his new position in Stockholm with a fresh outlook and a regained sense of confidence. He would forever argue that science thrived in diverse urban centers, and that it was hindered in small, inward-looking college towns. This was a recurring debate in Sweden that intensified whenever examination and other academic rights were offered to new institutions, such as Stockholm Högskola, which received such rights in 1904, two years after the end of Arrhenius' presidency.[26]

TRANSNATIONAL CLIMATE CHANGE

Rossby, the prodigal son going west, was inspired by both sides of Arrhenius. The bold scientist who explained planetary climate change and ice ages and launched spectacular ideas of industrial use of fossil fuels, and the academic entrepreneur who believed in close collaboration between science and society. He left for the United States in a spirit not totally unlike that of Arrhenius when he had broken up from Uppsala, in disappointment of the ungenerous and narrow-minded attitudes of his colleagues. For Rossby, it became second nature to break up and start afresh. He left Bergen once his ideas were established there. He also left the SMHI, a bureaucracy, he thought, controlled more by

[25] Elisabeth T. Crawford, *Arrhenius: From Ionic Theory to the Greenhouse Effect* (Canton, MA: Science History Publications, 1995). Sverker Sörlin, "Arrhenius, Svante August," In: Noretta Koertge, ed. *New Dictionary of Scientific Biography*, vol. 1 (of 8) (New York: Charles Scribner's Sons, 2007), 111–113.

[26] Elisabeth Crawford, *The Beginnings of the Nobel Institution: The Science Prizes, 1901–1915* (Cambridge: Cambridge University Press, 1987). The argument would come up again in the interwar years when the KTH Royal Institute of Technology in Stockholm received the right to confer the doctoral degree, much to Uppsala's chagrain. Henrik Björk, *Teknikens art och teknikernas grad: Föreställningar om teknik, vetenskap och kultur speglade i debatterna kring en teknisk doktorsgrad, 1900–1927*. Stockholm papers in history and philosophy of technology (1992).

rules and routines than by imaginative new thinking. A fellowship from the newly started (1919) Sweden–America Foundation allowed him to spend part of 1926 at the US Weather Bureau, where he studied dynamic meteorology.

Rossby soon left the Weather Bureau as well, again disappointed, but also enticed by new prospects. In 1926, the Daniel Guggenheim Fund for the Promotion of Aeronautics announced that it intended to allocate USD 2.5 million for a program to promote education and research in aviation, "particularly in its use as a regular means of transportation of both goods and people." Aviation safety needed to be ensured. The same year federal legislation was prepared pertaining to airports, aeroplanes, and rules for air traffic and its supervision. It became clear that the aviation sector and the US Weather Bureau should work together. This led to contacts between the Guggenheim Fund and Rossby, whose ideas and enthusiasm made such a strong impression that the Fund employed him to run its newly formed aviation program. Military applications were in the picture from an early stage. Chemical engineer and naval lieutenant Francis W. Reichelderfer was an intermediary between Rossby and Guggenheim. Reichelderfer and Rossby were both members of the Guggenheim Interdepartmental Committee on Aeronautical Meteorology, which began its work in 1927. Other members were representatives for the military, the weather service, and the ministry of trade.

In the years that followed, intensive efforts were made to establish what would be called "aerial roads," in other words the air space equivalent of the motorways being built for rapidly expanding motor vehicle traffic. These flight corridors were based on the work of a network of weather observers in cooperation with airlines, telecommunications operators (to communicate weather information), and government authorities. California was selected as a test case, partly because there was a great need there, with the fast changes in weather where the Pacific met the Sierra Nevada mountain range, and partly because the Guggenheim Fund was at the time building up aeronautical engineering research capacity at the California Institute of Technology in Pasadena. Rossby was appointed to head the project, and in the unrestricted, experimental space of the Guggenheim Fund, with his natural leadership skills, his capacity for work, and his endearing, diplomatic disposition, he was able to thrive and his star rose quickly on the US meteorological firmament. The most important outcome of his fellowship period was something he had not planned on at all: for the first time, Rossby felt that he was part of something where he could really play a role. He would stay for a year, but remained in the United States for more than twenty years, as an American citizen from 1939. He was essentially equal parts American and Swedish for the remainder of his life.[27]

Rossby became the golden boy of American meteorology, heading the US war effort in his field and serving as a close aid to the War Office, founding institutes first at MIT, then at Chicago, where hundreds of wartime meteorologists were trained.[28] Aerial and maritime warfare, but also nuclear weapons testing, drove demand for improved weather predictions in the US Navy and the US Air Force, and also by the atomic warfare authorities, including the Atomic Energy Commission.[29] Computers held out a promise that the enormous calculation work could be done by machines, and in the early days of the Cold War, Rossby was wooed to lead the Princeton Computer Project housed at the Institute for Advanced Study in Princeton, New Jersey. Insisting invitations came both from John von Neumann, the Hungarian born Cold Warrior mathematician, and J. Robert Oppenheimer, who headed the Institute.[30] Instead, in 1947, at the height of his career in the United States and with top security clearance in the Pentagon, where he continued to serve into the 1950s, Rossby suddenly returned to Sweden and Stockholm, where a new, remarkable career took off. One that would with time give him a crucial role in the critical early phase of Stockholm's rise as a site of global environmental governance. The move was blessed by the prime minister himself, Tage Erlander, a signal that able leadership of Swedish

[27] For Rossby's early years in the United States, see Fleming, *Inventing Atmospheric Science.*

[28] Douglas R. Allen, "The Genesis of Meteorology at the University of Chicago," *Bulletin of the American Meteorological Society* 82(2001):9, 1905–1910. Doel, "Constituting the Postwar Earth Sciences." Fleming, *Inventing Atmospheric Science.*

[29] Generally on the development of numerical meteorology in the United States in Kristine Harper, *Weather by the Numbers: The Genesis of Modern Meteorology* (Cambridge, MA: MIT Press, 2008), which also presents some of the key Scandinavian connections. More detail, not least on the latter, can be found in Kristine Harper, *Boundaries of Research: Civilian Leadership, Military Funding, and the International Network Surrounding the Development of Numerical Weather Prediction in the United States*, unpublished Ph.D. diss. (Corvallis, OR: Oregon State University, 2003).

[30] Jule Charney tells Rossby in a letter dated "Electronic Computer Project, Princeton New Jersey, March 16, 1949" that he had spoken to John von Neuman who was "immediately taken with the idea and expressed a desire to know more about the possibilities." Rossby collection, Stockholm Meteorological Institute Archives, Stockholm University. This may have been the first of what became a long string of invites and cajoles to Rossby. See multiple letters from von Neumann and Robert Oppenheimer from 1949 through 1953. Institute for Advanced Study Archives, Princeton, New Jersey.

Cold War meteorology was an issue at the highest political level.[31] Right after the most devastating war that the world had seen, this was a small but significant move to improve Swedish meteorology for times of peace and war. Nobody could imagine what the future would hold instead.

A COMPUTER IN STOCKHOLM

In the following years, Rossby gathered members of his large international network in the Swedish capital at the Meteorological Institute of Stockholm Högskola (later acronym: MISU), which he had founded in 1947 upon his return to Sweden. The natural born institution builder, Rossby, in 1955 established within MISU the International Meteorological Institute, which with government funding has since helped solidify Stockholm's standing as a hub for climate science by facilitating visits of top scientists for seminars, sabbaticals, and the exchange of ideas. His new journal, *Tellus*, was also published by MISU/IMI. In the United States, he had already founded the *Journal of Meteorology* (later changed to the *Journal of the Atmospheric Sciences*), which had quickly become a leading forum. *Tellus* was equally successful. In its very first year (1949), it published articles by some of the biggest names in meteorology from all around the world. Many papers in the early years were devoted to the intense work in both Stockholm and Princeton on numerical weather prediction by the likes of John von Neumann, and Rossby's Chicago protégé, Jule Charney. In the course of the 1950s, it was in *Tellus* that ground-breaking studies of the new emerging orthodoxy of *anthropogenic* climate change were published. Canada-born physicist Gilbert Plass, who was the first to scientifically re-establish the Arrhenius theory of the effect of carbon dioxide on global warming in a famous article in 1956, and Roger Revelle at Scripps were visiting scientists who published in the journal, along with a growing band of Rossby's own students and colleagues in Stockholm, including Bolin.[32]

31 The Prime Minister also blessed the proposal for public funding of Rossby's IMI in 1951 (see later). Sverker Sörlin, "Narratives and Counter Narratives of Climate Change: North Atlantic Glaciology and Meteorology, ca 1930–1955," *Journal of Historical Geography* 35(2009):2, 237–255.

32 Gilbert N. Plass, "The Carbon Dioxide Theory of Climatic Change," *Tellus* 8(1956):2, 140–154; Roger Revelle & Hans E. Suess, "Carbon Dioxide Exchange between Atmosphere and Ocean and the Question of an Increase of Atmospheric CO_2 during the Past Decades," *Tellus* 9(1957):1, 18–27, argued along similar lines, reinforcing the revival of Callendar's hypothesis. Bert Bolin, "On the Exchange of Carbon Dioxide between the Atmosphere and the Sea," *Tellus* 12(1960):3, 274–281.

In comparison with the big Princeton project, Stockholm may at first have seemed like a dead end, but in practice it was almost the opposite. The Swedes were determined to build their own computer, the so-called BESK, *Binär Elektronisk Sekvenskalkylator*, plans for which were fully known in Princeton, especially since the computer there, which took a full five years to complete, was used as a model for BESK.[33] The project found a home at the Royal Institute of Technology's former location on Drottninggatan in Stockholm's old "Latin Quarter," near the eighteenth-century Observatory on the top of a hill overlooking the entire city center of the rapidly growing Stockholm conurbation. One of the main stakeholders in Swedish computers was the Swedish Air Force, which had an interest in computer-based support for airplane construction at SAAB industries, but also hoped for better and automated forecasts. Rossby understood (even better now that the Princeton project had started) the enormous potential of computers and wrote to von Neumann to make sure that the Swedes traveling to the United States to learn about them would receive the best possible help. What he meant was of course that they would get what they needed to build a computer in Sweden. Erik Stemme was one of the several Swedes who went to the United States to learn more about computers. His destination was Princeton, and when he returned in 1950, he was assigned the task of developing BESK, which was put into operation in 1953.[34]

Meanwhile, Bert Bolin (Figure 3.3), then an Air Force meteorologist, also spent a year in the United States, 1950–1951, first in Chicago, then six months in Princeton with Charney. When Bolin was there, the computer was not sufficiently developed, so in order to test weather forecasts, the Princeton team had to go down to the US military base in Aberdeen, Maryland, where the already famous ENIAC computer, programmed by a group of women, was already in operation.[35] Bert Bolin acted as a bridge between the Air Force and Rossby's department at Stockholm University, where he became a research student and started

33 A. Wiin-Nielsen, "The Birth of Numerical Weather Prediction," *Tellus A: Dynamic Meteorology and Oceanography* 43(1991):4, 36–52, on 46. Kristine Harper, *Weather and Climate: Decade by Decade* (New York: Facts on File, 2007), 109–110. Anders Persson, "Early Operational Numerical Weather Prediction Outside the USA: An Historical Introduction: Part 1: Internationalism and Engineering NWP in Sweden, 1952–69," *Meteorological Applications* 12(2005):2, 135–159.

34 Bert Bolin, "Carl-Gustaf Rossby: The Stockholm Years, 1947–57," *Tellus A* 51(1999), 1: 9.

35 Thomas Haigh, Mark Priestley & Crispin Rope, *ENIAC in Action: Making and Remaking the Modern Computer* (Cambridge, MA: MIT Press, 2016).

FIGURE 3.3 Bert Bolin early in his career. Trained as a meteorologist, Bolin became a PhD student of Carl-Gustaf Rossby in the late 1940s and succeeded him as leader of the Meteorological Institute on Rossby's death in 1957. Photo: Sam Stadener/Scanpix.

to teach. He was also essential in the daily operational leadership in the work on numerical weather predictions, especially since Rossby spent long periods every year in the United States.[36]

With Norman Phillips, who Rossby had met in Chicago, also joining the team, a project to determine how a computer could be used for weather forecasts supported by the Swedish Board of Computing Machinery (*Matematikmaskinnämnden*) was launched in the autumn of 1953. Phillips' role was to rewrite the code that existed for ENIAC for BESK.[37] Rossby's brother, Åke, also had a small role to play. The Swedish National Defence Radio Establishment (FRA), where Åke was bureau chief from the time the agency was established in 1942, was a big

[36] Wiin-Nielsen, "The Birth." Henning Rodhe, "Bert Bolin (1925–2007) – A World Leading Scientist and Science Organiser," *Tellus B: Chemical and Physical Meterology* 65(2013): 1, 1–9.

[37] Bolin, "Rossby," 9. Persson, "Early Operational Numerical Weather Prediction," 145.

stakeholder in a Swedish computing device, but, for reasons of secrecy, could not be at the forefront. The two brothers were, however, in touch with each other. Thus, early Swedish computer development was able to benefit from two Rossby brothers collaborating.

The weather project was an experimental workshop for what had been called the early mainframe computers' "elite users," that is, public authorities, researchers, and the military. They worked around the clock for two months as the autumn nights grew longer. Bert Bolin had described how the Swedish Air Force created weather maps that were sent by military couriers to the computer on Drottninggatan in Stockholm, where observational data was extracted and fed into the computer. The computer worked throughout the night, and early in the morning, the results were couriered back to the Air Force, which could then produce a weather forecast for the following day and two days ahead. This was autumn 1953, and it would take a year and a half before the Americans were able to produce their first forecast, in July 1955.

It could be regarded as a race between the BESK team and the Princeton team to see who would be first to generate computer-supported numerical forecasts. In reality, the opposite was the case; it was a close cooperation, led by Rossby and von Neumann and with large security policy interests in the balance. The two teams consisted of a long series of personal connections. The results and information were exchanged on an ongoing basis that was facilitated by Rossby serving as the common denominator for the projects, building relationships between individuals, groups, and authorities, and ultimately between nations, a diplomatic scientist role of a kind that his PhD student Bert Bolin would adopt and develop even further. Rossby himself, commuting between the two continents, had numerous opportunities to communicate information to the American side. The Americans thought highly of the BESK computer at KTH. It performed at least as well, if not better, than the one in Princeton.[38]

"A UNIQUE EXPERIMENT OF PLANETARY DIMENSIONS"

The meteorological world that was taking shape in the 1950s focused not only on civil aviation and military strategy needs. A world with a growing population and dramatically increasing consumption of

[38] Harper, *Boundaries of Research*, 419.

natural resources and energy was also a world in which the composition of the atmosphere was changing. In an article on "Current Problems in Meteorology" (1956/1959), Rossby took up the new lines for the science he had shaped in a fundamental way. The composition of air, not least "air pollution" as people were then starting to say, impacted both humans and their environment. This is also how Rossby was presented in the *Time Magazine* article in 1956, "Man's Milieu" – literally the same name as that of the Stockholm Conference, "The Human Environment" – for which he was the cover image.[39]

Rossby would not be Rossby if he had not, in his typical style, started to take an interest in entirely different problems at the same time. Air pollution was one of his interests. Several of his research colleagues in Stockholm had gathered data on the chemical composition of the atmosphere and it was clear that large quantities of acidifying substances were entering Swedish air space from Denmark, Germany, and the British Isles. Over the next few decades, this would be an important environmental issue – a term that in the mid-1950s was not yet used – and would lead to negotiations between the countries involved. Rossby was also interested in nitrogen and how it could be carried into the atmosphere affecting the nutritional content of the ground and water, leading to eutrophication. Immediate inspiration came from agricultural scientists in Sweden, but he had been sensitized to large-scale issues through work he was engaged in on the Dust Bowl, the erosion disaster that hit the American mid-west in the 1930s.[40]

There was also the related issue of climate. Could climate also change as a result of human impact on the composition of the atmosphere? The idea was not a new one. Theoretically, the greenhouse effect had been known about since the 1800s, but after Arrhenius' article in 1896, ideas about the impact of carbon dioxide were largely forgotten. The general conception for more than half a century had been that humans did not have the power to impact anything as large as the planet or its climate. Only a few voices expressed a different view. One belonged to the British engineer Guy Stewart Callendar, who in 1938 made detailed calculations of the Earth's increasing

[39] "Man's Milieu," *Time* 68 (December 17, 1956): 68–79. Sverker Sörlin & Maria Bohn, "Commentary on C. G. Rossby," In: Libby Robin, Sverker Sörlin & Paul Warde, eds., *The Future of Nature: Documents of Global Change* (New Haven, CT: Yale University Press, 2013), 451–153.

[40] His Dust Bowl work at MIT was supported by the US Department of Agriculture. See "Man's Milieu."

average temperature since the 1800s, and related them to human use of fossil fuels. Few, including meteorologists and atmospheric scientists, agreed with Callendar's ideas.[41] But in the first half of the 1950s, they started to circulate. Rossby did pay attention to them, and in his 1957 "Current Problems" article, he discussed Callendar's ideas, and his data, at length, including a diagram that showed the long-term increase of CO_2 content in the atmosphere.

He had already shown occasional signs of interest in the issue. One of his first PhD students, Chaim Pekeris, had been interested in the CO_2 problem as early as the 1930s. Rossby had also established contact with Finnish chemist Kurt Buch, who had been gathering data for many years on CO_2 uptake in the ocean and had found that amounts were constantly increasing. At the beginning of the 1950s, Buch contacted Rossby again and explained that he was thinking along the same lines as Callendar. Receptive and curious as always, Rossby picked up on the idea. He did not launch a comprehensive research program at first, but gradually allowed more space for the study of atmospheric chemistry in his department, including carbon dioxide. His department at Stockholm University took the initiative to build a Scandinavian network tasked, among other things, with measuring CO_2 levels in the air.[42] In that process, the Scandinavians learned a lot, especially that there were major differences between sites.

The obvious remedy, thought Rossby, was to find locations of measurement in the "free atmosphere," far from large concentrations of people and industry. He laid this out in his "Current Problems" piece: the Arctic, Antarctica, Andean mountain tops, and Pacific islands. He also said that such a "rather extensive observational program will be conducted during the International Geophysical Year."[43] As Chairman of the Swedish IGY Committee and with Bert Bolin as his secretary and right-hand man, Rossby made sure that CO_2 measurements were included in the Swedish program. On the completion of Bolin's PhD in 1956, he also advised his student to now turn his attention to

[41] James R. Fleming, *The Callendar Effect: The Life and Work of Guy Stewart Callendar (1898–1964)* (Boston: American Meteorological Society, 2007).

[42] Maria Bohn, "Concentrating on CO_2: The Scandinavian and Arctic Measurements," In: J. R. Fleming & Vladimir Jankovich, eds., *Osiris 26: Klima* (Chicago, IL: The University of Chicago Press, 2011), 165–179. See also, Stig Fonselius & Folke Koroleff, with a preface by Kurt Buch, "Microdetermination of CO_2 in the Air, with Current Data for Scandinavia," *Tellus* 7(1955):2, 258–265.

[43] Rossby, "Current Problems in Meteorology" (orig. in Swedish 1957), English transl., in Bolin, *The Atmosphere*, 9–50, on 15.

atmospheric chemistry and problems of pollution in the air, including "biogeochemical cycles," which to a large extent came to mean greenhouse gases.[44] Now things started to move swiftly. Already in 1959, Bolin published (with Erik Eriksson) a seminal article on "Changes in the carbon dioxide content of the atmosphere and sea due to fossil fuel combustion" in the large Rossby memorial volume that Bolin also edited.[45]

Rossby was directly involved in the discussions about where and how Charles David Keeling's measurements of greenhouse gases – an IGY project – should be pursued, just as he indicated in the "Current Problems" article.[46] Few contributions to climate change science have meant more than the iconic rising curve of CO_2 that Keeling's instrument has provided since 1958, when he first installed it at 3,200 meters above sea level on the top of the Mauna Loa mountain in Hawaii. Al Gore has reported in his film and book *An Inconvenient Truth* (2006) how he as a student at Harvard encountered the curve in Roger Revelle's science classes, and how he became mesmerized and convinced it had profound implications. The curve itself is probably also the single most important piece of evidence to demonstrate the anthropogenic element in rising CO_2 levels in the atmosphere. There is simply no other reasonable explanation available, and Mauna Loa has turned into what sociologist of science Thomas Gieryn has called a "truth spot," a site where evidence accumulates to underwrite a complex claim, which is not otherwise easily confirmed.[47]

It is also worth noting how this curve, with its seemingly unstoppable upward trend, put geophysics into the growing category of upwardly slanting curves that signified various forms of growth: economic, demographic, and social. These environmental visuals with time grew into a genre of their own, but not until the early 2000s were they assembled to demonstrate how they all hung together, not just chronologically, all moving rapidly upward from around 1950. But also causally in the sense that it became evident that it was the economic and social curves, reflecting the enormous rise in human activity, that drove the steep rise in environmental indicators. In 2004, the phenomena that the curves reflected

44 Rodhe, "Bert Bolin," 2.

45 Bert Bolin & Erik Eriksson, "Changes in the Carbon Dioxide Content of the Atmosphere and Sea Due to Fossil Fuel Combustion" in Bolin, *The Atmosphere*, 130–142.

46 Correspondence between Rossby and Keeling in 1956 and 1957 in MISU archives.

47 Thomas F. Gieryn, *Truth-spots: How Places Make People Believe* (Chicago: University of Chicago Press, 2018).

were given a name, "the Great Acceleration."[48] Among all these curves, the Keeling curve has been the most iconic.

In the Swedish IGY program, CO_2 measurements were made from a base on Spitsbergen. Several articles were published in *Tellus* presenting then current ideas about climate change, at a time when these were regarded as new and perceived by most people as wild guesswork. Rossby also refuted the more common idea of a new ice age. Hence, in the mid-1950s, which in this context was very early, he was on the side of those who kept an open mind about whether or not humans were responsible for large-scale climate change:

> It has been pointed out frequently that mankind is now performing a unique experiment of impressive planetary dimensions by now consuming during a few hundred years all the fossil fuel deposited during millions of years. The meteorological consequences of this experiment are as yet by no means clarified, but there is no doubt that an increase of carbon-dioxide content in the atmosphere would lead to an … increase of the mean temperature of the atmosphere.[49]

Thus spoke Rossby. His voice in *Time Magazine*, just like that of Greta Thunberg, on the cover of the same magazine 63 years later, drew on the widest possible international networks and could certainly be linked to the United States. The voice now came from Stockholm, an emerging global "truth spot" where a few people had started to believe in the emergence of a new, and perhaps very different human–Earth relationship.

ARRIVING AT THE RIGHT TIME

Rossby was in the middle of a workshop at his alma mater, Stockholm Högskola, when he died suddenly and unexpectedly of a heart attack on August 19, 1957. Restlessly active to the very end, he had just returned from yet another of his many trips to the United States. As a visionary, he was typically focused on the future and on opportunities, rather than

[48] Will Steffen, Angelina Sanderson, Peter Tyson, Jill Jäger, Pamela Matson, Berrien Moore III, Frank Oldfield, Katherine Richardson, John Schellnhuber, B. L. Turner II & Robert Wasson, *Global Change and the Earth System: A Planet under Pressure*, IGBP Book Series (Berlin, Heidelberg, New York: Springer, 2004). Robert Costanza, Lisa J. Graumlich & Will Steffen, eds., *Sustainability or Collapse? An Integrated History and Future of People on Earth*, Dahlem Workshop Report 96 (Cambridge, MA: MIT Press, 2007). Will Steffen, Wendy Broadgate, Lisa Deutsch, Owen Gaffney & Cornelia Ludwig, "The Trajectory of the Anthropocene: The Great Acceleration," *The Anthropocene Review* 2(2015):1, 81–98.

[49] Rossby, "Current Problems" (1959), 14.

what he had already achieved. As a research director, he had qualities that sociological research has identified as fundamental for successful research environments. He was charismatic, a visionary with a strategic focus and good relationships with funding sources, and a leader able to incentivize and generate enthusiasm for the projects he wanted to implement. Those around him felt as if they were taking part in something big and important.

He nurtured talent in research students, including Horace Byers from Berkeley who came to MIT, Jule Charney who received his PhD from the then less enticing meteorology environment at UCLA and was brought into Rossby's magic circle, and Bert Bolin who was the wise and multitalented meteorologist whom he could entrust with anything, knowing that he would always be useful. Their careers all followed a similar pattern. They met Rossby as a teacher or mentor, then, through dialogue or mutual understanding, adopted his research agenda and worked faithfully on it within the network. Rossby retained his leadership role through all turns in the road and regardless of the subsequent successes of the individual members of his network. In many ways, they outshone him in individual achievements, not least in mathematics or research methods. But there was never any doubt about which roles his collaborators had. Few of them seemed to question this arrangement, although some would occasionally find him hard to get into dialogue with, unsurprisingly given his frequent travel, extensive networks, and dense portfolio of activities. Similarly, Rossby was never envious; the more successful his acolytes, the better it was for him.

Personal qualities, however eminent, can only explain part of Rossby's significance. Another key explanation was that he was active in Stockholm – one of the research environments in the world where geophysical issues had long had a high status, and where links to Bergen and Bjerknes were already established. He was catapulted from the Nordic region out into the world and like a boomerang, he came back. The ties between the Nordic environments had been established since the end of the 1800s. Bjerknes' chair at Stockholm University was the most important one. They were strengthened after World War I by geographer and glaciologist Hans Ahlmann's work in Western Norway with Bergen as the base, and by Tor Bergeron, who became the Bergen School's unofficial emissar in Sweden.[50] It was Ahlmann who in a

[50] Tor Bergeron, "The Young Carl-Gustaf Rossby," In: *The Atmosphere and the Sea in Motion*, 51–55.

public inquiry with Uppsala professor Harald Norinder had first suggested that Rossby should be invited to return to Sweden.[51] He was also the one to first suggest that the Rossby family should be allowed to take over the Ahlmann family's flat at the top of the Observatory on "the Hill" when Ahlmann, since 1930 Professor of Geography at Stockholm University, became Ambassador to Oslo in 1950. The cupola overlooking the Swedish capital became their home until 1954.[52] Although its core activity was astronomy, Observation Hill in Stockholm had always been a site of climate- and weather-related science. It kept a temperature record from 1756 (in Sweden only preceded by a record from Uppsala since 1722) and during Ahlmann's time as the leading glaciologist and climate scientist in the country, Stockholm Högskola's Geography Department moved into the building in 1934. Civil weather services started to emerge in several countries during the second half of the 1800s, and what is now called the Swedish Meteorological and Hydrological Institute was instigated in 1873.

It was also a matter of timing. Rossby returned to Stockholm just as the very first phases of "the environment" were taking shape in the post-World War II years, with his first full year in Sweden, 1948, being the breakthrough moment for the new understanding of the concept.[53] He was a top US researcher with unique institutional resources and was recruited to come to Sweden to work in a field and during a period when Sweden and the United States had extensive security policy collaboration, and where Swedish research was also greatly motivated by military needs, especially when the technically advanced Swedish Air Force – in need of regularly updated and accurate weather forecasts – was expanding rapidly.

Rossby's visions were, perhaps paradoxically, easier to realize in Sweden, and in Stockholm, where he managed to assemble all the institutional, academic, and political networks that he needed into a small workable space of operation. As we have seen, the Prime Minister

[51] Their report, "Sakkunnige rörande organisationen av den meteorologiska forskningen och undervisningen" (1946), Riksarkivet/Swedish National Archives, SE/RA/321102, was submitted to the Department of Ecclesiastical affairs January 26, 1946. Further details in Sverker Sörlin, "Ice Diplomacy and Climate Change: Hans Ahlmann and the Quest for a Nordic Region beyond Borders," In: Sörlin, ed., *Science, Geopolitics and Culture in the Polar Region: Norden beyond Borders* (Farnham: Ashgate, 2013), 23–54, on 43–46.

[52] Personal information by Staffan Helmfrid, Ahlmann's student, later President of Stockholm University. Interviewed by Sverker Sörlin, February 22, 2011.

[53] Warde et al., *The Environment*, ch. 1.

seconded, and funded, his move from the United States to Stockholm and then his expansion to an international institute. The Air Force supported him, and the university employed him and hosted him. KTH and the Swedish state, through the Board of Computing Machinery, provided world-class computing capacity, and several public agencies and funding councils also stepped in. However, he would have gotten nowhere with his visions if the additional funding he needed could not be found in American military sources, which were not limited in their geographical distribution. From this massive institutional backing, just as much as from his personal charisma and creativity, rose the convening power that he then used skillfully. Rossby brought to Stockholm ever-growing networks of scientists from several countries – not just from the trans-Atlantic sphere but also from the developing world that was starting to rise after the partition of India in 1947 and the following decade.

He translated his visions to Stockholm and started a tradition to build Earth System science institutions on a grand scale. One of the early elements of such an evolution, *avant la lettre*, was Arrhenius' Nobel Institute, which then followed an orientation to nuclear physics with the Stockholm accelerator, just across the yard on the extended Stockholm University campus north of the city (now the main campus). Later would follow massive growth in the geosciences and in environmental science, and the neighboring Academy of Sciences would transform with the work of, among others, Bolin and his student in turn, the young Paul Crutzen. Decades later, the Stockholm Environment Institute was formed, followed by the Stockholm Resilience Center, the latter on the same campus.

In retrospect, this was all a legacy ultimately derived from Rossby. His passion and curiosity for the unknown included the entire Earth. What these non-linear forces had in store for us could ultimately not be figured out in the middle of the 1950s. There was an element of idealism and internationalism in his project. His vision was that it would serve the common good if there was more collaboration, as "an altruistic post-war idealistic idea, as a way for Sweden to extend its resources and influence." In the same spirit, he founded *Tellus*, as an explicit attempt to de-nationalize the circulation of ideas, hitherto separated into journals for the French and the British, and to mix Europeans with Americans.[54] In the 1956 issue of *Time Magazine* with Rossby on the cover, he concluded

[54] Persson, "Early Operational Numerical Weather Prediction," quote on 140.

by saying: "We should have a great deal of respect for the planet on which we live."[55] His words were prophetic in more than one sense: they inadvertently also heralded the forging of institutions in Stockholm to turn these words into deeds.

A PART OF "HIGHER DIPLOMACY"

One major question remains unanswered – why did Rossby return to Sweden at all? How should this remarkable start of what proved to be a central role for Sweden as a player in the emerging environmental and Earth system sciences be explained? It was certainly not planned. In the early post-war years, nobody could imagine what would come: the rise of "the environment," the understanding of anthropogenic climate change, and the entire idea that humanity must manage its relationship with "the planet" in entirely new ways. Zooming out, a broader view of the events leading up to the peak moment in the summer of 1957, we find that much is explained by institutions, and by path dependencies from earlier work that is taken up by individual system actors who can see opportunities and act boldly and convincingly.

In early 1945, when World War II was coming to an end, the Swedish Government commissioned Stockholm University geography professor Hans Ahlmann and Uppsala professor in atmospheric electricity Harald Norinder to explore how to strengthen meteorology in Sweden. These men would propose a new infrastructure involving new professorships, better education programs, and other necessary measures. One of the first measures proposed in their report was to contact Rossby. This happened even before there was time to consider anything else. Such was Rossby's status.

Such was also the status of meteorology. Ahlmann wrote to his friend Harald Ulrik Sverdrup, the Norwegian oceanographer who headed the Scripps Institution in La Jolla, California, in April 1945, that he considered the assignment to be part of "the higher diplomacy."[56] A statement which, at first glance, may seem strange – a simple, academic inquiry – if it wasn't already clear what significance atmospheric sciences and geophysics were beginning to have at that time. World War II had been a high-tech war with air and naval forces playing important roles, and it revealed the need for knowledge of what would soon be called "the

55 "Man's Milieu," 79.

56 Ahlmann to Sverdrup, April 12, 1945, Ahlmann Collection, KVA Archives, vol. 8.

environment" – the US armed forces were among the first to talk seriously about "the environment" in the modern sense.[57] Warfare had verticality, from the sea floor to the stratosphere; the atmosphere was key. The status of the most prominent experts in the field rose quickly. Sverdrup wrote to Ahlmann in October 1945 about how Vilhelm Bjerknes would be an obvious candidate for the Nobel Prize in Physics. "[W]eather forecasting has played an important part in winning the war." But even "men trained by him" had played a very significant role, claimed Sverdrup.[58] One of them was Rossby.

A main recommendation in the Ahlmann-Norinder report was to establish Rossby in Sweden and build a program with him at the helm.[59] A meeting was organized in Stockholm in January 1946 when Rossby met with Ahlmann and Norinder as well as Minister of Ecclesiastical Affairs, Tage Erlander. Another suggestion was to strengthen the meteorology program in Uppsala. This would change later on, partly because Rossby had more faith in locating his activities in Stockholm, where ties could be created with the weather service whose staff also needed to be trained. In Uppsala, the shift toward Stockholm was of course unpopular. A letter in the winter of 1946 confirms the fact that Ahlmann kept Rossby informed about the obstacles that existed in establishing the program at Stockholm University as well as SMHI, which was expecting the new meteorological capacity to be built up there.[60] Tage Erlander, on the other hand, was on board, both as Minister for Ecclesiastical Affairs when Rossby's professorship was granted and again in 1951 when, as Prime Minister, he supported Rossby's plans for a department of meteorology at the university with the diligent Lund physicist Torsten Gustafson, Erlander's trusted science advisor, providing the connection.[61]

Exactly how and when the idea of bringing Rossby back to Sweden first came about is unclear. There is evidence that Rossby had wanted to come back for some time.[62] Among other things, he wanted to apply for a position in Lund in the mid-1930s. In 1935, Jacob Bjerknes wrote to his Swedish colleague Tor Bergeron (who had spent a long time in Bergen)

57 Doel, "Constituting the Postwar Earth Sciences."
58 Sverdrup to Ahlmann, October 26, 1945. Ahlmann Collection, KVA Archives, vol. 19.
59 Ahlmann and Norinder, Report to the Department of Education (January 1946).
60 Ahlmann to Rossby, February 20, 1946, one undated (February 1946). Ahlmann Collection, KVA Archives, vol. 9.
61 Tage Erlander, *Dagböcker 1950–1951*, ed. Sven Erlander (Stockholm: Gidlunds, 2001), on 310. Gustafson to Rossby, March 7, 1951. Rossby collection, MISU.
62 Bolin, "Carl-Gustaf Rossby" (1999).

to say that Rossby was interested in the position as Director General of the Meteorological and Hydrological Service following Axel Wallén, who had once been Rossby's boss when he received his fellowship to go to the United States a decade earlier. How serious he was about this is hard to judge, but it is clear at least that he visited Sweden during several summers in the 1930s and on a few occasions Bergen as well. He arranged for his Stockholm relatives to visit him in Boston in the late 1930s. Whatever he felt during these years, his growing commitments and the new, tempting opportunities that always came in between made it difficult for him to leave the United States; and when the war broke out, it became impossible.

That Rossby and Ahlmann knew about each other was clear; they both had a strong connection to Bergen and both were renowned internationally in their fields. Ahlmann also had expertise in polar climate change, which proved to be important to the Americans during the war. He had already confirmed that warming was happening in the Arctic. At a lecture for the Swedish Society for Anthropology and Geography in March 1943, Ahlmann suggested that melting was also taking place in Antarctica, which would explain the rising sea levels that had been observed since the 1800s.[63] This lecture would be translated (we do not know exactly when) into English by Swedish-born Karin A. Gleim, who worked in Rossby's department in Chicago and was sent from there in February 1947 to Captain Howard Hutchinson at the Office of Naval Research (ONR). Everything to do with knowledge of ice and cold had gained strategic significance as the military plans for the Arctic entered a more intense phase during the Cold War.[64]

Another motive for Ahlmann could be that he saw the potential in a scientific collaboration with Rossby. Ahlmann, who was already convinced that climate change was happening on a large scale in the polar regions, needed the support of meteorological research to further his hypothesis, and Rossby could provide him with this. Rossby could also open doors for him in the US. There appeared to be a mutual interest. Rossby invited Ahlmann to a meeting attended by leading European climate scientists in

[63] Ahlmann, "Is och hav i Arktis" [Ice and Sea in the Arctic], *Kungl. Svenska Vetenskapsakademiens årsbok* 41(1943), 327–336.

[64] Doel, "Constituting the Postwar Earth Sciences." Janet Martin-Nielsen, "'An Orgy of Hypothesizing': The Construction of Glaciological Knowledge in Cold War America," In: Julia Hertzberg, Christian Kehrt & Franziska Toma & Julia Lajus, eds., *Ice and Snow in the Cold War* (New York: Berghahn, 2019), 69–88. Stephen Bocking & Daniel Heidt, eds., *Cold Science: Environmental Knowledge in the North American Arctic during the Cold War* (New York: Routledge, 2019).

Stockholm in 1948. This was a typical move by Rossby, one that would immediately give him a key role in his new position and place Stockholm meteorology on the map. Rossby, in turn, was kept informed by Ahlmann about the upcoming Norwegian–British–Swedish Antarctic Expedition (NBSX), and Rossby published an essay by Ahlmann in *Tellus* in which Ahlmann launched his plans. Rossby worked with Ahlmann's students in preparation of a Festschrift entitled *Glaciers and Climate* honoring their professor on his sixtieth birthday in November 1949.[65]

During the same period, in 1945–1946, when contacts with Rossby intensified, and by all accounts helped by these, Ahlmann forged his own relationships with leading figures in the military meteorological institutions in the United States, including the military. His request to be on board a US flight over Greenland was finally confirmed, after months of correspondence, by Colonel Dan Yates, who had a leading position in the Air Weather Service of the US Army and was part of Rossby's personal wartime network. Yates made flights available to him and at the same time took the opportunity to ask Ahlmann to inform him about the results of the visit.[66] Ahlmann's interest could have been motivated by the trip he made in the summer of 1945 to Leningrad and Moscow, where he obtained plausible information on the strong Soviet development in technology and science in the Arctic, with obvious military implications.[67] His interest in Soviet Arctic science, glaciology in particular, had always been strong, both linked to his connections in Bergen and to his own professional network, which now came in handy.[68]

After the war, and as the Soviet Union also had its own nuclear weapons from 1948, and when the United States started testing nuclear weapons at home in Nevada in 1950, scientific intelligence was becoming increasingly significant, not least in cold regions. It was clear that Rossby had maintained his contacts with those in political and military circles in the US, even after he moved back to Sweden. He was a member of the powerful Research Development Board and was directly involved during the legendary Vannevar Bush's leadership of strategic US research planning

65 Letters from Rossby to Ahlmann February 7, March 4, and April 1, 1946, Ahlmann collection, vol. 19, KVA Archives.

66 Yates to Ahlmann June 24, 1946, Ahlmann Collection, KVA Archives, vol. 19. For the full context, see Sörlin, "Ice Diplomacy," on 35–38, esp. footnote 43.

67 Ahlmann reported on his Soviet trip, "Geographer in the Soviet Union," *Geographical Journal* 106(1945), 217–221. Sörlin, "Ice Diplomacy," 35–36.

68 Julia Lajus & Sverker Sörlin, "Melting the Glacial Curtain: The Politics of Scandinavian-Soviet Networks in the Geophysical Field Sciences between Two Polar Years, 1932/33-1957/58," *Journal of Historical Geography* 44(2014), 44–59.

in the post-war years. Rossby's good connections came in handy when Ahlmann, in May and June 1947, went on an extended trip to the United States to lecture at some thirty universities and institutions around the country. During the trip, Ahlmann and Rossby visited the Pentagon, where Ahlmann held a lecture about the impact of climate change on the conditions for warfare in the Arctic. The conclusion was that the Americans needed to strengthen their ties with Swedish research, and that they should send officers and students to Stockholm. There were many indications that the planning for the US trip was the work of Rossby in cooperation with Harald Ulrik Sverdrup in La Jolla. It was Rossby's, and to some extent Sverdrup's, network of individual contacts and institutions that were visited and it was Rossby's questions, with a focus on the implications of the climate change studied by Ahlmann, that were on the agenda.

The contacts continued when Rossby came to Stockholm. He was part of an arrangement with the US Air Force's research department in Cambridge, Massachusetts, involving sending officers from there for advanced training in Stockholm. From 1947 up to the mid-1950s, there were always several US military personnel there at any given time. They were able to take the knowledge and methods they learned about back with them to the United States. Rossby also developed a proposal for the ONR in March 1949 about acting as a central "screening" venue for young, gifted meteorologists so that the best would be selected and enticed to the United States where they could be useful to the US Navy. The idea was based on the way things had worked at his international seminar in Chicago, where there was a strong inflow of foreign talent. Rossby now thought the same thing could happen in Stockholm, but with an emphasis on Europe and Great Britain. He created a budget for twenty "visitors" a year, each of whom would spend a couple of weeks in Stockholm. In other words, Rossby and his colleagues would separate the wheat from the chaff for the US Navy. "The weakest factor in any system of direct importation of scientific talent into the U.S.A. is the lack of real knowledge relative to the actual or potential capabilities of suggested or possible candidates." At the same time, he took it upon himself, with the help of his colleagues, to send reports, seminar papers, and other materials that could be of interest to the Americans. For this service, the ONR would pay USD 4,800 a year, which was equivalent to more than Rossby's full grant for his professorship, including "assistants and expenses" (a total of SEK 21,100).[69]

[69] Rossby collection. MISU archives, Stockholm University, Department of Meteorology. Thanks to PhD student Maria Bohn for identifying these documents.

Could the Americans gain anything from Rossby's move to Sweden? Regardless of the initial intentions, the answer was apparently yes. Although the United States had unparalleled research capacity domestically, they could not do everything on their own. At the same time, they needed access to all possible kinds of intelligence because their geopolitical role had gone global. The Scandinavian countries were already prominent in meteorology and climate research. Rossby knew it, and the United States knew it too. Ahlmann was the undisputed authority on polar climate change. The Scandinavians had also been outstanding in weather forecasting ever since the early days of the Bergen School. Under Rossby, Stockholm could be developed into a European center of excellence. This would be entirely in line with the official US post-war strategy, which, as historian John Krige has shown, was to allow various centers of excellence in Europe to work uninhibited but with financial support from US sources so that the Americans could get access to the results.[70] An American military man, Flight Lieutenant and meteorologist Philip D. Thompson, who was the first director of the computer project in Princeton in 1948–1949 and who had very good insight into research in both the United States and Europe, considered Rossby's Department of Meteorology in Stockholm to be a perfect "listening post" for meteorological intelligence.[71]

This is also how we can understand the sympathetic stand that the American military and research establishment took in relation to Sweden. Rossby, and other scientists working in a similar trans-Atlantic setting, gained considerable respect not only for the scientific quality of their work but also for their collaborative capacity. Rossby's MISU institute was soon to receive, through sometimes winding institutional roads, substantial funds from American military sources. The CIA's Office of Scientific Intelligence (OSI) had in its August 1949 report "The State of Science in Sweden" concluded that Swedish science was of a high standard and thus contributed to countering the earlier notion that neutral Sweden could be a threat to US ambitions.[72] The US Air Force also

[70] Krige, *American Hegemony*.

[71] Kristine C. Harper, "Research from the Boundary Layer: Civilian Leadership, Military Funding and the Development of Numerical Weather Prediction (1946–1955)," *Social Studies of Science* 33(2003):5, 667–696, on 686.

[72] Ronald E. Doel & Allan A. Needell, "Science, Scientists, and the CIA: Balancing International Ideals, National Needs, and Professional Opportunities," In: Rhodri Jeffreys-Jones & Christopher Andrew, eds., *Eternal Vigilance?: 50 Years of the CIA* (London & Portland, 1997), 70–71.

concluded that involvement, including funding, in Rossby's new institute in Stockholm would benefit the Princeton Project. The Princeton Project and the new program at MISU were thus essentially linked, institutionally and through their US military funding, and through a steady stream of military staff from the United States visiting Rossby's institute.[73]

Rossby mobilized military support in Sweden, connecting with the Air Force in particular.[74] In many organizational ways, the Swedish development resembled the American. In 1950, when the Swedes started building their own computer, which was crucial for the work on numerical weather predictions, the Swedish Air Force played an important role.[75] The military dimension of meteorology was both an everyday thing and somewhat of a red herring. First of all, neutral Sweden did in fact maintain clandestine links with the United States all throughout the Cold War to an extent that was not disclosed until public inquiries studied this in the 1990s (e.g., the Neutrality Commission).[76] Nonetheless, Rossby was cautious to make sure that the funds were "decontaminated" via a civilian institution, for example, the Woods Hole Oceanographic Institution.[77] Even after the demise of the idea of a Nordic Defense Union in the winter of 1949, when Denmark and Norway joined NATO, Norway and Sweden maintained their cooperation in selected areas of strategic interest. One of these areas was meteorology.[78]

73 This paragraph and the following are based on Harper, *Boundaries of Research*, 299–300.

74 Bert Bolin interview (Sörlin), June 2, 2005; Harper, *Boundaries of Research*. Sörlin, "Narratives," documents evidence of MISU funding from the Swedish Air Force as well as from the US Air Force and from the ONR.

75 Per Lundin, ed., *Att arbeta med 1950-talets matematikmaskiner: Transkript av ett vittnesseminarium vid Tekniska museet i Stockholm den 12 september 2005* [Working with the Computing Machines of the 1950s: Transcript of a Witness Seminar] (Stockholm: Filosofi och teknikhistoria, 2006), 21–23. In his statement at the witness seminar, Bolin underscores the close cooperation with the Swedish Air Force, which provided funding and, perhaps more importantly, provided the necessary weather maps and staff for the drawing of maps and for tests of the computer-based models in regular prognostication activities. This statement is also corroborated by numerous documents in the Military Archives of Sweden (Stockholm), for example, a document dated March 1, 1956, regarding "running of BESK [acronym for the Swedish computer] during military training operation," ref. 25(578: 5). See also Germund Dahlquist, "Väderleksberäkningar på BESK" [Weather Computations on BESK], *Teknisk tidskrift*, 1955, 293–296.

76 Thomas Jonter, "Det amerikanska spåret: En undersökning av IB:s bildande och eventuella kopplingar till USA," SOU 2002:95 *Forskarrapporter till Säkerhetstjänstkommissionen*, 135–191.

77 Thompson to Chief, Atmospheric Analysis Laboratory, c. November 1953 (Thompson papers, Correspondence 1953–1954); see Harper, *Boundaries of Research*, 420.

78 "Plan for coordination of military meteorological service" was one of five, out of a total of twelve, points in a memo from the Head of the Swedish Defence Staff

CONVENING AND CONVINCING – INSTITUTIONS AND NETWORKS

This brings the story back to where we started – in Kiruna on July 2, 1957, on the second day of the IGY. Another research environment that benefited from US military links was the Kiruna Geophysical Observatory KGO, founded in 1957 as part of the Swedish effort in the International Geophysical Year and renamed twice to become in 1987 the Swedish Institute for Space Physics (under the Swedish acronym IRF). The Institute was formed under the Swedish Academy of Sciences and was initially part of the Academy's network of northern research stations.[79] Public funding became available since the Institute was a centerpiece of Swedish IGY planning.[80] The Institute's first director Bengt Hultqvist at Stockholm Högskola prepared during his break between his PhD exam and his move to Kiruna by following Hannes Alfvén's seminars at KTH and by making an extended trip to the United States in the autumn of 1956. Among his stops was the Geophysical Institute in College, Alaska, near Fairbanks, and the Cambridge Air Force Research Center in Massachusetts, where Rossby had connections.[81]

In the following years, the Kiruna Institute undertook contract research for the US Air Force according to a constantly growing

[Försvarsstabschefen] that were approved by the Swedish Government and were also followed up by operational contacts in the following years. Memo, September 3, 1949, and details of coordination quoted by the Neutrality Commission, SOU 1994:11 *Had there been a war--: preparations for the reception of military assistance 1949–1969: Report of the Commission on Neutrality Policy*, chapter 6, and by Magnus Petersson, *"Brödrafolkens väl": Svensk-norska säkerhetspolitiska relationer 1949–1969* ["The Welfare of Our Sister Nation": Swedish Security Relations 1949–1969] (Stockholm: Santerus, 2003), 234–235. However, the politics of the weather-computing revolution were not easy, and the activities in Stockholm should not be equated with the political agenda on the American side, nor, in particular, the one represented by von Neumann, who was a zealous anti-Communist. Rossby seems rather to have fostered an internationalism that went far beyond the demarcation lines of the Cold War, a spirit that was to be continued in the department. Persson, "Early Operational Numerical Weather Prediction," on 140.

79 These also comprised the Abisko research station, founded in 1906, the Tarfala station for glaciological research, founded in 1945, and a small observatory on the Pårte Mountain in the alpine area of Sarek.

80 Bengt Hultqvist interview (Sörlin), March 13, 2007. Ingrid Sandahl, "Rymden" [Space], in *Kunskapsarena Norrland* [Knowledge Arena Norrland] (Umeå, 2007).

81 Bengt Hultqvist, "Reserapport rörande studiebesök i U.S.A.," [Report from a Study Tour in the U.S.], Report to the Board of the Royal Swedish Academy of Sciences' Research Stations in northern Norrland, December 7, 1956, Mimeo, Institutet för rymdfysik [Archives of the Swedish Institute of Space Physics, hereafter IRF], Kiruna.

program on "HF- and VHF-Auroral zone propagation." The first quarterly report comprised the period October 15, 1957, to January 15, 1958, and reports were issued regularly until 1967, when the program ended. By and large, the Americans funded basic research, and they encouraged the publication of results in peer-reviewed journals. Nonetheless, they also secured, in Kiruna as at many other research sites in Europe, observations and data that were collected by their funded partners. The first equipment to be furnished by the Air Research and Development Command was delivered to Kiruna on January 13, 1958.[82] In 1958, after Sputnik, a new task was added to the program: to track satellites and provide data concerning the time for the closest approach, by telegraphic means, to the Space Group of the Air Force Cambridge Research Center, during the first days after the launch of a new satellite. Likewise, the Swedish Institute was to provide the American military with Doppler data of every new Sputnik "for ten days to two weeks," and the scientists in Kiruna delivered.[83] The obvious reason for the Americans to be interested in the Kiruna Institute was its geographical location, far north of the Arctic Circle and in the vicinity of Soviet territory. The close proximity to the northern Soviet Union also came in handy when the Americans wanted information on upper atmosphere effects of the nuclear weapons tests in Novaya Zemlya between September and November 1961 and Semipalatinsk in October 1962 (twice).[84]

The US military funds were crucial for the fledgling institute, which could probably do little but applaud the booster they received. In 1960, a full 55 percent of all funds at the Institute came from the US Air Force, in 1959 and 1961, the contribution was at 40 percent, and still in 1963, a third of the income came from the American military.[85] With increasing criticism of the United States in Swedish politics and the media, the Air Force program attracted negative attention.

[82] "Studies on HF- and VHF-Auroral zone propagation: Quarterly Technical Status Report No. 1, October 15, 1957–January 15, 1958," by Bengt Hultqvist, Contractor & Johannes Ortner PI., Contract No. AF 61(514)-1314, IRF Archives.

[83] *Quarterly Report*, January 5, 1959, on 8.

[84] Terence Elkins and Alv Egeland, *Ionospheric Effects Associated with Nuclear Weapon Tests July–December 1962*. Scientific Report KGO 63 2, March 1, 1963; B[engt] Hultqvist et al., *Report on Observations Made at Kiruna Geophysical Observatory during the Series of Nuclear Weapon Tests Carried Out at Novaja Zemlja between 10 Sept. and 4 Nov. 1961*, Scientific Report KGO 611, December 11, 1961.

[85] Swedish Institute for Space Physics, Report to the Space Commission [Synpunkter på rymdutredningen], memo, January 4, 1995, Table 6. IRF Archives.

Ultimately, it was also questioned by the Swedish Government and subsequently terminated in 1967. Thus closed a special Cold War chapter of science-politics relations between Sweden and the United States. However, as we have seen, another development was still only in an early phase at this point and had received a tremendous boost by the trans-Atlantic networks. The initial rise of Stockholm as a global environmental hub actually occurred as an effect of events taking place in Washington, DC, at MIT in Cambridge and the nearby Air Force base in Massachusetts, in Princeton, New Jersey, in Chicago, in Fairbanks, Alaska, and at the Scripps Institution in La Jolla, California. Not only in the United States, but also in Svalbard, Norway where some of the most essential Swedish research was conducted, and in Oslo, where some of the best mathematicians and physicists in the computer project were based, and in Kiruna, where the new geophysical observatory took shape. Rossby's world-leading hub of atmospheric science, with profound implications for environmental and climate knowledge, was a fermenting element. The work that took place there was evidence of a capacity to collaborate – connect, convene, conceptualize, and contribute to what was now becoming central developments in human–Earth understanding.

Nothing comes from nothing. There was already something, and that something already had a wide geography spatially and a wide reach conceptually. Yet, its energies were concentrated in Stockholm, and especially in a set of institutions and buildings scattered in available public buildings and lands on the northern edge of the city. This is where the Observatory was located, neighbor with university premises and a range of institutions, including the KTH Royal Institute of Technology's early buildings where the BESK computer was hosted. A step further away were both the new KTH campus, opened in 1917, and the Karolinska Institute, where Rolf Sievert had his laboratory. Yet another step in the same direction, along the tramline to the leafy northeastern suburbs, was an area formerly called *Vetenskapsstaden*, Science City. This is now where the Stockholm University campus is located, but its roots go back to the eighteenth century, when, on royal grounds, experimental agriculture was initiated. In the nineteenth century, the Experimental Field, as it was called, grew in size. In the early part of the twentieth century, both the Royal Swedish Academy of Sciences and the copious Museum of National History with all its collections and comprehensive research moved out from the city to be located here, as well as Svante Arrhenius' Nobel Laboratory of

Physical Chemistry.[86] In the 1960s and 1970s, most of the university followed, breaking up from its inner city addresses, where Rossby had worked on his return to Sweden and also had once been a mathematics and physics student. The Arrhenius Laboratory, the new home to the Stockholm University meteorologists, opened in 1973. Forces were coalescing to shape a center and these forces found a spatial manifestation in the form and planning of the city itself.

86 On the spatial evolution of Stockholm science and scholarship, see Beckman, *Naturens palats* (1999), which discusses the notion of a Stockholm "Science City" external to the city center with references to London (Kensington) and Berlin (Dahlem; "ein deutsches Oxford"), 119–137, quote on 127; Fredric Bedoire & Per Thullberg, *Stockholms universitet 1878–1978* (Uppsala: Almqvist & Wiksell, 1978); Inga Elmqvist et al., eds., *Kunskapens kvarter: Husen runt Observatorielunden* [The Streets of Science: Houses around the Observatory] (Stockholm: Stadsbiblioteket, 2000); Julia Lindqvist, *Att bygga kunskapsstaden: En studie av högskolornas framväxt i Stockholm 1850–1960* [To Build a Scientific City: A Study of the Institution of Science in Stockholm 1850–1960] (Stockholm: KTH, 2008).

4

The Swedish UN Initiative

A major step in the institutionalization of climate science took place in the summer of 1967 at an otherwise obscure venue on the outskirts of Stockholm. The site had a few years earlier been set aside for the construction of an oil storage terminal, but the too-porous ground on a hillside overlooking the entrance to Stockholm's inner harbor instead became the location of an education and training facility for trade unions and other civic organizations associated with the Social Democratic Party. The compound's scenic setting stood in stark contrast to the austere modernist architecture of the Skepparholmen conference center, where around seventy world-leading climate scientists gathered in late June for two weeks of intensive discussions on scientific as well as organizational issues. This seminal event in international climate science would take place just months before a major societal breakthrough for environmental issues in Sweden and the launch of an initiative in the United Nations by Swedish diplomats that would lead to the first-ever global environmental conference.

A decade after the untimely death of Carl-Gustaf Rossby, the conference would confirm that Bert Bolin – Rossby's disciple and successor as head of the International Meteorological Institute (IMI) at Stockholm University – had become one of the leading figures in the advancement and organization of atmospheric science. His career path had in a sense been set twenty-five years earlier when his father Richard, a school teacher who was himself fascinated by weather phenomena, brought his teenage son to the Swedish Meteorological and Hydrological Institute to meet the renowned meteorologist Anders Ångström. Even as a young boy, Bolin had, like his father, demonstrated a keen interest in weather, keeping temperature records in his home town of Nyköping, 100 km south of Stockholm. A year after his career coaching encounter with Ångström, Bolin began his undergraduate studies in mathematics and physics at his father's alma mater, Uppsala University. He subsequently moved to Stockholm for military service, where, as fate would have it, he came into contact with the recently returned Rossby who was in the process of establishing IMI. Over the next few years, Bolin completed his Master's degree and began his PhD training under Rossby at Stockholm Högskola's Meteorological Institute, with 1950–1951 as a year in-between spent working with numerical weather modeling in Chicago and Princeton, where he would draw upon his mentor's extensive contact network.[1] In the years leading up to the 1967 Global Atmospheric Research Programme (GARP) meeting, during which he helped launch the Swedish space program and served as scientific director of the European Space Research Organization, Bolin would continue to expand his engagement with cutting-edge technologies, such as satellites and sounding rockets, that were revolutionizing climate science and becoming another field where Cold War Sweden showed ambitions.[2]

The long legacy of cutting-edge climate research at Stockholm University and Bolin's skills as a science organizer ensured that the Swedish capital would be situated at the center of an emerging international movement of pioneering meteorologists and atmospheric scientists that included, among others, Jule Charney, Thomas Malone, Verner Suomi, Morris Tepper, and Rolando V. Garcia. The GARP Study Conference, named after a project its planners had been calling the Global Atmospheric Research Program, was organized by Bolin on behalf of the Committee on Atmospheric Sciences. Chaired by Bolin, the Committee was established in 1964 by the International Council of Scientific Unions (ICSU), together with the International Union of Geodesy and Geophysics. This was the opening phase of the large international collaborative science programs, and the Swedes were among the early adopters taking leading roles.

[1] Rodhe, "Bert Bolin." B. Bolin, *A History of the Science and Politics of Climate Change: The Role of the Intergovernmental Panel on Climate Change* (Cambridge: Cambridge University Press, 2007). Bolin maintained a continuous correspondence with Rossby during his time in Princeton, a practice that was to continue – by necessity since it was his mentor's habit to spend long periods in the US – until Rossby's passing in 1957. Bolin papers, KVA Archives, Stockholm, and Rossby papers, MISU Archives, Stockholm University.

[2] Chapter 3 above. Sörlin & Wormbs, "Rockets and Reindeer."

GARP 1967: DEPARTURE POINT FOR THE INSTITUTIONALIZATION OF CLIMATE SCIENCE

The primary purpose of the GARP Conference was to plan a major international initiative, first proposed by Jule Charney (see Chapter 3), to study the global circulation of the troposphere and lower stratosphere. GARP had initially been conceived of as a limited-time experiment, coordinated by ICSU and World Meteorological Organization (WMO), that would, among other methodologies, involve the deployment of helium-filled balloons to conduct atmospheric measurements in different world regions.[3] Like the International Geophysical Year (IGY) a decade earlier, the GARP project was planned to last one to two years. It was, however, decided at Skepparholmen to institutionalize the GARP project by establishing it as a long-term program that would encompass a series of global experiments and observational activities over time. Based on his scientific credentials and – perhaps even more importantly – his personal integrity, communication skills, and ability to inspire others,[4] Bolin was appointed to chair the GARP Joint Organizing Committee by its sponsors, ICSU and WMO. Chairing GARP marked a milestone in Bolin's ascent as a science diplomat acting on the world stage, a career trajectory that paralleled the rise of climate change as a major political as well as scientific issue.

Over the next thirty years, during which he remained closely connected to Stockholm University as Professor of Meteorology and director of the IMI, Bolin became the central figure in the institutionalization of global climate change science. He was deeply engaged in shaping GARP successors such as the World Climate Programme (established in 1979), the World Climate Research Programme (1980), and the Intergovernmental Panel on Climate Change (IPCC) (1988), which he would help establish and chair for the first decade. The evolution of the

[3] World Meteorological Organization, International Council of Scientific Unions, International Union of Geodesy and Geophysics, *Report on the study conference held at Stockholm, June 28 to July 11, 1967, on the Global Atmospheric Research Programme (GARP)* (Geneva: World Meteorological Organization, 1967). Walter Orr Roberts, "The Global Atmospheric Research Program," *Bulletin of the American Meteorological Society* 48(1967):2, 85–88. Erik M. Conway, *Atmospheric Science at NASA: A History* (Baltimore: Johns Hopkins University Press, 2008). Henning Rodhe, "Bert Bolin and his Scientific Career," *Tellus B: Chemical and Physical Meteorology* 43(1991):4, 3–7. Henning Rodhe, "Bert Bolin (1925–2007) – A World Leading Climate Scientist and Science Organizer," *Tellus B: Chemical and Physical Meteorology* 65 (2013).

[4] Spencer Weart, "The Evolution of International Cooperation in Climate Science." B. Watson, "Bert Bolin (1925–2008)," *Nature* 451, 642 (2008).

intertwined science and politics of climate change as embodied by Bolin was ultimately reflected in the awarding of the Nobel Peace Prize to both Al Gore and the IPCC in 2007, shortly before the passing of the Panel's founding chairman at the end of that year.

In the summer of 1967, climate change was still only an incipient scientific concern limited to an international network of atmospheric specialists such as those gathered at Skepparholmen, and environmental issues were not yet entrenched on the international political agenda. Within eighteen months, however, the issue of climate change would be raised, for the first time, before the United Nations General Assembly (UNGA), by a Swedish diplomat, Sverker Åström. Sweden's UN ambassador, Åström, had been briefed on environmental issues by a scientifically well-read biochemist, Hans Palmstierna, who had recently risen to prominence after publishing a concerned scientist's environmental exposé, tellingly entitled *Plunder, Famine, Poisoning*.[5] Although quite different in style, and not written with quite the same literary prowess, Palmstierna's book can for its influence and its landmark status be considered Sweden's answer to Rachel Carson's *Silent Spring*.[6] In a similar wide-ranging indictment of humankind's growing impact on the global environment and the reciprocal effect ecological degradation was inflicting upon human well-being, Åström urged the Assembly to adopt a Swedish proposal to convene a major international conference on the human environment under UN auspices. Following Åström's speech on December 3, 1968, the UNGA unanimously adopted resolution 2398, *Problems of the human environment*, a document that had been drafted by Sweden's UN delegation with support from members of the Swedish scientific community. The successful outcome of "the Swedish initiative," one year in the making at that point, set in motion an intensive four-year preparatory period that would culminate in the 1972 United Nations Conference on the Human Environment.[7]

After the adoption of UNGA resolution 2398, the small cadre of Swedish diplomats and scientists that leveraged their respective skills to insert the human environment onto the UN agenda continued to collaborate, while also engaging a wider range of experts in support of the Stockholm Conference. Bert Bolin played a leading role in the

[5] Hans Palmstierna, *Plundring, svält, förgiftning* (Stockholm: Rabén & Sjögren, 1967).

[6] Eric Paglia, "Not a Proper Crisis," *The Anthropocene Review* 2(2015):3, 247–261.

[7] Eric Paglia, "The Swedish Initiative and the 1972 Stockholm Conference: The Decisive Role of Science Diplomacy in the Emergence of Global Environmental Governance," *Humanities and Social Sciences Communications* 8(2021):1.

Swedish preparations, but climate change was not the atmospheric issue that would represent his and Sweden's main contribution to the Conference's substantial scientific foundation. Rather, another previously opaque side effect of fossil fuel combustion – acid rain – had surfaced as a significant national issue in Sweden in autumn 1967. It was selected by Bolin as the topic of a Swedish case study in order to illustrate and draw political attention to the transboundary nature of contemporary environmental problems. As such, they required international cooperation to combat.[8]

SUMMER AND FALL 1967: BREAKTHROUGH MONTHS

In his seminal history of the environmental movement that also covers the 1972 Stockholm Conference, political scientist John McCormick suggests that the pernicious impact of acid pollution on fish and lakes in Sweden was a motivating factor behind the Swedish Initiative. Although a direct link between the discovery of acid rain and Sweden's intervention at the UN several months later is disputed by some of the primary participants,[9] the Swedish UN Mission was keenly aware of the explosive nature of the acid rain issue when drafting Sverker Åström's December 1968 address to the UNGA and did consider it a catalyst for the Swedish Initiative.[10] The issue had emerged within limited scientific circles in the spring of 1967,[11] before its major public splash that autumn. The key figures included Bert Bolin, atmospheric chemist Göran Persson of the Swedish EPA, and Bengt Lundholm, a zoologist and senior figure within Sweden's scientific bureaucracy who had led a 1964 government inquiry into natural resources. Given their scientific credentials and professional status, these three and a cadre of other experts that straddled the realms of science and policy were well positioned to sway political elites on questions related to the environment.

[8] Henning Rodhe, "Bert Bolin (1925–2007) – A World Leading Climate Scientist and Science Organizer," *Tellus B: Chemical and Physical Meteorology* 65(2013).

[9] John McCormick, *Reclaiming Paradise: The Global Environmental Movement.* (Bloomington & Indianapolis: Indiana University Press, 1989). McCormick's claim is countered by Lars-Göran Engfeldt (2009) based on his review of Swedish foreign ministry documents.

[10] It was ultimately decided not to include acid rain in Åström's speech due to the scientific uncertainty that still remained in the autumn of 1968. Lars-Göran Engfeldt, personal communication, July 23, 2022.

[11] Lars J. Lundgren, *Acid Rain on the Agenda: A Picture of a Chain of Events in Sweden, 1966–1968* (Lund: Lund University Press, 1998).

Thus, while the summer of 1967 GARP meeting laid the groundwork for the long-term institutionalization of climate science, the gathering storm of acid rain was poised to become a major societal issue as part of the environmental turn in Sweden that year.[12] It would soon spill over to become an issue of international politics.

The breakthrough was precipitated by yet another Swedish scientific expert, this time a soil specialist, Svante Odén, who would publish his consequential findings in an unconventional outlet for cutting-edge science: the culture section of Sweden's largest daily newspaper. The rapid rise of acid rain on the political agenda and the increasing level of alarm over the consequences of industrialized economic expansion were in fact the result of a pair of scientifically informed polemics that were published in October 1967. Odén's newspaper article and the aforementioned book by Hans Palmstierna together thrust the environment into the center of Swedish societal debate.[13] Public awareness of environmental issues was, however, far from nonexistent at that point. The environmental awakening that had begun several years earlier in other parts of the world, notably the United States, had soon spread to Sweden. The 1963 translation of *Silent Spring* made a powerful impact, and the discovery that mercury – used in some Swedish pesticides – was accumulating in fish and birds, and perhaps people, generated significant public concern. Research on the history of environmental concepts, with "environmental" (*miljö-*) as the prefix, and their appearance in Swedish media and other printed sources confirms very clearly that 1963 and 1964 were the years when a more widespread "environmental awareness" (*miljömedvetande*) – one of the words that entered circulation – had started its ascent.[14] Environment as a prefix had already circulated internationally for some time. For example, "environmental sciences," was used by the South African-British primatologist and science advisor Solly Zuckerman in 1959. It would continue to spread in the 1960s as a prefix in the titles of books and conferences, and eventually in university departments as well as other areas of society.[15]

[12] David Larsson Heidenblad, *The Environmental Turn in Postwar Sweden: A New History of Knowledge* (Lund: Lund University Press, 2021).

[13] Larsson Heidenblad, *The Environmental Turn.*

[14] Ann-Mari Sellerberg, *Miljöns sociala dynamik – om ambivalens, skepsis, utpekanden, avslöjanden m.m.* Research Reports from the Department of Sociology (Lund University, 1994). Paul Warde, "The Environment," In: Peter Coates, David Moon & Paul Warde, eds., *Local Places, Global Processes* (Oxford: Windgather Press, 2016), 32–46.

[15] Sverker Sörlin, "Environment," In: Noel Castree, Mike Hulme & James D. Proctor, eds., *Companion to Environmental Studies* (London & New York: Routledge, 2018), 27–32.

In Swedish, the word *miljö* itself became widely used in the 1960s. Following an international pattern, it quickly shifted meaning around that time from its previous usage in pedagogy and on social issues to its current connotation with nature, typically under pressure or in some state of precarious decay. "Environment" in the 1960s signaled pollution, chemical contamination, and the spread of toxic elements in air and water; most of these were common concerns. Public interest was considerable and carried over into politics and public discourse, as well as arts and literature. In 1966, *Spillran av ett moln* (On the Shred of a Cloud) by Rolf Edberg became a national bestseller. Not bad for a secular, quasi-philosophical sermon *qua* novel on the relationship of humans and nature by a Swedish diplomat, Social Democratic politician, journalist, and, with that book, a novelist.[16] On the applied side, the Swedish government together with industrial interests had earlier that year created the Swedish Environmental Research Institute to develop technological solutions to environmental problems associated with manufacturing and resource extraction.[17] And on July 1, 1967, Sweden established its Environmental Protection Agency, the world's first.[18]

It was a Swedish moment of momentum on environmental matters. The growing interest and concern over pollution also precipitated political action on a pan-Nordic level at the end of 1966. Preparations were at that point initiated by the Nordic Council for a high-level conference that would be held in Stockholm in May 1967. The Nordic Conference on Soil and Water Pollution, held in Lidingö outside Stockholm, involved 170 government officials, scientific experts, environmental activists, and industry representatives from Denmark, Finland, Norway, and Sweden. The conference marked the start of a multi-year intergovernmental process of institutionalizing political cooperation and scientific exchange on environmental issues between the Nordic countries. It would culminate with a first-of-its-kind treaty, the Nordic Environmental Protection Convention, which was signed in Stockholm in February 1974. The convention was the first international agreement to apply aspects of the 1972 UN Conference on the Human Environment, in particular Principle 21 of the Stockholm Declaration, which made countries

[16] Rolf Edberg, *Spillran av ett moln: Anteckningar i färdaboken* (Stockholm: Norstedts, 1966).

[17] Eva Bingel et al., *Från vetenskap till verklighet i 50 år* (Stockholm: IVL Svenska Miljöinstitutet, 2016).

[18] John McNeill, *Something New under the Sun: An Environmental History of the Twentieth-Century World* (New York: W. W. Norton & Company, 2000), 350.

accountable for any environmental damage they cause beyond their national borders. Even more than institutionalizing this principle within the Nordic context, the signatories of the 1974 Convention had their sights set on eventually expanding the idea of a transboundary pollution treaty to parts of Europe that were the source of acid rain in the North.[19]

The highly visible October 1967 interventions that would play out in the popular media and captivate the Swedish public were thus made under favorable preconditions for a receptive population and political class in Sweden. The first was Palmstierna's broad-based indictment of industrialized society. With his short book, Palmstierna would instantly become Sweden's leading environmentalist while also coming to serve as a government official, Social Democratic Party advisor, and, as will be explained later in this chapter, a key contributor to the Swedish UN initiative.[20] The other science-based critique that captured the attention of policymakers and the general public that month was targeted specifically at the atmospheric issue of acidification, the chemical process that was surreptitiously devastating soils, lakes, and forests in Sweden. As acid rain had not previously been identified as an area of acute environmental concern, it was not yet an issue of domestic or international politics. This would change literally overnight with a newspaper article published on the morning of October 24, 1967.

SVANTE ODÉN'S ACIDIFICATION INTERVENTION

The discovery of acid rain as a transboundary ecological menace was delivered to the Swedish public through what amounted to a direct action by Svante Odén (Figure 4.1), a soil chemist at the Royal Agricultural College in Uppsala, an hour's train ride north of Stockholm. Odén, who had spent several years at Bolin's Meteorological Institute at Stockholm

[19] Melina Antonia Buns 2021a and 2021b. "The Emergence of Nordic Environmental Cooperation, 1967–1974" https://nordics.info/show/artikel/the-emergence-of-nordic-environmental-cooperation-1967-1974/ and "The 1974 Nordic Environmental Protection Convention" https://nordics.info/show/artikel/the-1974-nordic-environmental-protection-convention/. Melina Antonia Buns, "Making a Model: The 1974 Nordic Environmental Protection Convention and Nordic Attempts to Form International Environmental Law." *Scandinavian Journal of History*, 2022. https://doi.org/10.1080/03468755.2022.2069151.

[20] Eric Paglia, *The Northward Course of the Anthropocene: Transformation, Temporality and Telecoupling in a Time of Environmental Crisis*, PhD diss. (Stockholm: KTH Royal Institute of Technology, 2016).

FIGURE 4.1 Soil chemist Svante Odén, whose scientific discovery and October 1967 exposé of acid rain in the Swedish press brought transboundary environmental problems to the attention of the public and policymakers in Sweden and beyond. Photo: Sture Foto/Scanpix.

University in the early 1960s,[21] had for the previous decade been managing the European Air Chemistry Network (EACN). The EACN was a series of monitoring sites in Sweden and other parts of Europe that measured the chemical composition of precipitation, providing data over extended durations and large spatial scales. The network had been established in the late 1940s by soil scientist Hans Egnér and Erik Eriksson, a meteorologist and later professor of hydrology at Uppsala University who over the course of the 1950s employed EACN data in developing a general theory of biogeochemical circulation.[22]

An international expansion of the network was facilitated by Carl-Gustaf Rossby. In EACN, Rossby saw an opportunity to improve his understanding of long-range nutrient cycles and the relationship between

[21] Henning Rodhe, "Bert Bolin."

[22] Ellis B. Cowling, "Acid Precipitation in Historical Perspective," *Environmental Science & Technology* 16(1982):2, 110A–123A.

atmospheric and biogeochemical processes. Rossby and Eriksson, who joined Stockholm University's Meteorological Institute in 1954, were at the time leading the rapid development of the integrative field of atmospheric chemistry.[23] They convened a series of international conferences in Stockholm in 1954, 1955, 1956, and 1957 that brought together experts from a range of disciplines, including meteorology, oceanography, geography, and agriculture. The conferences' proceedings were presented in the Meteorological Institute's in-house scientific journal *Tellus*,[24] where some of the first articles on atmospheric chemistry and "Precipitation Chemistry" had been published.[25] Among the conference participants was Odén, who would take over the management of EACN after Rossby's death in 1957,[26] and Eriksson's departure for Yale that same year. Odén's immersion in the massive amount of data generated by EACN would a decade later lead to his discovery of the destructive environmental effects of acid precipitation.

Twenty years of EACN data analyzed by Odén demonstrated that, by 1967, acid rain was in fact intensifying across time and space. Rather than first publishing the results of his research on the likely cause of increasing acidification of Scandinavian ecosystems in a peer-reviewed academic journal, Odén went to the popular press with his explosive findings. His article "Nederbördens försurning" (The Acidification of Precipitation), which evoked "an insidious chemical warfare among the nations of Europe," appeared in *Dagens Nyheter* on October 24, 1967, provoking the public outcry and political response that Odén had intended.[27] In the article, appearing in the newspaper's culture

[23] Cowling, "Acid Precipitation."

[24] Erik Eriksson, "Report on an Informal Conference in Atmospheric Chemistry Held at the Meteorological Institute, University of Stockholm, May 24–26, 1954," *Tellus* 6(1954):3, 302–307. Erik Eriksson, "Report on the Second Informal Conference on Atmospheric Chemistry, held at the Meteorological Institute, University of Stockholm, May 31 – June 4, 1955," *Tellus* 7(1955):3, 388–394. W. A. Mordy, "Report on the Third Annual Conference on Atmospheric Chemistry, May 28–30, 1956," *Tellus* 9(1957):1, 127–134. Georg Hugo Neumann. "The Fourth Annual Conference in Atmospheric Chemistry May 20–22, 1957," *Tellus* 10(1958):1, 165–169.

[25] American Meteorological Society University Corporation for Atmospheric Research, Tape Recorded Interview Project. Interview with C. C. [Carl Christian] Wallén conducted by Gordon Cartwright. November 21, 1995, December 9, 1995. Transcript retrieved November 18, 2021 from https://opensky.ucar.edu/islandora/object/archives%3A7653/datastream/OBJ/view

[26] Rachel Emma Rothschild, *Poisonous Skies: Acid Rain and the Globalization of Pollution* (Chicago: University of Chicago Press, 2019).

[27] Peringe Grennfelt et al., "Acid Rain and Air Pollution: 50 Years of Progress in Environmental Science and Policy," *Ambio* 49(2020): 849–864.

section – in and of itself a sign of how widespread interest in the environment had become – he attributed declining pH levels in waters and soils, and the consequent ecological degradation in Sweden, to the long-range transport of sulfur dioxide from the burning of fuel oil and coal in continental Europe and the United Kingdom.[28] Odén's acidification hypothesis was further amplified in an address by Sven Brohult, director of the respected Swedish Academy of Engineering Sciences, delivered the same night as the article's publication in *Dagens Nyheter*. Within days, the article was picked up by European and North American media, and the question of acid rain received additional treatment in Swedish scientific committees and government reports in the weeks and months that followed. With backing from figures such as Bengt Lundholm, a leading member of the Air Protection Council at the Swedish EPA among other official responsibilities, acid rain generated an immediate reaction by the Swedish government.[29] The issue was brought to the Organisation for Economic Co-operation and Development (OECD), where it initially received a tepid response. In the years following Odén's scientific exposé, however, transboundary acid rain would become one of the first, most prominent, and truly international environmental issues. It would also become the central scientific focus of Sweden's UN Stockholm Conference preparations.

HANS PALMSTIERNA'S POPULAR SCIENCE POLEMIC

The other October 1967 media event that brought the environment to the forefront of Swedish societal debate was the publication and extensive press coverage of Hans Palmstierna's (Figure 4.2) bestseller. In *Plunder, Famine, Poisoning*, Palmstierna situated the ongoing despoilment of the natural environment and its reciprocal impact on human well-being within a long-term narrative of the technological advance and industrialization of societies in the Global North. The slim paperback consisted of a series of short chapters that catalogued a selection of environmental problems such as soil erosion, overexploitation of the oceans, and the chemical contamination of ecosystems and human bodies resulting from the use of pesticides. Foreshadowing a

[28] Arne Kaijser, "Combatting 'Acid Rain': Protecting the Common European Sky," In: Anna-Katharina Wöbse & Patrick Kupper, eds., *Greening Europe: Environmental Protection in the Long Twentieth Century – A Handbook* (Berlin & Boston: De Gruyter Oldenbourg, 2021), 363–388.

[29] Lundgren, *Acid Rain*, 78–94.

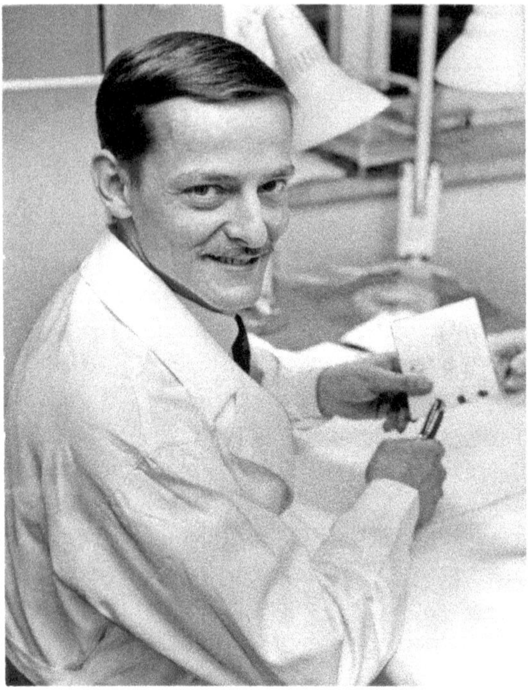

FIGURE 4.2 Biochemist Hans Palmstierna at work at a government bacteriological lab in 1965. His bestselling polemic *Plunder, Famine, Poisoning* helped spark the environmental awakening in Sweden in autumn 1967. He also played a critical scientific role in the Swedish initiative at the United Nations that led to the 1972 Stockholm Conference. Photo: Owe Sjöblom/Svenska Dagbladet.

fundamental theme of the Stockholm Conference, *Plunder* also condemned global inequalities and endemic poverty, incorporating a proto-sustainable development and environmental justice outlook grounded in Palmstierna's and Sweden's sense of solidarity with the Global South. Without radical reform and coordinated international action, he concluded, population growth and modern society's widespread waste of natural resources would jeopardize the survival of human civilization.

In the weeks following its publication, reviews and even editorials praising *Plunder* appeared in most major newspapers in Sweden, and Palmstierna himself became a media phenomenon – a "meta-specialist" that could speak broadly and with scientific authority on the complex and multifaceted environmental crisis that seemed to be spreading and accelerating.[30] He was interviewed on prominent news programs, penned newspaper articles, and his book became, in a Swedish popular movement tradition with roots going back to the beginning of the century, the object of discussion in "study circles" across the country.[31] Palmstierna's general critique can be seen in contrast, and perhaps as a complement, to the singular focus on the ecologically destructive sulfur dioxide – acid rain relationship exposed by Odén.[32] The pair of high-profile interventions, together with similar yet less sensational contributions to the public debate in the autumn of 1967, initiated a turning point in Swedish environmental consciousness. Rather than just a passing moment in the environmental history of a small northern European nation, the environmental awakening in Sweden set the stage for a diplomatic gambit at the UN several weeks later that would lead to the establishment of the human environment as an issue of international concern.[33]

THE SWEDISH INITIATIVE: SCIENCE DIPLOMACY AT THE UNITED NATIONS 1967–1968

The catalyst for the diplomatic process that would culminate in the Stockholm Conference came from Inga Thorsson (Figure 4.3), a Swedish politician and diplomat who in 1967 had become Director of the Social Development Division at the United Nations. A dedicated nuclear disarmament and women's rights advocate, Thorsson and her career trajectory drew inspiration from a 1941 book by the Swedish author and journalist Elin Wägner (see Chapter 1). In *Väckarklocka* (Alarm Clock), Wägner had put forward a feminist and ecological perspective, somewhat ironically called "functional poverty." This was a program which in its

30 David Larsson Heidenblad, "Mapping a New History of the Ecological Turn: The Circulation of Environmental Knowledge in Sweden 1967," *Environment and History* 24(2018):2, 265–284. Paul Warde, Libby Robin & Sverker Sörlin, *The Environment – A History of the Idea* (Baltimore: Johns Hopkins University Press, 2018).

31 Larsson Heidenblad, *Environmental Turn.* Staffan Larsson & Henrik Nordblad, *Study Circles in Sweden: An Overview with a Bibliography of Existing Literature* (Linköping: Linköping University Electronic Press, 2010). http://urn.kb.se/resolve?urn=urn:nbn:se:liu:diva-57887

32 The first edition of *Plunder* did not make reference to acid rain, but following Odén's revelation in October 1967, future editions of Palmstierna's book, as well as many of his media interventions, mentioned the acid rain issue (Lundgren 1998). The scientific material Palmstierna produced in 1968 in support of the Swedish Initiative also included references to acid rain.

33 Paglia, "The Swedish Initiative."

FIGURE 4.3 Politician and diplomat Inga Thorsson, a strong advocate of women's rights, nuclear disarmament, and international development, who sparked the Swedish initiative at the United Nations in 1967, leading to the convening of the Stockholm Conference five years later. Photo: Freddy Lindström/Scanpix.

main thrusts (though perhaps not the acclaim of poverty) was shared by Thorsson, who also questioned aspects of patriarchal industrial society and its reliance on the unconstrained advance of technology.[34] Wägner, a study circle supporter teaching at a women's "folk high school" for rural women with little formal education and also a member of the Swedish Academy that awards the literary Nobel Prize, had become a controversial figure for her 1941 book. Not everyone liked its ideas of small-scale economics. It was heavily critiqued by Karin Koch, an up-and-coming economist who became the first female member of the Swedish government in 1949. "Does Elin Wägner want to take us back to a preindustrial society based on natural self-sufficiency, or not?"[35]

[34] Elin Wägner, *Väckarklocka* (Stockholm: Bonnier, 1941).
[35] Karin Koch, review in *Social-Demokraten*, cited in Boel Hackman, *Elin Wägner* (Stockholm: Bonniers, 2005).

Koch represented the mainstream Social Democrat opinion of industrial modernization. For a long time, voices such as Wägner's had a hard time finding their way as the party was leading the country into a future based on increasing wealth and equal welfare, and with a limited understanding of nature preservation. Wägner and other green feminists were torn between industrial modernism and retrospective utopianism, along the lines of the anarchist Peter Kropotkin, the agrarian reformist Henry George, and other rural radicals. This was a far cry from Palmstierna's science and technology approach to radicalism, and it was he who became the government's environmental advisor, rather than a representative of the green camp.[36] A few decades later, the fault lines of this conflict were reshuffled as a first wave of green transformation shook Swedish political and public life. It did not in any major way change the ideological ball game, but the discussion broadened, and it certainly brought with it a whole new level of agency and activism.

Thorsson's environmental concerns were coupled with a keen interest in international development issues, which she worked on as an expert at the Swedish Foreign Ministry prior to her UN assignment.[37] Not long after her arrival in New York, Thorsson became involved in discussions on organizing a meeting on environmental issues under the UN Economic Commission for Europe – one of the five regional commissions under the United Nations Economic and Social Council (ECOSOC). This experience encouraged her to approach Under Secretary-General Phillipe de Seynes with the idea of convening a major international conference on the human environment at the highest level of the United Nations system. Such a conference, she suggested, could possibly supplant the fourth International Conference on the Peaceful Uses of Atomic Energy, which had recently been proposed by the UN Scientific Advisory Committee. The idea appealed to de Seynes, who was skeptical of yet another nuclear energy conference, and was strongly supported by other officials at the UN Secretariat. Thorsson's proposal was also firmly backed by Alva Myrdal, the politically influential head of Sweden's UN delegation, whose blessing provided the Swedish diplomats in New York the operating space to develop the initiative as they saw fit.[38]

36 Jonas Anshelm, *Socialdemokraterna och miljöfrågan: Om framstegstankens paradoxer* (Stehag: Symposion, 1995).

37 Gunnel Karlsson, "Inga Margarethe Thorsson," *Svenskt kvinnobiografiskt lexikon*, www.skbl.se/sv/artikel/IngaThorsson (retrieved November 16, 2021).

38 This section draws upon the experiences and insights of Ambassador Lars-Göran Engfeldt, including his 2009 and 2019 books, the latter his diplomatic memoirs, and

FIGURE 4.4 Swedish Prime Minister Tage Erlander (left), future Prime Minister Olof Palme (middle), and ambassador Sverker Åström (right), Sweden's permanent representative at the United Nations, in 1965. A political insider and major figure in shaping Swedish foreign policy during the post-war decades, Åström devised and implemented the diplomatic strategy that ensured the successful outcome of the Swedish initiative for an environmental conference under UN auspices. Photo: Sture Waldö/Aftonbladet.

The diplomatic strategy that was to unfold over the next twelve months was devised by Sweden's UN ambassador Sverker Åström (Figure 4.4), one of the most influential diplomats and shapers of

a series of interviews, conversations, and correspondences between Engfeldt and Eric Paglia between 2014 and 2020. Lars-Göran Engfeldt, *From Stockholm to Johannesburg and Beyond: The Evolution of the International System for Sustainable Development Governance and Its Implications* (Stockholm: Government Offices of Sweden, 2009); Lars-Göran Engfeldt, *Diplomati i förändring: I UD:s tjänst 1965–2009* (Stockholm: Ekerlids Förlag, 2019). Sverker Åström's 1992 memoirs were also a key resource for this section, Sverker Åström, *Ögonblick: Från ett halvsekel i UD-tjänst* (Stockholm: Bonnier Alba, 1992). Göran Bäckstrand of the Swedish Foreign Ministry was another important participant in the processes described in this section and a further source of first-hand insight gained through several interviews and email exchanges with Eric Paglia. Documents from the Swedish Foreign Ministry archives (in Sweden's National Archives, https://sok.riksarkivet.se/) also underpinned much of the empirical analysis in this section. Additional analysis of archival documents can be found in Paglia, "The Swedish Initiative."

Swedish foreign policy of his generation. Before becoming permanent representative at the Swedish UN mission in 1964, Åström held senior positions at the foreign ministry, where he was the head of the political department from 1956, and served at Sweden's embassies in Moscow, Washington, and London. In Åström's estimation, government leaders had been too slow in recognizing the global significance of the environmental issue. Emboldened by the environmental awakening back home in Sweden, he was convinced that the time had become ripe for international action and that the diplomatic initiative would receive domestic political support.[39] His strategy centered on working through ECOSOC to bring a formal motion before the General Assembly for adoption.[40] This would entail a mobilization of Swedish bureaucratic and scientific resources to help formulate a convincing case that Åström and his colleagues could use to persuade other, possibly skeptical or even hostile, delegations on the imperative for convening an environmental conference under UN auspices. Ironically, the intellectual basis of the conference imagined by Thorsson, Myrdal, and Åström represented a de facto rebuke of the model of industrialized economic growth that had made Sweden's welfare state the envy of much of the world. Demonstrating his commitment to internationalism, and despite his position as a government official, even the institution of national sovereignty was seen by Åström as an obstacle in tackling transnational threats to the human environment. The UN, in the eyes of Åström, represented the best hope for humanity in its global struggle with overpopulation, resource depletion, and the spread of toxic substances.[41]

The first step was to swiftly inject the idea of an environmental conference into circulation at the United Nations by way of a short statement before the General Assembly. Delivered on December 13, 1967, by Börje Billner, the deputy head of Sweden's UN mission, the main points of the Swedish proposal were reminiscent of the radical reflexivity contained in the writings of Elin Wägner and Hans Palmstierna, despite all their differences. In calling for a conference on the human

39 Åström, *Ögonblick*; Engfeldt, *From Stockholm to Johannesburg*.

40 Based on UN procedure, ECOSOC was a necessary first step for the UNGA to take up a matter like the Swedish Initiative.

41 Sverker Åström, "Global Consensus or Global Catastrophe?" *Ambio* 1(1972):1, 2–5. www.jstor.org/stable/4311934. Sverker Sörlin, "The Environment as Seen through the Life of a Journal: *Ambio* 1972–2022." *Ambio* 50(2021): 10–30, on 15. https://doi.org/10.1007/s13280-020-01421-w

environment, Billner identified technology-driven industrial development as the principal driver of various forms of pollution and their subsequent effect on humans:

> The impact of the technological revolution that is taking place around us is felt by all peoples, irrespective of their present technological level. It has far-reaching effects on the environment of man. The human body and the human mind are subjected to serious and ever-increasing inconveniences and dangers. These are caused by air pollution, water pollution, sulfur fall-out waste, etc. – in short by all the secondary effects related to the process of industrialization and urbanization.[42]

The statement makes clear that even at this early stage of the "Swedish initiative," as Sweden's diplomatic intervention was referred to in UN circles, human wellbeing would be a primary focus, perhaps even more so than the environment itself. For the purposes of the conference, as imagined by the diplomats at Sweden's UN mission, the environment was a medium through which technology, industrialization, and urbanization impacted people, rather than an object explicitly valued in and of itself, separate from its influence on humans. This anthropocentric conception of the environment – the *human* environment – was a 1960s novelty that complemented and to some extent supplanted conservationist and preservationist ethics that saw nature as something that should be set aside to be managed or protected against human encroachment. The Swedish initiative and the Stockholm Conference were decisive in elaborating and elevating the new human environment concept to the level of international politics. Although "human environment" today may sound anachronistic, the human dimension is implied and almost always present – if not predominant – in discussions on the environment, which are often integrated or interchangeable with the concept of sustainable development.

The overarching themes of Billner's brief statement – preliminary or even tentative as it was at that point – would become further developed and scientifically substantiated as the Swedish initiative progressed over the course of 1968. In a memorandum telexed to the foreign ministry in Stockholm on March 13, 1968, Åström outlined his diplomatic strategy and general objectives for the conference. These included increasing worldwide environmental awareness, engagement, and cooperation

42 "Statement by Mr. Börje Billner before the General Assembly on December 13, 1967, on agenda item 27, Question of holding further conferences on the peaceful sues of atomic energy," cited in Paglia, "The Swedish Initiative."

among actors from government, science, and civil society. He also identified potential diplomatic pitfalls, such as opposition from financial interests or from countries that might see the environment as potentially sidetracking the development agenda, and targeted two upcoming ECOSOC sessions in May and July 1968 as opportunities for advancing the idea of an environmental conference and securing support from other UN member states.[43] The first stage of diplomatic strategizing took place almost exclusively in New York, where Åström consulted with other UN delegations as well as experts such as the American cultural anthropologist and public intellectual Margaret Mead.[44] Moving forward with the initiative, however, demanded greater consultation and coordination with Stockholm, and the participation of relevant government ministries and specialized agencies. Moreover, preparing for the ECOSOC sessions, where Åström would make the case for a conference based on the state of the global environment, required a level of scientific expertise that the Swedish UN mission did not have in house. The Foreign Ministry therefore convened a pair of interagency meetings in April 1968 that were pivotal in formulating the next phase of the Swedish initiative.

STOCKHOLM, 1968: FORTIFYING THE SWEDISH INITIATIVE WITH ENVIRONMENTAL EXPERTISE

Attending the first meeting, held on April 10, were senior officials from government authorities that represented sectors such as environmental protection, public health, scientific and defense-related research, international development, and foreign affairs.[45] Sverker Åström, chairing the meeting, laid out the overarching objectives of the conference: to raise environmental awareness internationally and align action between governments, scientists, and industry. He emphasized that the environment should be interpreted in the widest possible manner to encompass its range of relevant social and economic development dimensions. Related to this broad interpretation of the issue, Åström and the other officials agreed that Sweden's call for universal engagement on the environment needed to be carefully motivated in anticipation of potential pushback by some developing countries. This concern proved justified.

43 Sverker Åström, Memorandum to Swedish Foreign Ministry," March 13, 1968.

44 Engfeldt, *From Stockholm to Johannesburg*.

45 Kanslisekreterare Göran Bäckstrand, Swedish Foreign Ministry, Promemoria April 11, 1968, "FN-konferens för miljövård."

While India and especially Brazil, as well as other countries of the Global South, would later express degrees of skepticism, it was France and the United Kingdom that would be the first countries to question the need of convening an environmental conference under United Nations auspices.

Åström and Börje Billner also foresaw resistance on another front: established UN bodies that could very well perceive an environment conference, and new institutions that might follow from it, as an intrusion on their organizational turf. UNESCO, for example, considered the environment to be within its remit and was convening the Man in the Biosphere conference just a few months later. Although, as Billner noted at the meeting, the Biosphere conference, scheduled for September 1968, was narrowly focused on conservation issues as compared to the much more expansive human environment conference envisioned by the architects of the Swedish initiative. Navigating the intricacies and sensitivities of the UN system presented a formidable challenge that would command Åström's attention in the months ahead. The more immediate concern that he brought up at the interagency meeting in Stockholm was the drafting of a thirty to forty-page scientific document that could underpin a two-hour presentation at the United Nations. Åström considered the report he called for to be of the utmost importance since, from his extensive diplomatic experience, he expected it would serve as a point of reference and framework for conference preparations going forward.[46]

Two weeks later, the second Foreign Ministry meeting addressed the imperative of infusing Sweden's incipient environmental diplomacy with scientific substance.[47] On the agenda was the key statements Åström would deliver at the upcoming ECOSOC sessions. These needed to be scientifically grounded in order to convince other delegations of the rationale for convening a major environmental conference. Representing the Swedish EPA, Hans Palmstierna was tasked with preparing a memorandum that would provide a scientific perspective on the declining state of the global environment. One year after he had penned *Plunder*, the report Palmstierna would rapidly produce in support of the Swedish initiative was not limited to scientific facts and

[46] Göran Bäckstrand, Swedish Foreign Ministry, Promemoria April 16, 1968, Sammanträde om miljövårdsproblem.

[47] Kanslisekreterare Bäckstrand, Swedish Foreign Ministry, Promemoria April 25, 1968, "Kort resumé av diskussioner och beslut vid sammanträde med den förberedande arbetsgruppen för eventuell FN-konferens kring miljövårdsproblem."

analysis. Similar to his popular science writing, Palmstierna's assessment included strong normative positions, historical comparisons, and political prescriptions. The report condemned unequal access to natural resources over time and space, as well as the recurring pattern of small groups enriching themselves at the expense of the natural environment and wider swaths of humanity. Since modern environmental problems and associated development issues were qualitatively different from those of the past, often transcending political boundaries or driven by transnational processes, Palmstierna asserted that measures taken at the local or national level would in many cases be futile. He therefore called for solutions based on international solidarity:

> As will be evident, many of the problems are international and must be solved internationally. The very big problems, common to all nations, cannot be solved by any individual country. A prerequisite to the solution of these problems is an intimate and trustful cooperation between all the nations of this globe.[48]

The two-part memorandum covered a wide range of specific aspects of the environmental crisis, and in turn, infused Åström's ECOSOC statements with a degree of scientifically informed dramaturgy and a sweeping historical narrative on the relationship of humans to the environment.

Situating the Swedish initiative within the context of the UN Development Decade, Åström's July 19 ECOSOC speech drew upon Palmstierna in outlining the main components of the physical environment that supported human survival: fertile soil, fresh water, living oceans, and clean air, and how human activity had undermined their quality and integrity. Åström evoked what would today be called ecological tipping points and cited the emergence of transboundary environmental problems such as acid rain, eutrophication, and even human-driven climate destabilization as motivating factors for international cooperation. The lion's share of the blame for the deteriorating condition of the global environment was assigned to accelerating urbanization and unrestrained industrialization – a cautionary tale from the Global North that developing countries should avoid repeating. A pillar of the Swedish initiative was precisely this idea that societies at earlier stages of industrialization could draw lessons from the myriad mistakes made by wealthier nations in the course of their own development, which came at great

[48] Letter containing a translation of Palmstierna's text (Swedish to English) sent by Lennart Eckerberg at the Swedish Foreign Ministry to Sverker Åström on July 5, 1968, "Översättning av svensk miljövårdsdokument till ECOSOC/ Anförande i FN:s Ekonomiska och sociala råd (ECOSOC)."

human and environmental cost. Learning from these historical experiences, according to Åström and Palmstierna, developing countries could embark upon pathways toward prosperity that were more sustainable than the kind of unreflexive economic growth paradigm that had devastated the biosphere.

Internalizing the environment at an early phase of economic planning would also have the added benefit of reducing costs associated with ecological degradation later on. Furthermore, despite differing levels of development, the environment had become a common concern for all of humanity, impacting people and countries in all parts of the world. The Swedes at the center of the initiative were convinced that a major international conference on the human environment could serve as a catalyst for mobilizing the resources, knowledge, and political will necessary to rethink, reform, and redirect this deeply flawed trajectory of industrialized economic development. Particularly in places where its roots did not yet run deep. To this end, the synthesis of Sverker Åström's diplomatic acumen and Hans Palmstierna's scientific expertise and dramatic flair in framing the environmental crisis provided the Swedish initiative with a strong science diplomacy foundation upon which a compelling case could be made.[49]

Both of Åström's ECOSOC statements,[50] as well as his December 1968 UNGA speech,[51] mirrored many of the points contained in Palmstierna's memorandum, which reiterated themes such as overconsumption and chemical poisoning from his popular science bestseller *Plunder*. Recent advances in science and technology, according to Palmstierna, had led to excessive exploitation of natural resources, rendering environmental degradation a structural condition in industrial societies and a rupture in humanity's long-term relationship with nature. On what he considered a fully settled planet characterized by exponential population growth, migration was no longer an option in the face of localized resource depletion. International cooperation, assistance, and solidarity, as well as the sharing of appropriate technologies, were critical necessities to

[49] Paglia, "The Swedish Initiative."

[50] "Statement by the Representative of Sweden, Ambassador Sverker Åström, on agenda items 5 (c) and (d) in the Economic Committee of the Economic and Social Council, May 24, 1968"; and, "Statement by Ambassador Sverker Åström, on agenda item 12, the question of convening an international Conference on the problems of human environment, in the ECOSOC meeting in Geneva, on July 19, 1968." Swedish Ministry for Foreign Affairs.

[51] "Statement by the Swedish Representative, Ambassador Sverker Åström, in the General Assembly on Tuesday, 3 December, 1968."

secure less destructive development pathways for countries of the Global South while also reforming the economies of the Global North. A crucial aspect of this narrative was that a healthy environment was a prerequisite for development, rather than something that could be sacrificed on the altar of economic growth. This relationship between environment and development would prove to be an essential source of tension later in the conference's preparation period (see Chapter 5) and in the decades since Stockholm.

The core arguments and main objectives of Sweden's science diplomacy carried through to the decisive UN resolutions to which the emergence of global environmental governance, as associated with the Stockholm Conference, can be traced. Even after undergoing scrutiny from UN bodies and member states, the texts of both foundational documents – UNGA resolution 2398 *Problems of the human environment*, and ECOSOC resolution 1346 (XLV), which recommended the General Assembly take up the idea of an environmental conference – retained indelible Swedish fingerprints. The ECOSOC resolution strongly resembled the Explanatory Memorandum that Sverker Åström had submitted to UN Secretary General U Thant to formally inscribe the Swedish initiative on the ECOSOC agenda.[52] And the seminal UNGA 2398 was written by a young diplomat at Sweden's UN mission, Lars-Göran Engfeldt (Figure 4.5), who also drafted Åström's pivotal speech before the General Assembly on December 3, 1968, where the resolution was adopted unanimously.[53]

The Swedish initiative embodied by Åström's statements and the UN resolutions was a formative moment in the globalization of environmental politics. Further, over the course of the almost four-year preparation period that followed, international development and its links to the environment would not only become an intrinsic part of the proposed conference but also the basis of the future concept of sustainable development. In front of the full Assembly in December 1968, Åström revisited many of the points he had made before ECOSOC earlier in the year, articulating his vision of an action-oriented event lasting two to three weeks with broad civil society participation. To avoid complications

[52] The Explanatory Memorandum included a cover letter dated May 20, 1968 written by Sverker Åström and addressed to U.N. Secretary-General U Thant requesting that an item entitled "The question of convening an international conference on the problems of human environment" be inscribed on the agenda of the 45th Session of the Economic and Social Council.

[53] Lars-Göran Engfeldt, personal communication (Paglia), March 8, 2016.

FIGURE 4.5 Swedish UN delegation and Crown Prince Carl-Gustaf in September 1970 at the 25th anniversary commemoration of the founding of the United Nations. First row from left: Ambassador Lennart Petri, Crown Prince Carl-Gustaf, Sweden's UN Ambassador Olof Rydbeck. Second row from left: Secretary of Embassy Lars-Göran Engfeldt, First Secretary of Embassy Torsten Örn, and Minister Kaj Sundberg. Photo: United Nations.

inside the UN system, he also emphasized that he did not foresee any new institutions resulting from the conference. The scientific case made by Åström also followed his previous statements, with some small but significant modifications. Global warming caused by carbon dioxide emissions was according to Åström now the more likely climate change scenario, compared to his earlier ECOSOC statements that had given equal weight to the possibility of global cooling from particle pollution. Perhaps surprisingly, acid rain was not mentioned in Åström's UNGA speech. The issue was, however, of great concern for Sweden and would soon become a central aspect of Swedish activity during the preparation period set in motion with the adoption of UNGA 2398.

5

Twelve Days in Stockholm, June 1972

Even after the Swedish initiative evolved into an official item on the UN agenda, Sverker Åström and the small circle of diplomats at Sweden's UN mission remained indispensable in leading preparations for the conference. Soon after the adoption of UNGA 2398, Åström communicated to the Foreign Ministry that other UN delegations expected that Sweden would maintain its de facto conference leadership position during the preparation process. Swedish officials would continue to play such a role until autumn 1970. In a diplomatic maneuver orchestrated by Åström, Swedish Prime Minister Olof Palme then reached out to his Canadian counterpart Pierre Trudeau for the purpose of recruiting Maurice Strong (Figure 5.1), a former energy company executive and head of the Canadian International Development Agency, to become the conference's secretary general. Strong's eventual appointment filled an increasingly apparent leadership vacuum at the executive level and brought coherence as well as an injection of personal energy and flavor to the preparation process.[1]

Up until the time of Strong's appointment, Sweden's imprint on conference preparations, organizationally as well as scientifically, remained predominant. This was both by design and practical necessity. Exercising the political influence he enjoyed back in Stockholm, Åström in February 1969 convinced the Swedish government to extend an invitation to the United Nations to host the conference, providing him with additional leverage in shaping the agenda at an early stage of the

1 Åström, *Ögonblick*. Engfeldt, *From Stockholm to Johannesburg*. Wade Rowland, *The Plot to Save the World* (Toronto: Clarke, Irwin & Company, 1973). Canadian Prime Minister Justin Trudeau is the son of Pierre Trudeau.

FIGURE 5.1 The recruitment of Maurice Strong in early 1970 to become Secretary General of the upcoming Stockholm Conference was a diplomatic masterstroke that would help alleviate the emerging North-South tensions that threatened the conference. Strong's personal energy and background as a businessman and head of the Canadian International Development Agency were indispensable in convincing countries of the Global South that the conference would not subordinate their development interests to environmental protection. Twenty years later, Strong would be called upon to serve in the same capacity for the 1992 Rio Earth Summit. Photo: UN Photo/Yutaka Nagata.

preparation period. He demonstrated considerable diplomatic finesse in enrolling other UN delegations in supporting the Swedish-led planning process and orchestrated negotiations surrounding another UNGA resolution, which reaffirmed the aims of the conference and provided greater structure for preparations going forward, including the creation of a twenty-seven-country Preparatory Committee. Åström's persistent engagement brought much-needed coherence to the arrangement of the Committee's critical first gathering in March 1970, which took place amid mounting uncertainty over how conference preparations should proceed.[2]

[2] Engfeldt, *From Stockholm to Johannesburg*.

The first PrepCom meeting, as well as a series of informal consultations orchestrated by Åström in advance, represented an important milestone on the road to Stockholm. The meeting took place several months before Strong signed on as secretary general and can be seen as the climax of the two years of Swedish conference leadership. In his opening statement on March 10, Åström put forward a set of fundamental principles and primary objectives for the conference that could guide the efforts of the Preparatory Committee. These included the crafting of a United Nations Declaration on the Human Environment and the drafting of conventions on particular environmental problems, such as pesticide toxification of ecosystems. They also encompassed the creation of a comprehensive global system for environmental monitoring that would collect scientific data on issues ranging from marine pollution to rising atmospheric CO_2 levels in support of national and international policy action. An overarching principle for the entire conference, according to Åström, was that "action in the environmental field must be seen as an integral part of the whole process of economic and social development everywhere ... to provide people with the kind of environment which is necessary for man's material prosperity, physical and mental well-being and spiritual life."[3]

Åström's statement, as well as documentation on environmental issues circulated in advance of the PrepCom meeting, was drafted by Sweden's UN mission with substantial support from Swedish scientists – several of whom traveled to New York to help prepare for and take part in the meeting as part of the Swedish delegation. Supporting the efforts of Åström, Engfeldt and Billner, the Swedish science diplomacy contingent included Hans Palmstierna, at that point secretary of the National Environmental Advisory Committee, and Arne Engström, professor of medical biophysics at the Karolinska Institute and secretary of the influential Scientific Advisory Council of the Swedish Government.[4] Palmstierna and Engström were members of a group of hand-picked experts that the Swedish government had established for the purpose of supporting Sweden's conference preparation efforts. With Engström as chair, the Committee for Research and Factual Issues (*Kommittén for forskning och andra sakfrågor*, CRF) would not only oversee Sweden's

[3] Statement by Ambassador Sverker Åström, Representative of Sweden, on Tuesday, March 10, 1970. *Preparatory Committee for the United Nations Conference on the Human Environment*, first session, March 10–20, 1970.

[4] United Nations General Assembly, *Report of the Preparatory Committee for the United Nations Conference on the Human Environment.* April 6, 1970. A/CONF.48/PC/6.

scientific output in advance of the conference but also contribute to the preparatory work within the UN Secretariat through diplomatic channels that included a junior member of the Swedish mission who was situated at the center of conference preparations.[5]

Swedish experts had already been providing scientific advice to conference leadership for over a year by the time of the first PrepCom meeting in March 1970. It had become apparent as early as January 1969 that the UN Secretariat and its understaffed Office for Science and Technology lacked the in-house expertise and organizational capacity required to carry out the task of preparing an event on the scale and complexity of the human environment conference. This led to an informal arrangement in which the young diplomat Lars-Göran Engfeldt, who had recently arrived in New York to serve as second secretary at the Swedish UN mission, was discretely attached to the UN Secretariat.[6] Inside the Secretariat, he would serve as a diplomatic advisor as well as a conduit through which experts back in Stockholm could convey much-needed scientific information on the environment to conference planners. Engfeldt's strategic placement would also provide an additional avenue for Swedish influence over the preparatory process.

One direct result of Engfeldt and Åström's sway within the Secretariat was a call for participating countries to produce national environmental reports. Although the conference was conceived as a forum and catalyst for the internationalization of the environment, a primary ambition of the Swedish initiative was to mobilize societies, scientific institutions, and public authorities at the national level. If the conference and its preparations were to only involve UN delegations without deeper domestic engagement in home countries, the Swedes feared that the lasting societal impacts they had envisioned would never come to fruition. They therefore developed a strategy that would promote greater societal participation in conference preparations while also producing a global inventory of environmental knowledge. Operating through the UN Secretariat, Åström and Engfeldt arranged for a circular note to be distributed to member states mandating them to produce national environmental reports and issue-specific case studies as part of their conference preparation efforts.

In the late 1960s, the level of scientific knowledge on environmental issues was relatively poor in many countries, particularly in the

[5] Engfeldt, *From Stockholm to Johannesburg*. Jan Mårtenson. *Att kyssa ett träd: Memoarer* [To Kiss a Tree: Memoirs] (Stockholm: Wahlström & Widstrand, 2000).

[6] Engfeldt, *From Stockholm to Johannesburg*.

developing world. Moreover, links between the government and other societal sectors where research on the environment could take place were tenuous or nonexistent. Hence, in order to stimulate interaction between public authorities and experts from civil society, the Secretariat provided detailed instructions for drafting reports that, by design, compelled governments to coordinate with their national scientific establishments. This unheralded yet significant institution-building aspect of the Swedish initiative resulted in over eighty national reports and a range of environmental case studies that provided an unprecedented overview of the state of the global environment.[7] Preparing Sweden's suite of reports was the primary purpose of the Committee for Research and Factual Issues. The scientific documentation produced under CRF auspices, including three case studies and the national environmental report – the English version of which included summaries in Arabic, Chinese, French, Russian, and Spanish – would also serve as instruments of Swedish science diplomacy. In cooperation with the foreign ministry, the reports were widely distributed through Swedish embassies to be used as templates by developing countries in the drafting of their own national reports. In some cases, Swedish scientists were deployed as informal envoys to advise and assist government officials and local experts in the writing of national reports.

One such science diplomacy mission involved dispatching systems analyst Lars Ingelstam to Brazil in 1970 at a particularly sensitive moment in conference preparations. Scientists at MIT were at the time developing the World Systems computer model that would underpin the Club of Rome's *Limits to Growth* report. Some developing countries, including Brazil, were becoming concerned that environmental protection could entail restrictions on exploiting natural resources and other measures inhibiting economic growth. In this increasingly fraught political context, Ingelstam's dual-purpose assignment was to not only advise on drafting the national report but to also employ his expertise in systems theory to convince outspokenly suspicious Brazilian officials of the merits of the conference and international environmental cooperation on the environment and its compatibility with economic development.[8] Although Brazil continued to criticize the conference, the boycott that some Swedish officials feared did not in the end take place.

[7] Paglia, "The Swedish Initiative." Engfeldt, *From Stockholm to Johannesburg*. McCormick, *Reclaiming Paradise*. Maurice F. Strong, "The Stockholm Conference: Where Science and Politics Meet," *Ambio* 1(1972):3, 73–78. www.jstor.org/stable/4311954.

[8] Paglia, "The Swedish Initiative." Lars Ingelstam, personal communication (Paglia), April 21, 2016.

SWEDEN'S CASE STUDY ON ACID RAIN

Ingelstam's contribution to Sweden's extensive conference preparation effort also included co-authoring the country's case study on acid rain, an issue that had become a significant concern for Swedish policymakers in the wake of Svante Odén's autumn 1967 exposé and a scientific article he published in 1968.[9] While the majority of the report's eight co-authors were natural scientists, Ingelstam's expertise as an engineer, mathematician, and systems specialist enabled the application of a systems analysis approach to the acid rain problem in what was likely the first environmental study of its kind.[10] The lead author of the case study was CRF member Bert Bolin, who had suggested acid rain, rather than his signature issue of climate change, as the subject of the Swedish study. Bolin's backing of acid rain was based on the recommendation and lobbying efforts of Odén, who pointed to the likely long-term societal costs of acidification and the inevitability that developing countries would also have to confront the issue as they industrialized.[11]

Given the seminal research on atmospheric chemistry that had been conducted at Stockholm University since the 1950s, acid rain was a topic well suited for Bolin's Department of Meteorology. Bringing in colleagues Lennart Granat and Henning Rodhe, who was in the process of completing his dissertation on the long-range transport of sulfur,[12] Bolin led an eight-expert team that also included Odén, Ingelstam, and forest ecologist Carl Olof Tamm in producing a comprehensive case study on the causes, effects, and costs of acid rain.[13] As Bolin's PhD student and in many ways his scientific protégé in the study of biogeochemical cycles and climate change over the course of his long career at Stockholm University, Rodhe played a key role in coordinating the project.[14]

[9] Svante Odén, "The Acidification of Air Precipitation and Its Consequences in the Natural Environment," *Ecological Bulletins* vol. 1 (Stockholm: Swedish Natural Science Research Council, 1968).

[10] Peringe Grennfelt et al., "Acid Rain and Air Pollution: 50 Years of Progress in Environmental Science and Policy," *Ambio* 49(2020): 849–864.

[11] Rothschild, *Poisonous Skies*.

[12] Henning Rodhe, *Long Range Transport through the Atmosphere: A Study of Sulfur as an Air Pollutant* (Stockholm: Stockholm University, 1972).

[13] B. Bolin, L. Granat, L. Ingelstam, M. Johannesson, E. Mattsson, S. Oden, H. Rodhe & C. O. Tamm, *Air Pollution across National Boundaries: The Impact on the Environment of Sulfur in Air and Precipitation. Sweden's Case Study for the United Nations Conference on the Human Environment* (Stockholm: Royal Ministry for Foreign Affairs and Royal Ministry of Agriculture, 1971).

[14] At the beginning of his Ph.D. training, Rodhe also assisted with preparations for the 1967 GARP meeting in Stockholm. By the late 1980s, he would become one of Sweden's

By emphasizing the transboundary nature of the problem, *Air pollution across national boundaries: The impact on the environment of sulfur in air and precipitation* reinforced the political case for international action that Sweden advocated for in the OECD and, later, within the United Nations system.[15] The case study would also serve as an instrument of Sweden's Conference-related science diplomacy. As part of Engfeldt and Åström's efforts to stimulate scientific report writing in developing countries, the Swedish foreign ministry distributed the acid rain study through its embassies around the world. The demonstration of scientific ambition that the Swedish study represented would provide governments, expert agencies, and scientific establishments a template of sorts for their own contributions to the Stockholm Conference.[16] Sweden's acid rain case study would be one of three major reports published in 1971 that were closely linked to Stockholm and the conference preparation process. Each would become a milestone in its respective scientific context and contribute to the formation of knowledge networks and environmental governance institutions in the years following the Stockholm Conference. The three reports would also inform conference participants on emerging scientific issues related to the human environment and contribute directly to key Stockholm outcomes.[17]

SMIC: REPORT FROM AN INFLUENTIAL 1971 CLIMATE SUMMIT

The report *Inadvertent Climate Modification* was the result of a three-week workshop Study of Man's Impact on Climate, which was held in the summer of 1971 at the Wijk conference center on the island of Lidingö in the Stockholm archipelago.[18] Sponsored by MIT and jointly hosted by the Royal Swedish Academy of Sciences and the Royal Swedish Academy of Engineering Sciences, the SMIC workshop was

leading experts on climate change and was deeply involved with the first three IPCC assessment reports (1990, 1992, and 1992) published during Bolin's chairmanship of the Panel. Henning Rodhe, personal communication, July 24, 2022.

15 Grennfelt et al., "Acid Rain."

16 Paglia, "The Swedish Initiative."

17 Eric Paglia & Sverker Sörlin, "Greening Our Common Fate: Stockholm as a Node of Global Environmental Memory," In: G. Sluga, K. Darian-Smith & M. Herren, eds., *Sites of International Memory: A Century of Commemoration and Internationalization* (Philadelphia, PA: University of Pennsylvania Press, 2023).

18 Study of Man's Impact on Climate (SMIC). *Inadvertent Climate Modification: Report of the Study of Man's Impact on Climate* (Cambridge, MA: The MIT Press, 1971).

explicitly intended to contribute to conference preparations by assembling many of the world's leading atmospheric experts. Their remit was to reach a consensus and draft a report on the state of the global climate based on the latest scientific knowledge in the field.[19]

The workshop organizer was MIT management professor Carroll Wilson, a senior advisor to Stockholm Conference leadership, who as part of conference preparations had arranged the Study of Critical Environmental Problems workshop the previous summer in Williamstown, Massachusetts. While that workshop attracted mostly American experts and encompassed a range of global-scale environmental issues, SMIC focused exclusively on climate change and was by design highly international.[20] Among the thirty prominent atmospheric scientists from fourteen countries attending the workshop were luminaries like Russian geographer-glaciologist Mikhail Budyko, German climatologist Hermann Flohn, US climate expert William Kellogg, and Princeton climate modeler Syukuro Manabe (a half-century later receiving the 2021 Nobel Prize in physics).[21] Also present was a young Stephen Schneider, who later became a prominent public voice on climate change, serving as rapporteur.[22] All but one participant, the Indian physicist Pisharoth Rama Pisharoty, were from industrialized countries. Curiously, Bert Bolin did not participate in the workshop, and only one Swedish scientist, Erik Eriksson, took part in the role of consultant. Opening with a Sanskrit prayer that asked Mother Earth to forgive human transgressions, the 308-page SMIC report was rapidly published by the MIT Press just a few months after the workshop so it could be widely circulated in advance of the Stockholm Conference.

A draft of the manuscript was delivered to the UN Secretariat by September 1971 in order to influence the third meeting of the conference's Preparatory Committee and was further distributed to selected

[19] David M. Hart & David G. Victor, "Scientific Elites and the Making of US Policy for Climate Change Research, 1957–74," *Social Studies of Science* 23(1993):4, 643–680. www.jstor.org/stable/285728.

[20] Spencer R. Weart, "The Evolution of International Cooperation in Climate Science," *Journal of International Organization Studies* 3 (2012):1, 41–59.

[21] Syukuro Manabe & Richard T. Wetherald, "*Thermal Equilibrium* of the Atmosphere with a Given Distribution of Relative Humidity," *Journal of Atmospheric Science* 24(1967): 241–259. Manabe's work in the 1960s and after was instrumental in the evolution of atmospheric general circulation modeling; see: Paul N. Edwards, "A Brief History of Atmospheric General Circulation Modeling," *International Geophysics* 70(2000): 67–90.

[22] Stephen, Schneider, *Science as a Contact Sport: Inside the Battle to Save Earth's Climate* (Washington, DC: National Geographic, 2009).

scientific organizations, including WMO, ICSU and GARP.[23] The report was structured in such a way as to be of interest both to scientists and relevant officials and institutions, such as national science-funding authorities, international development organizations, and Stockholm Conference delegates, for whom it was considered "required reading."[24] The first and second sections of the report were pedagogic, providing a lay audience with an overview of the past and possible future of the global climate, including societal impacts and anthropogenic drivers of change. The third and fourth sections were geared toward the scientific community and went into far greater technical detail on the dynamics of the climate system. All sections and subsections included sets of specific recommendations.

Even more important than the scientific aspects of the workshop and report, SMIC represented a turning point in the promotion of climate change as an issue of global political relevance. It also served as a social bonding opportunity that strengthened the growing international network of climate experts,[25] which would in the years ahead play an increasingly decisive agenda-setting role in the realm of climate politics. From a practical scientific standpoint, a key SMIC objective of greatly expanding data collection capabilities was realized a year later with recommendation 79 of the Stockholm Conference Action Plan, which called for the creation of a global climate monitoring network.[26] The latter reflected converging interests between communities of atmospheric and biological scientists that skillfully leveraged the Stockholm Conference to attract new resources and expand their respective research infrastructures.

SCOPE 1: SEMINAL REPORT ON GLOBAL ENVIRONMENTAL MONITORING

While the SMIC participants were on Lidingö discussing mankind's impact on climate and the potential consequences for the human environment, members of another international scientific network spent the

23 SMIC, *Inadvertent Climate Modification.*
24 Spencer Weart, *The Discovery of Global Warming* (Cambridge, MA: Harvard University Press, 2008).
25 Paul N. Edwards, *A Vast Machine: Computer Models, Climate Data, and the Politics of Global Warming* (Cambridge, MA: MIT Press, 2010). Paul Edwards & Myanna Lahsen, "Climate Science and Politics in the United States," Unpublished manuscript (1999). http://pne.people.si.umich.edu/PDF/PMNPC/USA.pdf (accessed July 18, 2024).
26 Edwards, *A Vast Machine.*

summer of 1971 in Stockholm drafting a report advocating the establishment of a Global Environmental Monitoring System (GEMS). It would be the first in a long series of influential reports published by SCOPE, the Scientific Committee on Problems of the Environment, an organization founded by ICSU in 1969 for the purpose of producing interdisciplinary scientific assessments on complex issues at the intersection of environment and society. The series of numbered SCOPE reports would run for almost five decades, involving hundreds of natural and social scientists writing on a broad range of environmental issues. Many of the widely distributed volumes would influence governance initiatives and research trajectories within and across a range of scientific disciplines.

Among the founders of SCOPE was Bengt Lundholm,[27] secretary of the Ecological Research Committee at Sweden's Natural Science Research Council and a member of the Committee for Research and Factual Issues that would oversee the Swedish scientific reports for the Stockholm Conference. He was an early critic of the use of biocides and a strong promoter of ecology as a field of knowledge that in his view ought to influence policy in Sweden,[28] where, like in other industrialized countries in the 1960s and 1970s, many prominent environmental activists were ecologists or biologists.[29] Lundholm's outlook on the policy relevance of ecology also extended to his work with international development. With financial support from the Swedish International Development Agency, Lundholm and the Ecological Research Committee hosted the international symposium "Ecology and the Less Developed Countries" in April 1971 during the lead-up to the Stockholm Conference. The symposium featured papers – published in an issue of the Ecological Research Committee's *Ecological Bulletins* – by experts that, like Lundholm, straddled the realms of science and policy. Participants, primarily from the Global North, included James Lee, the first environmental advisor at the World Bank; Lee Talbot, an ecologist, senior official in several US presidential administrations, and architect of

[27] Gilbert F. White, "SCOPE: The First Sixteen Years," *Environmental Conservation* 14(1987):1, 7–13.

[28] Lundholm edited the volume *Därför ekologi: Vetenskapen om vår överlevnad* [Therefore Ecology: The Science of Our Survival] (Stockholm: Askild & Kärnekull, 1971), with a foreword by former Swedish prime minister Tage Erlander.

[29] Thomas Söderqvist, *The Ecologists: From Merry Naturalists to Saviours of the Nation: A Sociologically Informed Narrative Survey of the Ecologization of Sweden 1895–1975* (Stockholm: Almqvist & Wiksell, 1986). Anna Kaijser & David Larsson Heidenblad, "Young Activists in Muddy Boots: Fältbiologerna and the Ecological Turn in Sweden, 1959–1974," *Scandinavian Journal of History* 43(2018):3, 301–323.

major environmental legislation like the Endangered Species Act; Dale Jenkins, director of the ecology program at the Smithsonian Institution; and Anders Forsse, a career civil servant who from the mid-1960s to the mid-1980s was one of the most influential figures in shaping Swedish development policy. Lundholm himself contributed papers on environmental monitoring and "the DDT Dilemma."[30]

As a highly active science administrator with a PhD in zoology and a zeal for international cooperation, Lundholm had also been deeply engaged with the International Biological Program (IBP), where he had since 1968 promoted the idea of a Global Network of Environmental Monitoring (GNEM). Lundholm's intellectual and organizational leadership of international efforts to institutionalize environmental monitoring would result in a major report and, in the wake of the Stockholm Conference, a widely dispersed monitoring network managed by the United Nations' new environment program. On his initiative, Sweden's Ecological Research Committee hosted a series of GNEM planning meetings in Stockholm, with Lundholm chairing a team of Swedish natural scientists, including Erik Eriksson, that proposed an international monitoring network of "Global Baseline Stations."[31] The founding of SCOPE and the approach of the Stockholm Conference provided Lundholm and like-minded scientific elites with a platform and political context for further advancing the idea of an environmental monitoring network.[32] An initiative to this end was supported by the conference's Secretary General Maurice Strong, upon whose request a SCOPE report on environmental monitoring was produced,[33] with work being initiated in Stockholm in March 1971.[34]

Lundholm served as chairman of the Commission on Monitoring, a body that had been effectively transferred to SCOPE from the IBP. He led a trio of scientists that also included Swedish biologist Sören Svensson and Welsh ecologist Gordon Goodman – another SCOPE

30 "Ecology and the Less Developed Countries," *Bulletins from the Ecological Research Committee/NFR* Issue 13 (Stockholm, 1971).

31 Elena Aronova, "Environmental Monitoring in the Making: From Surveying Nature's Resources to Monitoring Nature's Change," *Historical Social Research* 40(2015):2, 222–245.

32 Selcer, *The Postwar Origins*.

33 Bengt Lundholm, "Remote Sensing and International Affairs," *Ambio* 1(1972):5, 166–173. www.jstor.org/stable/4311974.

34 Enora Javaudin, "Environmental Problem-Solvers?: Scientists and the Stockholm Conference," In: Wolfram Kaiser & Jan-Henrik Meyer, eds., *International Organizations and Environmental Protection: Conservatoin and Globalization in the Twentieth Century* (New York: Berghahn Books, 2017), 74–102.

architect who would several years later relocate to Stockholm to become founding director of the Beijer Institute – in drafting the first report in the SCOPE series. The intention behind SCOPE 1, *Global Environmental Monitoring: A Report submitted to the United Nations Conference on the Human Environment, Stockholm 1972*, was to convince conference participants on the merits and feasibility of establishing a widely dispersed network of monitoring stations that could collect and process large quantities of environmental data to facilitate the management of the biosphere. The report, which served as a background paper on environmental monitoring for Stockholm Conference delegates,[35] put forward eighteen detailed recommendations on how such a network could be organized and what specific environmental indicators should be monitored.[36] The essence of the system elaborated in SCOPE 1 would be endorsed in the Stockholm Conference Action Plan and institutionalized as Earthwatch and the Global Environmental Monitoring System within the United Nations Environment Programme.[37] Earthwatch, formally established at the first UNEP Governing Council meeting in 1973, functioned as a catalyst and clearing house for environmental data in support of policy formation, although the system proved difficult to maintain over time for financial, technical, and political reasons.[38] Overall, twenty three of the Stockholm Action Plan's 109 recommendations involved environmental monitoring.[39]

TWELVE DAYS IN STOCKHOLM: THE UNITED NATIONS CONFERENCE ON THE HUMAN ENVIRONMENT

Less detailed yet more influential than the Action Plan in articulating a normative outlook on the human environment and establishing a basis for international environmental law, the Declaration of the United Nations Conference on the Human Environment was the other landmark document adopted at the 1972 Stockholm Conference. An intergovernmental

35 C. C. Wallén, "History of 'Earthwatch' 1972–1995" (1997), available at: http://yabaha.net/dahl/earthw/History.htm (accessed April 29, 2022).

36 Commission on Monitoring of the Scientific Committee on Problems of the Environment (SCOPE) of the International Council of Scientific Unions (ICSU). *Global Environmental Monitoring: A Report Submitted to the United Nations Conference on the Human Environment, Stockholm 1972* (Stockholm: ICSU and SCOPE, 1971).

37 Aronova, "Environmental Monitoring."

38 Maria Ivanova, *The Untold Story of the World's Leading Environmental Institution: UNEP at Fifty* (Cambridge, MA: The MIT Press, 2021).

39 Lundholm, "Remote Sensing."

working group began drafting the eventual Stockholm Declaration in May 1971 and had essentially reached a consensus on its main messages by early 1972. At the start of the Conference in June that year, however, objections by the People's Republic of China – which had recently replaced Taiwan in representing China in the United Nations – complicated the Declaration's adoption by demanding that certain aspects, specifically wording on population and weapons of mass destruction, be modified or removed.[40] With Swedish legal expert and future foreign minister Hans Blix playing a leading role, and with strong engagement from Conference president Ingemund Bengtsson and secretary general Maurice Strong,[41] intensive discussions and drawn-out drafting work continued until the final hours of the Conference on June 16, when a version of the text was ultimately adopted by acclamation, with the nonparticipation of China.[42] In its final form, officially adopted on December 15 by the UN General Assembly, the Stockholm Declaration consisted of a preamble outlining both the importance of environmental protection and the imperative of economic development and a set of twenty-six principles encompassing related aspects ranging from intergenerational equity and environmental education to pollution prevention and the rational use of natural resources. Most significantly, Principle 21 consolidated the idea of environmental harm prevention across national borders – a key legal concept – that would be reiterated twenty years later as Principle 2 of the 1992 Rio Declaration. Throughout, the Stockholm Declaration emphasized, even elevated above environmental concerns, the priority of development and the sovereignty of states, particularly over the exploitation of natural resources.

The Conference had begun on June 5 amid great expectations on the part of the over 1,200 delegates from 113 countries, the thousands of environmental and other activists, and the large media contingent that had descended upon Stockholm for the first-ever United Nations conference of its kind. Notably absent were government representatives of the Soviet Union and most of the Eastern Bloc countries, which had boycotted the Conference over a dispute in the UN over the participation

[40] Hans Blix, "History of the Stockholm Declaration," In: Myron H. Nordquist, John Norton Moore & Said Mahmoudi, eds., *The Stockholm Declaration and Law of the Marine Environment* (Leiden: Brill Nijhoff, 2003), 15–24. Louis B. Sohn, "The Stockholm Declaration on the Human Environment," *The Harvard International Law Journal* 14(1973):3, 423–515.

[41] Peter Stone, *Did We Save the Earth at Stockholm?: The People and Politics in the Conference on the Human Environment* (London: Earth Island, 1973).

[42] McCormick, *Reclaiming Paradise*; Engfeldt, *From Stockholm to Johannesburg*; Blix, "History."

of East Germany. Yet the potential boycott by states of the Global South that had concerned conference planners did not in the end take place. This was largely due to the painstaking preparation process and the shuttle diplomacy of Maurice Strong, which had to some extent assuaged, although certainly did not dispel,[43] the grievances and suspicions of developing countries surrounding the specter of the environment being used as a pretense for imposing limits to economic development. Conference preparations included distilling some 700 pages of official documentation from hundreds of scientific reports drafted by UN member states and international organizations. Four two-week meetings were held between March 1970 and March 1972 of the twenty-seven-country Preparatory Committee, chaired by Jamaican diplomat Keith Johnson. Regional seminars took place in Bangkok, Addis Ababa, Mexico City, and Beirut, and two gatherings of leading development economists at Columbia University in New York in 1970 and in the Swiss resort town of Founex in 1971.[44]

The Founex seminar and the report that resulted from it were watershed moments in the effort to reconcile economic growth with environmental protection.[45] It foreshadowed the idea of sustainable development, which would be articulated some fifteen years later in the Brundtland Report.[46] Along with Maurice Strong, who lobbied developing countries extensively during the lead-up to the Conference, the initiator of both the Columbia and Founex seminars was the British economist Barbara Ward (Lady Jackson), a strong advocate for the development prerogatives of the Global South. Capping off the preparatory process, Ward and French microbiologist and environmental thinker René Dubos were commissioned by Strong to author the Conference's scientific and symbolic centerpiece, *Only One Earth: The Care and Maintenance of a Small Planet*. The report, which Ward

[43] Macekura, *Of Limits and Growth*.

[44] Engfeldt, *From Stockholm to Johannesburg*; Macekura, *Of Limits and Growth*; Rowland, *The Plot to Save the World*.

[45] P. Rambach, "Founex Report on Development and Environment," *International Conciliation* 586(1972): 7–36. Michael W. Manulak, "Developing World Environmental Cooperation: The Founex Seminar and the Stockholm Conference," In: Wolfram Kaiser & Jan-Henrik Meyer, eds., *International Organizations and Environmental Protection: Conservatoin and Globalization in the Twentieth Century* (New York: Berghahn Books, 2017), 103–127.

[46] Björn-Ola Linnér & Henrik Selin. "The United Nations Conference on Sustainable Development: Forty Years in the Making," *Environment and Planning C: Government and Policy* 31(2013):6, 971–987.

and Dubos drafted after collecting contributions from a Committee of Corresponding Consultants consisting of 152 experts from fifty-eight countries representing all world regions, provided Stockholm with its slogan and would become the "officially unofficial" Conference bible.[47]

Sweden's Prime Minister Olof Palme opened the Conference with an internationally broadcast speech at the Stockholm Opera House in which he indicted industrialized countries for the vast majority of environmental degradation to date,[48] and condemned the United States for conducting in Vietnam a campaign of what he called "ecocide," setting a tone that elicited accolades from developing country delegates.[49] The idea of ecocide, which would decades later inspire the End Ecocide movement founded by Scottish barrister Polly Higgins,[50] became one of the many themes of the Environment Forum, a parallel series of events facilitated by the Swedish government and the Conference's preparatory committee to provide scientists, environmental NGOs, and others a place to discuss issues related to the human environment.[51] Although separate from official UNCHE proceedings, the at times cacophonous debates and conclusions of the Environment Forum were reported in the *Stockholm Conference Eco*, a newsletter produced by Friends of the Earth and The Ecologist magazine that was distributed to Conference participants.[52]

The other "outer conference" was less connected to the official UN event and encompassed an even wider political agenda that also included indigenous peoples' rights and opposition to the Vietnam War.[53] The People's Forum (Folkets Forum) was a gathering of a broad spectrum of radical groups as well as counterculture constellations and celebrities such as Hog Farm, Wavy Gravy, the legendary psychedelic rock band The Grateful Dead, Stuart Brand, publisher of the *Whole Earth Catalogue*, and the somewhat unlikely Walter Hickel, President Nixon's former Minister of the Interior.[54] Organized by the Swedish

47 Selcer, *The Postwar Origins*, 201.

48 Lynton Keith Caldwell, *International Environmental Policy: Emergence and Dimensions* (Durham & London: Duke University Press, 1990).

49 Selcer, *The Postwar Origins*.

50 Polly Higgins, *Eradicating Ecocide: Laws and Governance to Prevent the Destruction of Our Planet* (London: Shepheard-Walwyn, 2010).

51 Stone, *Did We Save the Earth*; Caldwell, *International Environmental Policy*.

52 McCormick, *Reclaiming Paradise*.

53 Rowland, *The Plot to Save the World*.

54 Leonidas Aretakis, *Extas i folkhemmet: Sveriges psykedeliska historia* [Ecstacy in the People's Home: Sweden's Psychedelic History] (Stockholm: Natur & Kultur, 2022), 425–427 (e-book pagination).

leftist organization Powwow,[55] "Woodstockholm," as some called it, was safely sequestered far away from the main Conference on the outskirts of the city at an abandoned airfield in Skarpnäck, where Swedish authorities had set up a tent city for the several thousand young activists that had traveled to Stockholm to protest or raise awareness for their particular cause and, as it were, consume large amounts of narcotics.[56] In the rather more posh setting of the Grand Ballroom of Stockholm's Grand Hotel, the International Institute of Environmental Affairs and the International Population Institute arranged a well-attended Distinguished Lecture Series for Conference delegates that featured talks by Club of Rome founder Aurelio Peccei, Swedish Nobel Prize-winning economist Gunnar Myrdal,[57] former science advisor to the British government Sir Solly Zuckerman, and Norwegian explorer and ethnographer Thor Heyerdahl, as well as lectures by René Dubos and Barbara Ward (Figure 5.2).[58] The *Only One Earth* authors, along with other Environment Forum intellectuals such as Egyptian ecologist Mohamed Kassas, American anthropologist Margaret Mead (Figure 5.2), and MIT professor Carroll Wilson, also participated in a big picture brainstorming event under the banner of "Where do we go from now" arranged by Olof Palme and Maurice Strong at the Swedish prime minister's retreat at Harpsund halfway through the conference.[59]

In one of her many contributions to the success of Stockholm, Barbara Ward, together with Margaret Mead (Figure 5.2), served as a liaison between the Environment Forum and the Conference plenary, providing a voice for the over 400 NGOs present in Stockholm.[60] Although there was a degree of overlap between topics addressed at UNCHE and the various side events of the "Stockholm circus,"[61] the Environment Forum

55 McCormick, *Reclaiming Paradise.*

56 Aretakis, *Extas.*

57 Myrdal was co-recipient together with Friedrich Hayek of the 1974 Sveriges Riksbank Prize in Economic Sciences in Memory of Alfred Nobel. His June 8, 1972 lecture during the Stockholm Conference, "Economics of an Improved Environment" was published, along with other speeches from the Distinguished Lecture Series in Maurice Strong (ed.), *Who Speaks for the Earth?: Seven Citizens of the World on Major Issues of the Global Environment* (New York: W. W. Norton & Company, 1973).

58 Sidney Hyman, *The Aspen Idea* (Norman: University of Oklahoma Press, 1975). Rowland, *The Plot to Save the World.*

59 Javaudin (2017) and Selcer (2018) devote several pages to this fascinating "backstage" event during the Stockholm Conference.

60 McCormick, *Reclaiming Paradise*; Caldwell, *International Environmental Policy*; Engfeldt, *From Stockholm to Johannesburg*; Macekura, *Of Limits and Growth.*

61 Selcer, *The Postwar Origins.*

FIGURE 5.2 Economist, author, and journalist Barbara Ward (left), a globally renowned champion of international development during the 1960s and 1970s, was a moral and intellectual leader of the Stockholm Conference, co-authoring its unofficial report *Only One Earth* and serving as a bridge between various stakeholder groups. Cultural anthropologist Margaret Mead (right) was one of the many prominent participants of the parallel civil society Environment Forum, for which she and Ward functioned as a communication channel with the official conference. Photo: UN Photo/ Yutaka Nagata.

encompassed exhibitions, films, lectures, and discussions on a wide variety of issues, such as workplace environments and supersonic transport, that were not otherwise part of the official Conference agenda.[62] These moreover included complex and controversial questions like limits to economic growth, socioeconomic justice, and the nexus of population, pollution, technology, and natural resources. This sparked heated debates between Paul Ehrlich and Barry Commoner and their respective followers as part of an ongoing feud between the two prominent

[62] Lars Emmelin. "The Stockholm Conferences," *Ambio* 1(1972):4, 135–140. Workplace environment was the topic of one of the reports Sweden submitted for the Conference and was mentioned by Olof Palme in his opening speech. It was also an area of concern and activism for Hans Palmstierna (see Heidenblad 2021). Supersonic travel, which some feared could damage the ozone layer, was a sensitive issue for France and the United Kingdom due to the Concorde program (Engfeldt 2009).

scientist-environmentalists.[63] The large number of activists in Stockholm also made their presence felt through demonstrations in the city center in the vicinity of the three main Conference venues of Folkets hus (the People's House), the Old Parliament building, and what was at the time known as the New Parliament building and is today Stockholm's Culture House. Several hundred custom-designed "environment bicycles" were provided by the Swedish organizing committee to facilitate sustainable mobility around downtown Stockholm, along with shuttle bus services and a fleet of UN-blue Volvos and Saabs for Conference delegations.[64]

By the time of the Stockholm Conference, Commoner – an American biologist and author of the bestselling 1971 ecological critique of technology-driven capitalism *The Closing Circle* – had come to consider the environment as "the world's most dangerous political issue" due to the substantial North-South divisions that had surfaced during the four years of preparations.[65] Despite the acrimony of the months leading up to the Conference, and the eventual Eastern Bloc boycott, the twelve days in Stockholm, abetted by a stretch of superb early-summer weather and a great deal of popular interest from around the world, went exceptionally well. There was no major controversy, aside from some of the theatrics of the outer conferences and the drama surrounding the redrafting and adoption of the Stockholm Declaration. The celebrated high point came toward the end, in a speech that encapsulated the tensions and tradeoffs at the core of the Conference. India's Prime Minister Indira Gandhi (Figure 5.3), the only foreign head of state present in Stockholm, embodied the concerns of developing countries. In her landmark address before a June 14 plenary session at Folkets Hus, Gandhi articulated the essential – some would argue ostensible – contradiction between environment and development at the heart of the human environment concept:

> On the one hand, the rich look askance at our continuing poverty – on the other, they warn us against their own methods. We do not wish to impoverish the environment any further and yet we cannot for a moment forget the grim poverty of large numbers of people. Are not poverty and need the greatest polluters?

[63] Andrew Feenberg, "The Commoner-Ehrlich Debate: Environmentalism and the Politics of Survival," In: David Macauley, ed., *Minding Nature: The Philosophers of Ecology* (New York: Guilford Press, 1996).

[64] Stone, *Did We Save the Earth*. Mårtenson, *Att kyssa ett träd*.

[65] Barry Commoner, *The Closing Circle: Nature, Man, and Technology* (New York: Random House, 1971). Barry Commoner, "Motherhood in Stockholm," *Harper's*, June (1972), 49–54. Macekura (2015) employs Commoner's formulation as the title for his chapter on the Stockholm Conference.

FIGURE 5.3 India's Prime Minister Indira Gandhi delivering the most celebrated speech of the Stockholm Conference in which she eloquently articulated the concerns of the Global South on the issue of the human environment and its implications for development. Photo: UN Photo/Yutaka Nagata.

The successful conclusion of the Conference with the adoption of the Stockholm Declaration on June 16 was reaffirmed six months later at the UN General Assembly in New York with Resolution 2997 of December 15, 1972, which formally established the United Nations Environment Programme and thus secured the long-term institutional legacy of Stockholm and its extensive preparation period. The resolution mandated a developing country majority in the UNEP Governing Council, consisting of sixteen seats for countries of Africa, thirteen for Asia, six for Eastern Europe, ten for Latin America, and thirteen for Western Europe and other states. Maurice Strong was appointed as the founding Executive Director of UNEP, which, in line with developing country demands,[66] and through skillful diplomatic maneuvering by Kenya's UN delegation, would be based in the Global South.[67] At a meeting of the Second Committee of the UN General Assembly in November 1972, the Kenyan capital of Nairobi was selected to host UNEP, rather

[66] Stone, *Did We Save the Earth.*
[67] Ivanova, *The Untold Story.*

than the established UN centers of Geneva or New York, the US preference of Vienna,[68] or even Stockholm, as some had suggested during the Conference.[69] Stockholm thus signaled not only the arrival of the environment on the international political agenda but also a relative increase in the prestige and influence of developing countries within the UN system. Yet the Stockholm Conference, and its ambitious Action Plan and precedent-setting Declaration, decidedly did not resolve the conflicts and complexities inherent in the world's most dangerous political issue. The several years leading up to June 1972 would, as Commoner perhaps feared, rather come to represent the onset of a polarization in global environmental diplomacy that has endured for half a century. As the world wrestled with the environment-development dilemma over the course of the 1970s and after, including developing country calls for a New International Economic Order,[70] and Strong set about establishing the secretariat for the first UN body situated in the Southern Hemisphere,[71] back in Stockholm, the concept of the human environment would serve as a catalyst. It helped bring about the creation of an international institute for environment and development and a new scholarly platform for the interaction of science and politics, while the convening power of the Swedish capital would continue to play a key role in advancing the organization of climate science.

AMBIO, THE ACADEMY OF SCIENCES, AND THE ENVIRONMENT

Not entirely content with the techniques that would be employed in implementing the Action Plan's recommendations, Bengt Lundholm, one of the architects of the initiatives that would become Earthwatch and GEMS at UNEP, delivered a keynote address at the October 1972 Eighth Symposium on Remote Sensing of the Environment in Ann Arbor, Michigan. In his address, he strongly advocated the use of satellites and other modern technologies in environmental monitoring. He also argued that the increased use of remote sensing would not only

[68] Macekura, *Of Limits and Growth*.

[69] Mårtenson, *Att kyssa ett träd*. Proposals for London, Madrid, Malta, Mexico City and New Delhi were also made during the Conference (Emmelin 1972).

[70] Linnér and Selin 2013; Carl Marklund, "Double Loyalties?: Small-State Solidarity and the Debates on New International Economic Order in Sweden during the Long 1970s," *Scandinavian Journal of History* 45(2020):3, 384–406.

[71] Macekura, *Of Limits and Growth*.

facilitate the collection of more comprehensive and higher-quality environmental data but that satellites connected to geographically dispersed ground stations would also foster greater international cooperation.[72] Lundholm's idealistic outlook largely reflected a basic conviction that laid behind the Swedish initiative – that the biosphere would benefit from multilateral cooperation and that the human environment could serve as a medium for enhancing international solidarity and strengthening global institutions. Further, Lundholm's call for greater use of remote sensing dovetailed with the Swedish position in the UN Committee on the Peaceful Uses of Outer Space, another emerging area of international cooperation that Sweden had taken part in establishing.[73]

To extend the reach of Lundholm's message, a transcript of his speech, "Remote Sensing and International Affairs," was published in *Ambio – A Journal of the Human Environment* (Figure 5.4). The journal was a new initiative of the Royal Swedish Academy of Sciences, which was in the midst of a modernization effort to make itself more relevant and attuned to the prevailing environmentalist sentiment of the early 1970s.[74] Launched in February 1972 during the lead-up up to Stockholm, *Ambio* would over the course of the year publish interviews and articles by scientists and officials involved with the Conference, including Sverker Åström, Inga Thorsson, Maurice Strong, and Hans Palmstierna. The inaugural issue featured a condensed version of Sweden's acid rain case study,[75] and the lead story of the following issue was in effect an early warning on stratospheric ozone depletion written by the Dutch expatriate Paul Crutzen,[76] at that point a research associate at Stockholm University's Department of Meteorology. Crutzen was soon after appointed to the editorial board of the journal, where he would in 1974 publish another lead article on the deleterious effects of human activity on the ozone layer.[77] *Ambio* embodied the shift at the Academy and other scientific institutions from studying nature to understanding the entangled complexities of the human environment.[78] Aimed

[72] Lundholm, *Remote Sensing*.

[73] "Sweden Urges UN Action on Remote Sensing." *Ambio* 1(1972):5, 175–175. www.jstor.org/stable/4311976.

[74] Sverker Sörlin, "The Environment as Seen through the Life of a Journal: *Ambio* 1972–2022," *Ambio* 50(2021):1, 10–30. https://doi.org/10.1007/s13280-020-01421-w

[75] "Sulphur Pollution across National Boundaries," *Ambio* 1(1972):1, 15–20.

[76] Paul Crutzen, "SST's: A Threat to the Earth's Ozone Shield," *Ambio* 1(1972):2, 41–51.

[77] Paul Crutzen, "Estimates of Possible Variations in Total Ozone Due to Natural Causes and Human Activities," *Ambio* 3(1974):6, 201–210.

[78] Sörlin, "The Environment as Seen through the Life of a Journal."

FIGURE 5.4 Front cover of the February 1972 inaugural issue of *Ambio – A Journal of the Human Environment*, which over fifty years later continues to be published under the auspices of the Royal Swedish Academy of Sciences as a platform for policy-relevant science. Cover illustration by Nils Petersson. Courtesy of Ambio.

at a broad international audience of scientists, policymakers, and the general public, the varied contributions from Swedish and international experts have encompassed issues of environment, society, politics, economics, law, and technology. It has also served as a forum for the development of concepts and fields of research with close ties to Stockholm, including resilience, ecological economics, Earth system science, and the Anthropocene. The latter was the topic of the most-cited article in the journal's fifty-year history, a 2007 transdisciplinary examination of the Anthropocene concept by Earth system scientists Paul Crutzen and Will Steffen and the environmental historian John McNeill.[79]

Moreover, *Ambio* has since its inception made a point of publishing pieces more reflective in nature. One example is a June 1972 article,

[79] Will Steffen, Paul J. Crutzen & John R. McNeill, "The Anthropocene: Are Humans Now Overwhelming the Great Forces of Nature?," *Ambio* 36(2007):8, 614–621.

published in conjunction with the Stockholm Conference, on the professional and ethical responsibilities of scientists in relation to the environment. It was authored by the American nuclear physicist Jack Hollander,[80] who would several years later be appointed as the international chairman of the Academy's newly established Beijer Institute.[81] Over the course of its half-century at the Academy, *Ambio* has also served as a communication platform for policy initiatives, scientific research, and the results of high-level events associated with Stockholm that advanced both environmental and foreign policy imperatives. An early example was a series of *Ambio* Special Reports presenting the scientific papers and policy recommendations of the three installments of the Soviet-Swedish Symposium on the Pollution of the Baltic that were held in Stockholm and Riga in 1971, 1973, and 1975.[82] The meetings were the initiative of the Soviet Academy of Sciences, the Royal Swedish Academy of Engineering Sciences, and the Swedish National Environmental Protection Agency and led to the creation of a Joint Soviet-Swedish Working Group on Problems of the Human Environment. From the second symposium, other states bordering the Baltic Sea also took part. In promoting scientific cooperation and environmental protection across Cold War boundaries, the symposia documented by *Ambio* contributed to the creation of the first Baltic Sea Convention in 1974.[83] This commitment to Baltic Sea science and governance would continue into the 1990s and beyond. In conjunction with the September 1990 Baltic Sea Prime Ministers Conference in Ronneby, *Ambio* would for instance publish a Special Report on the state of the Baltic Sea environment to serve as a scientific point of departure for the discussions held by the heads of government that gathered in the southern Swedish city during the waning days of the Cold War.[84] Initiated

[80] Jack Hollander, "Scientists and the Environment: New Responsibilities," *Ambio* 1(1972):3, 116–119.

[81] In the same issue of *Ambio*, Hans Palmstierna contributed an article on the responsibility of scientists to provide actionable knowledge on the safety of indoor working environments. Hans Palmstierna. "An Open Letter: The Scientist and the Working Man," *Ambio* 1(1972):3, 110–115. www.jstor.org/stable/4311958.

[82] 1st Soviet-Swedish Symposium on the Pollution of the Baltic / 1ый Советско-Шведский Симпозиум по Загрязнению Балтийского Моря. *Ambio Special Report*, no. 1(1972). 2nd Soviet-Swedish Symposium on the Pollution of the Baltic / 2ой Советско-Шведский Симпозиум по Загрязнению Балтийского Моря, *Ambio Special Report*, no. 4 (1976), 3rd Soviet-Swedish Symposium on the Pollution of the Baltic / 3ий Советско-Шведский Симпозиум по Загрязмемию Балтийского Моря. *Ambio Special Report*, no. 5(1977).

[83] Norström, Stig, and Стиг Нурстрём. "Preface / Предисловие." *Ambio Special Report* (1976):4, 5–5. www.jstor.org/stable/25099570.

[84] *Ambio*, "Special Report Number 7: Current Status of the Baltic Sea," September 1990.

by Sweden and Poland, the Ronneby Conference was a key moment in enhancing regional cooperation under rapidly changing political circumstances. Taking into account advances in scientific knowledge on transboundary environmental problems, the meeting updated the 1974 Helsinki Convention and resulted in a Baltic Sea Declaration and the Baltic Sea Joint Comprehensive Environmental Programme.[85]

Like with the various Special Reports on environmental concerns related to the Baltic, the editors of *Ambio* have since the early 1970s taken the opportunity of international events in Sweden to assemble thematic issues on some of the most significant issues of the day. In September–October 1973, Stockholm hosted the ten-day United Nations Symposium on Population, Resources and Environment as part of preparations for the UN World Population Conference the following year.[86] The Stockholm symposium attracted world-leading experts such as biologist Paul Ehrlich, physicist John Holdren, and economist Ignacy Sachs and inspired a special issue of *Ambio* on the population question. International contributors to the *Ambio* issue included Barry Commoner, Amory Lovins, and Stephen Schneider, who provided a climate perspective on population, with Swedes such as Inga Thorsson, hydrologist Malin Falkenmark, and plant physiologist Georg Borgström – among the most prominent demographic doomsday prophets of the 1960s – adding articles on the question of population from their own areas of expertise.[87] Borgström's article addressed the water, land use, and economic challenges of feeding the burgeoning global population by the year 2000.[88] Later in the autumn of 1973, Borgström also participated in another international symposium in Stockholm, for which he contributed a paper on the links between food and energy.[89] Convened by the Royal Swedish Academy of Sciences, the Energy in Society symposium, although planned earlier, coincided with the 1973 oil crisis that thrust energy to the top of the international agenda.

85 Helsinki Commission, *30 Years of Protecting the Baltic Sea: HELCOM 1974–2004* (Helsinki: Helsinki Commission, 2004). Björn Hassler, "Protecting the Baltic Sea – The Helsinki Convention and National Interests," In: Olav Schram Stokke & Öystein B. Thommesen, eds., *Yearbook of International Co-operation on Environment and Development (YBICED) 2003/2004* (Oxford: Earthscan, 2004), 33–41.

86 United Nations, *Report of the Symposium on Population Resources and Environment* (Stockholm, September 26–October 5, 1973). E/CONF.60/CBP/3.

87 "Population," *Ambio* 3(1974): 3–4 (special issue).

88 Georg Borgström, "The Food-Population Dilema," *Ambio* 3(1974):3/4, 109–113. www.jstor.org/stable/4312062.

89 Georg Borgström, "Food, Feed, and Energy," *Ambio* 2(1973):6, 214–219. www.jstor.org/stable/4312029.

The special issue of *Ambio* based on the symposium's papers and proceedings also featured contributions from scientific pioneers such as ecosystem ecologist Howard T. Odum, who in his theoretically groundbreaking paper modeled interlinkages between energy, ecological, and economic systems,[90] and economist E. F. Schumacher, who had earlier in 1973 published his highly influential book *Small is Beautiful: A Study of Economics as if People Mattered.* In his *Ambio* article, Schumacher elaborated on how both industrialized and developing countries could reduce their dependence on fossil fuels by promoting decentralized energy technologies and pursuing lifestyles that were closer to nature.[91] The lead article of *Ambio's* Energy in Society special issue was written by its guest editor, Lars Kristoferson, who had also served as secretary for the symposium.[92] A plasma physicist at the Royal Institute of Technology in Stockholm, Kristoferson had previously written extensively on energy and environmental issues and would in the months following the Energy in Society event become closely involved with the establishment of a new institute at the Academy focusing on precisely those topics. His association with what would become the Beijer Institute – the International Institute for Energy, Resources, and the Human Environment – would shape the trajectory of his career as a scientist, administrator, and environmental advocate.[93] During the next fifteen years, he would help build what would prove to be the common ancestor that three of Stockholm's leading sustainable development institutions – Stockholm Environment Institute, the Beijer Institute of Ecological Economics, and the Stockholm Resilience Centre – can directly trace their lineage to.

BIRTH OF THE BEIJER INSTITUTE

The series of events that would lead to the founding of the Beijer Institute started several weeks after the Energy in Society symposium was held at the Academy. The impetus came from the Swedish industrialist and financier

[90] Howard T. Odum, "Energy, Ecology, and Economics," *Ambio* 2(1973):6, 220–227. www.jstor.org/stable/4312030. James J. Zucchetto, "Reflections on Howard T. Odum's Paper: Energy, Ecology and Economics, *Ambio* (1973)," *Ecological Modelling* 178(2004):1–2, 195–198. www.sciencedirect.com/science/article/abs/pii/S0304380003005532.

[91] Ernst Friedrich Schumacher. "Western Europe's Energy Crisis: A Problem of Life-Styles," *Ambio* 2(1973):6, 228–232. www.jstor.org/stable/4312031.

[92] Lars Kristoferson. "Energy in Society" *Ambio* 2(1973):6, 178–185. www.jstor.org/stable/4312025.

[93] Kristoferson would later in his career become Secretary General of WWF Sweden.

Kjell Beijer, whose fortune was in part linked to investments in the energy sector. Inspired by the 1972 Stockholm Conference, Beijer was interested in providing funding for a new Nobel Prize that would stimulate research in the fields of energy and environment. Through the Nobel Foundation's executive director Stig Ramel, who was a board member of Beijer's investment company, the idea was communicated to Carl Gustaf Bernhard, permanent secretary of the Royal Swedish Academy of Sciences, the body responsible for awarding the Nobel prizes in chemistry and physics as well as the economics prize in the memory of Alfred Nobel. Over the course of several meetings and communications in January 1974, Bernhard was able to convince Beijer that his interest in advancing energy and environmental research would be better served by financing the establishment of a new institute at the Academy dedicated to those domains.[94]

Such an institute was well suited to the Academy's expanding interest in issues of the human environment. In addition to launching *Ambio* in 1972 and convening the Energy in Society symposium in 1973 – as well as other thematic conferences on environment-related issues in the years that followed – the Academy also established an Environment Protection Committee in 1974 in order to shape research policy on natural resources and the environment. The committee, initially chaired by the diplomat and environmental philosopher Rolf Edberg, was not only active nationally but also served as Sweden's representation in the ICSU body SCOPE.[95] Bernhard and a small group of advisors, including Lars Kristoferson, operated through Swedish and international networks during 1974 to determine the optimal organizational structure and specific areas of research that the Beijer Institute would be based upon. One productive forum was a Pugwash meeting in Vienna arranged by the organization's president, Stockholm physicist and Nobel Laureate Hannes Alfvén of the Royal Institute of Technology.[96]

[94] This section draws, among other sources, upon a history of the Beijer Institute by the permanent secretary of the Royal Swedish Academy of Sciences in the 1970s Carl Gustaf Bernhard, *The Beijer Institute: The International Institute for Energy, Resources and the Human Environment* (Stockholm: Royal Swedish Academy of Sciences, 1991), as well as personal communications with Lars Kristoferson and a speech delivered by him at the Gordon Goodman Memorial Seminar on April 29, 2009.

[95] Bosse Sundin, "Environmental Protection and the National Parks," In Tore Frängsmyr (ed.), *Science in Sweden: The Royal Swedish Academy of Sciences 1739–1989* (Canton, MA: Science History Publications, 1989), 199–226.

[96] Bernhard, *The Beijer Institute*. The Pugwash Conferences on Science and World Affairs is an international organization, founded in 1957, that promotes peace and nuclear disarmament. It was awarded the Nobel Peace Prize in 1995. Alfvén founded the Swedish Pugwash chapter in 1964.

Plans for creating the Beijer Institute were formally announced in January 1975, receiving a positive reception from the international scientific press, including the journal *Nature*. To further develop the research and operational agenda of the institute, the Academy in autumn 1975 convened a symposium chaired by Carroll Wilson of MIT, with representatives from a range of international institutes and organizations such as UNEP and the International Institute of Applied Systems Analysis taking part. The symposium's report contributed to the work of the Academy's planning group, which delivered recommendations on the structure, staffing, and other aspects of the institute that were adopted in January 1977. The Beijer Institute would be formally inaugurated later that year, moving into its offices in the new Beijer Foundation-financed wing that was added on to the Academy's existing premises in the Frescati section of the city, adjacent to the campus of Stockholm University.[97]

Gordon Goodman, the Welsh ecologist and energy expert with long-standing ties to Stockholm, was appointed to become the founding director of the institute. The selection would prove to be a resounding success in terms of the Beijer Institute's scientific output, international engagement and expansion, and propensity for attracting funding from a wide range of sources during the twelve years of Goodman's leadership. His appointment would also represent a significant moment in the evolution of global environmental governance. As will be explained in Chapter 6, Goodman and the Beijer Institute would become integral parts of the institutionalization and politicization process that transformed climate change into a major international issue by the end of the 1980s. The network that the institute and its director would operate within had at the time of Beijer's founding been expanding its efforts of coordinating scientific activity for well over a decade. In this context, Goodman and the climate science movement's de facto leader, Bert Bolin, would come to embody Stockholm's decisive contribution to the science diplomacy of climate change.

GARP 1974, STOCKHOLM: THE RISE OF CLIMATE MODELING

While the Royal Swedish Academy of Sciences was in the process of deciding the organizational mission of the Beijer Institute in 1974, Bert Bolin was preparing an event that would represent a major step

[97] During the existence of the original Beijer Institute from 1977 to 1989, its full title variously included "the International Institute for Energy, Resources and the Human Environment" and "the International Institute for Energy and Human Ecology."

in structuring and setting the agenda for international climate research. Held in the same conference center as the SMIC workshop three years earlier, the International Study Conference on the Physical Basis of Climate and Climate Modelling brought seventy scientists to Stockholm for two weeks of intensive discussions in July–August 1974. The conference was chaired by Bolin and arranged under the auspices of GARP, the global research program that was established at the 1967 atmospheric study conference in Stockholm. The recently established United Nations Environment Program, which would become a driver of international climate science cooperation leading up to the founding of the IPCC in 1988, co-sponsored the 1974 GARP conference along with WMO and ICSU, which had co-organized the 1967 event.[98]

From a scientific standpoint, the primary difference between the two GARP gatherings was that discussions in 1967 centered on understanding atmospheric fluctuations affecting weather, while the 1974 Conference took the significant step of studying longer-term changes in climate through the expanded use of quantitative data and computer modeling. Bolin elaborated on the enormous benefits of computer modeling as well as satellite monitoring for weather and climate prediction in an *Ambio* article that recapitulated the content and conclusions of the 1974 GARP conference, published months before the actual conference report was released. The popular science article also explained how the climate system functioned, the importance of climate in the development of human civilization, and how climate science – including its technological tools – had evolved over the course of decades. In motivating the expansion of international scientific cooperation and investments in modern monitoring and computational technologies, Bolin noted that despite the advances of recent years, climate remained poorly understood and difficult to predict.[99] The further development of the GARP initiative for the purpose of increasing understanding of the general circulation of the atmosphere and identifying the drivers of climate change – natural and anthropogenic – followed the scientific insights of the 1971 SMIC report as well as recommendation 79(d) of the Stockholm Conference Action

[98] World Meteorological Organization (WMO), International Council of Scientific Unions (ICSU). *The Physical Basis of Climate and Climate Modelling: Report of the International Study Conference in Stockholm, July 29–August 10, 1974, Organised by WMO and ICSU and Supported by UNEP* (Geneva: World Meteorological Organization, International Council of Scientific Unions, 1975).

[99] Bert Bolin, "Modelling the Climate and Its Variations," *Ambio* 3(1974):5, 180–188. www.jstor.org/stable/4312077.

Plan. Similar to SMIC, the report from the 1974 GARP Conference stressed the importance of national monitoring programs and international scientific collaboration. Moreover, Bolin's foreword to the report called for intergovernmental cooperation to tackle the "global problem" of climate change, suggesting a political component to what had to that point mostly been an international scientific endeavor.[100]

GARP 1974 would structure the ongoing WMO-ICSU cooperation on climate science going forward.[101] The conference also represented the debut of UNEP as a sponsor and catalyst of scientific research on climate change. The continued collaboration of the three international organizations would underpin the further institutionalization and advancement of climate science well into the 1980s, primarily under the banner of the World Climate Programme – the successor to GARP that was established in 1980 after the first World Climate Conference held in Geneva the previous year. What is more, the imprimatur of WMO, ICSU, and UNEP would be strategically employed by entrepreneurial scientists seeking to enhance the political status of climate change, with the involvement of UNEP – a creation of the Stockholm Conference – ensuring that the legacy of 1972 would be carried into the science and politics of climate change for years to come. Bert Bolin and Gordon Goodman, and an array of other actors associated with Stockholm, would make decisive contributions – as much organizational as scientific – during the 1980s, a decade that would see global environmental governance begin to come of age.

100 WMO and ICSU, 1975.

101 Eugene W. Bierly, "The World Climate Program: Collaboration and Communication on a Global Scale," *The Annals of the American Academy of Political and Social Science* 495(1988): 106–116. www.jstor.org/stable/1045877.

6

Enter the Earth System

A trio of Stockholm Conference-inspired events in 1982, although differing substantially in scope and focus, foreshadowed a remarkable period of activity of global environmental governance. The five-year time span that followed, from 1983 to 1988, would encompass the establishment of an international regime to combat ozone depletion, the rise of the revolutionary field of Earth System science, the creation of a UN commission that would popularize the concept of sustainable development, and the institutionalization of climate change science, a key step in the ascent of the issue on the international political agenda. Meetings would again serve as crucial cogs in the forward motion of scientific and political processes.

Acid rain, the original transboundary environmental issue that to some extent precipitated the Stockholm Conference, had by 1982 become the object of a regional governance regime. As one of the countries most affected by sulfur dioxide emissions from Europe, Sweden considered itself a primary stakeholder and, due to the pathbreaking work of Swedish scientists, a foremost authority on the issue. Acid rain was thus an obvious theme around which to convene another international conference ten years after Stockholm in June 1982. Further, in November of that year, the Royal Swedish Academy of Sciences organized a small yet influential meeting that brought together a select group of scientists to set forth an agenda for addressing environmental issues in the 1980s. Both of these meetings in Sweden were preceded by a May 1982 Conference in Nairobi, organized by the United Nations Environment Programme (UNEP), to commemorate the tenth anniversary of the 1972 event in Stockholm that led to its creation.

The latter half of the 1980s was in fact a moveable feast of scientific conferences, workshops, and meetings that paved the way for the establishment of what became the main pillars of international climate governance.[1] In the middle of the decade, climate change was an issue of increasing scientific concern, but still of only marginal political interest. Yet within just a few years, a newly formed Intergovernmental Panel on Climate Change, the IPCC, had released its first assessment report, and an Intergovernmental Negotiating Committee (INC) had been set up to prepare a framework convention on climate change under UN auspices.[2] At the center of this rapid rise of climate change on the international political agenda was a core group of activist climate experts who served as knowledge brokers in raising concern among decision-makers and laying the scientific foundation for governmental action.[3] A great deal of the coordinated activities of these scientific policy entrepreneurs was orchestrated in Stockholm.

The perhaps most groundbreaking change emerged more modestly at first, but would eventually reconfigure the entire playing field of global environmental governance: the Earth System idea. During the same period, a new institution – the International Geosphere-Biosphere Programme (IGBP) – that incorporated climate within a wider Earth System paradigm of global environmental change was established. Its international secretariat would, not coincidentally, be placed in the Swedish capital. In addition, a rebranding of the Beijer Institute initiated by the Swedish government would spawn two new Stockholm-based organizations dedicated to the study and implementation of the 1987 Brundtland Report's concept of sustainable development.

TEN YEARS AFTER STOCKHOLM

Several international events, mostly forgotten today, that were positioned at the intersection of environment, science, and politics had taken place earlier in the decade. The tenth anniversary of the 1972 UN Conference was celebrated with a "Stockholm plus 10" meeting in May

[1] Jill Jäger, "From Conference to Conference," *Climatic Change* 20(1992): iii–vii.

[2] Daniel Bodansky, "Prologue to the Climate Change Convention," In: I. M. Mintzer & J. Amber Leonard, eds., *Negotiating Climate Change: The Inside Story of the Rio Convention* (Cambridge: Cambridge University Press, 1994), 45–74.

[3] Daniel Bodansky, "The History of the Global Climate Change Regime," In: U. Luterbacher and D. F. Sprinz, eds., *International Relations and Global Climate Change* (Cambridge, MA: The MIT Press, 2001), 23–40.

1982 that was more stock-taking than ground-breaking in ambition.[4] Officially a "session of a special character" of the Governing Council of the UN Environment Program, the nine-day conference was held at UNEP headquarters in Nairobi. It was presided over by the Program's Executive Director, Mostafa Tolba, an Egyptian plant pathologist who had led his country's delegation to Stockholm in 1972 and would serve as head of UNEP for almost two decades.[5] The meeting's main task was to review the implementation of the Stockholm Action Plan and make recommendations for UNEP activities for the rest of the decade.[6] It resulted in a Nairobi Declaration that revisited the Stockholm themes of poverty and underdevelopment as drivers of environmental degradation, while reaffirming the imperatives of the Stockholm Declaration and Action Plan from a decade earlier. The 1982 Declaration also called attention to the emerging atmospheric issues of acid rain, ozone depletion, and increasing carbon dioxide concentrations that would become – in part through the efforts of UNEP and Tolba – prominent issues of international environmental politics over the course of the 1980s.

The ceremonial dimension of the Nairobi Conference featured the awarding of the newly instituted UNEP Gold Medal to Kenyan president Daniel Arap Moi, founding UNEP director Maurice Strong, and King Carl XVI Gustaf of Sweden. The Swedish King's interest in the environment dates back to the Stockholm Conference's preparatory period when, as crown prince, he served as an intern at the Swedish UN delegation in New York under the tutelage of Lars-Göran Engfeldt and attended meetings of the Conference's preparatory committee in Geneva.[7] Since the early 1970s, Carl Gustaf has consistently demonstrated his interest in the environment by supporting scientific research and advocacy organizations – the royal family are patrons of WWF Sweden and the

[4] George P. Smith II, "The United Nations and the Environment: Sometimes a Great Notion?," *Texas International Law Journal* 335(1984):19, 335–364; Philip Shabecoff, "The Environment Revisited," *New York Times*, May 23, 1982; Stephen J. Macekura, *Of Limits and Growth: The Rise of Global Sustainable Development in the Twentieth Century* (Cambridge: Cambridge University Press, 2015); United Nations Environment Program Distr. General, "United Nations Environment Programme: Nairobi Declaration on the State of the Worldwide Environment," *International Legal Materials* 21(1982):3, 676–678.

[5] "Green Giant: A Creator of the Successful Regime to Reduce Global Emissions Has Died," *The Economist*, March 16, 2016.

[6] Lars-Göran Engfeldt, *From Stockholm to Johannesburg and Beyond: The Evolution of the International System for Sustainable Development Governance and Its Implications* (Stockholm: Government Offices of Sweden, Ministry of Foreign Affairs, 2009).

[7] Engfeldt, *From Stockholm to Johannesburg and Beyond*, p. 54.

King is chairman of its advisory council – as well as through a series of speeches at major international meetings, including the Rio Earth Summit in 1992.[8] That year, he also launched the Royal Colloquium, an annual gathering of leading international experts convened by the King to discuss climate, environment, and sustainability issues.[9] Further, for his fiftieth birthday in 1996, King Carl Gustaf received a gift from Mistra, the Foundation for Strategic Environmental Research, and several academies toward establishing a Royal Professorship, which typically entailed funding invited researchers to spend time in Swedish universities and institutes. In the context of Swedish society, the King's patronage and personal interest in the environment, as well as that of his daughter Crown Princess Victoria, have served to help raise the issue to a level that to some extent transcends everyday domestic politics.

WWF Sweden has since 1983 been based at the royal palace of Ulriksdal, located at the northern end of the Royal National City Park – a natural and cultural heritage preservation area established in the early 1990s that covers a significant swath of greater Stockholm.[10] The prior occupant of Ulriksdal Palace was the International Federation of the Institutes of Advanced Studies (IFIAS), an NGO and a network of some twenty leading scientific institutes that was founded in 1972 with backing from the Swedish Riksbanken, the Nobel and Rockefeller foundations, and the King of Sweden, who provided the premises for the IFIAS secretariat. The idea for an organization dedicated to promoting research on the increasingly complex and interconnected problems of the modern world – the *problematique*, in the parlance of the Club of Rome – came from the Swedish Nobel-laureate chemist Arne Tiselius during the 1969 Nobel Symposium "The Place of Value in a World of Facts."[11]

[8] Erik Söderman, *Bernadotteprojektet*, 305. Available at: http://kungensbiografi.eriksoderman.se (accessed July 22, 2024).

[9] Many of the colloquia have been prepared in collaboration with Anders Karlqvist, head of the Swedish Polar Research Secretariat, until 2009 with offices in the Royal Swedish Academy of Sciences, and using the Academy's flagship journal *Ambio* – with offices in the same building – as an outlet for many of the ensuing papers, sometimes organized as special issues.

[10] Henrik Ernstson & Sverker Sörlin, "Weaving Protective Stories: Connective Practices to Articulate Holistic Values in the Stockholm National Urban Park," *Environment and Planning A: Economy and Space* 41(2009):6, 1460–1479. For a brief history of Ulriksdal Palace, see: www.kungligaslotten.se/english/articles-movies-360/ulriksdal-palace/2018-04-12-ulriksdal-palace-a-history.html

[11] Arne Tiselius & Sam Nilsson, eds., *The Place of Value in a World of Facts: Proceedings of the Fourteenth Nobel Symposium* (Stockholm: Almqvist & Wiksell, 1970). Sven Widmalm, "The Place of Humanities in a World of Science: Nobel Symposium 14

This intellectual affinity between IFIAS and the Club of Rome was not incidental. The executive director and co-founder of the Nobel Symposia series, nuclear physicist Sam Nilsson, was in fact the Club's first Swedish member, and would become the founding director of IFIAS, a position he held until 1986. What is more, the two co-founders of the Club of Rome, Aurelio Peccei and Alexander King, would serve as IFIAS trustees,[12] with King – the influential Director General for Scientific Affairs at the Organization for Economic Co-operation and Development – becoming chairman in 1974 after Swedish diplomat and former Nobel Foundation executive director Nils Ståhle.[13] The Nilsson-King era included, among other IFIAS activities, an Ulriksdal seminar series (with associated publications) and a comprehensive State of the Planet report, written by King and published in two editions in 1976 and 1980.[14]

The Swedish government marked the tenth anniversary of the UN Conference on the Human Environment by organizing a conference on acid rain, an issue that Sweden had brought to international attention in the years leading up to Stockholm 1972. Several hundred scientists and political representatives from twenty-one countries, as well as observers from a range of international NGOs, attended the Stockholm Conference on the Acidification of the Environment in June 1982.[15] The weeklong gathering resulted in consensus reports by experts and government ministers. It took place during the interim period between the signing, in 1979, and the coming into force, in 1983, of the Convention on Long-range Transboundary Air Pollution, an international agreement involving thirty-four countries that Sweden had played an important

and the Vanishing Humanist," In: Anders Ekström & Hampus Östh Gustafsson, eds., *The Humanities and the Modern Politics of Knowledge: The Impact and Organization of the Humanities in Sweden, 1850–2020* (Amsterdam: Amsterdam University Press, 2022), 179–204.

[12] Aurelio Peccei, *The Human Quality* (Oxford & New York: Pergamon Press, 1977). Wrote Peccei of IFIAS (p. 57): "My dear friend Alexander King is its current chairman and I have been one of its trustees since its foundation. This establishes a link with The Club of Rome; in fact, IFIAS can be regarded, in a way, as its extension into multidisciplinary scientific research."

[13] IFIAS also had links to the Aspen Institute through the American atmospheric physicist Walter Orr Roberts, who held positions and managed projects – including on the impacts of climate change – for both institutes.

[14] Alexander King, *The State of the Planet: A Report Prepared for the International Federation of Institutes for Advanced Study (IFIAS), Stockholm* (Oxford, New York, Toronto, Sydney, Paris, Frankfurt: Pergamon Press, 1980).

[15] Bette Hielman, "1982 Stockholm Conference on Acidification of the Environment," *Environmental Science and Technology* 17(1983):1, 15A–18A.

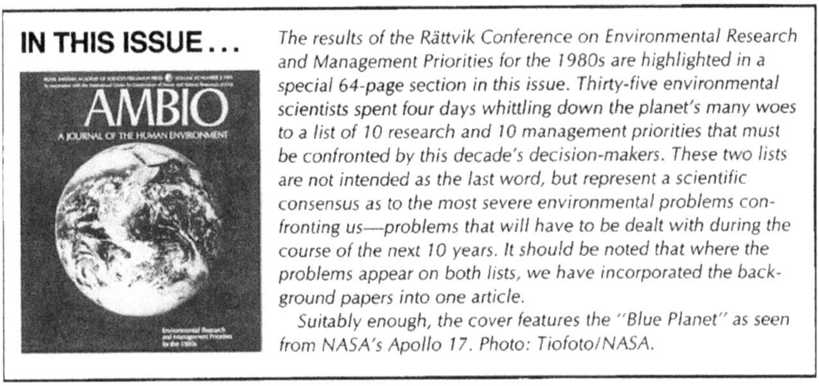

IN THIS ISSUE...

AMBIO

A JOURNAL OF THE HUMAN ENVIRONMENT

The results of the Rättvik Conference on Environmental Research and Management Priorities for the 1980s are highlighted in a special 64-page section in this issue. Thirty-five environmental scientists spent four days whittling down the planet's many woes to a list of 10 research and 10 management priorities that must be confronted by this decade's decision-makers. These two lists are not intended as the last word, but represent a scientific consensus as to the most severe environmental problems confronting us—problems that will have to be dealt with during the course of the next 10 years. It should be noted that where the problems appear on both lists, we have incorporated the background papers into one article.

Suitably enough, the cover features the "Blue Planet" as seen from NASA's Apollo 17. Photo: Tiofoto/NASA.

FIGURE 6.1 The 1982 Rättvik Conference convened by the Royal Swedish Academy of Sciences, and the special issue of the Academy's journal *Ambio* (Vol. 12, No. 2, 1983) demonstrated Sweden's continued interest in science organization and environmental agenda setting ten years after the Stockholm Conference. Courtesy of Ambio.

scientific and diplomatic role in bringing into existence.[16] The 1982 Acidification Conference in Stockholm served to accelerate the ratification of the LRTAP convention.[17]

Another 1982 event that explicitly invoked the spirit of the 1972 Stockholm Conference was the Rättvik Conference on Environmental Research and Management Priorities for the 1980s (Figure 6.1). The conference, organized by the Royal Swedish Academy of Sciences and held on the eastern shore of the magnificent Lake Siljan in central Sweden, further demonstrated Sweden's ambition to shape the scientific and political agenda on environmental issues. After a year spent selecting the scientists to invite and the specific topics to discuss – a curation process that the thirty-five hand-picked experts were involved with in advance of the conference – a total of ten research and ten management priorities were agreed upon.[18] Climate change, indicating the lukewarm level of concern that prevailed in the early 1980s, was selected by the participating scientific elites as a research priority but not as one of the ten most pressing

[16] Karin Bäckstrand & Henrik Selin, "Sweden – A Pioneer of Acidification Abatement," In: Arild Underdal & Kenneth Hanf, eds., *International Environmental Agreements and Domestic Politics: The Case of Acid Rain* (London & New York: Routledge, 2000).

[17] McCormick, *Reclaiming Paradise*, p. 185.

[18] Alf Johnels, "Conference on Environmental Research and Management Priorities for the 1980s," *Ambio* 12(1983):2, 58–59.

environmental management priorities. It was eclipsed in importance by issues like river basin management and the control of pathogens from human waste and their aquatic vectors.

The legacy of Stockholm 1972 was embodied in Rättvik by Maurice Strong, who delivered an expansive keynote address entitled "Towards a New Kind of Economic Growth." Engaging with the complex issue of growth, Strong suggested that industrialized countries should embrace a "new growth" economic paradigm, based on personal fulfillment rather than material production, while at the same time making a massive commitment to Third World development.[19] The environment-development dilemma was also addressed by the conference director Robert Munro. In a special issue of the journal *Ambio* dedicated to the Rättvik Conference, published the same year, 1983, as the launch of the Brundtland Commission, Munro sounded an optimistic tone on the possibility of, as it were, sustainable development:

> The decade (of the 1970s) began with a widely held view that one must choose environment or development, but ended with increasing acceptance of the view that environmental considerations were an essential and integral part of sustainable national and even global economic development.[20]

Munro was at the time an environmental consultant living in Stockholm with his wife Ingrid (née Mårtensson), an architect who had produced a report on urban environments for the Swedish government during the preparatory stage of the Stockholm Conference.[21] Robert Munro, a Canadian expert in environmental law, was an advisor to Maurice Strong for the Stockholm Conference and had been involved in a range of environmental and development projects and processes during the 1970s. Well-connected internationally, Munro was close to fellow Canadian Jim MacNeill, who was also a key participant in Stockholm Conference preparations on behalf of the Canadian government, and would from 1983 serve as secretary general of the Brundtland Commission. MacNeill, whose mother was Swedish, also had strong ties to Stockholm, having studied at Stockholm University in the early 1950s. Another one of Munro's contacts was the Yale-trained environmental lawyer James Gustave Speth, who had recently chaired the Council on Environmental

[19] Maurice Strong, "Towards a New Kind of Economic Growth," *Ambio* 12(1983):2, 64–66.

[20] R. D. Munro, "Environmental Research and Management Priorities for the 1980s," *Ambio* 12(1983):2, 60–61.

[21] Ingrid Munro would later found, in 1999, the microfinance organization Jamii Bora in Nairobi.

Quality under US President Jimmy Carter. Speth, who would also champion sustainability in the United Nations as Administrator of the UN Development Program during the 1990s, drew upon the Rättvik results while establishing and formulating the mission of the World Resources Institute, an influential environmental NGO he founded in 1982.[22] With its ties to Stockholm, Speth's institute would during the 1980s also become a partner in the publication of the Academy's journal *Ambio*.[23] Through Munro – a prime example of an expatriate contributor to Stockholm's outsized influence on environment and sustainability issues – the relatively small-scale Rättvik Conference had an impact on environmental governance well beyond the limited number of experts that participated in the meeting.

THE QUICKENING OF CLIMATE AS A POLITICAL ISSUE: THE PIVOTAL 1985 VILLACH CONFERENCE

The Rättvik Conference was another example of Sweden's long tradition of organizing international meetings, scientific as well as policy-oriented, on issues of the environment. Such conferences, and the individuals that organized and participated in them, have served as catalysts for advancing scientific knowledge and fostering policies, institutions, and networks. They have also constituted an important source of influence and leverage in promoting Stockholm's status as a leading hub of global environmental governance. Going back to at least the mid-1950s, Bert Bolin and many others had been arranging conferences in and around Stockholm, in settings ranging from stately nineteenth-century manor houses to the more austere functionalist facilities of 1960s modernism. In the middle of the 1980s, actors associated with Stockholm became instrumental in convening a series of events in locations outside Sweden that have since become synonymous with the emergence of climate change as a top-tier issue of international politics.

Well into the 1980s, climate change was largely outside the mainstream of environmentalism, and specialists in the field for the most part followed their own scientific track, separate from disciplines such as ecology that were often at the center of environmental politics. This would begin to shift sharply in the middle of the decade. A conference

[22] Robert Munro, personal communication (Paglia), March 8, 2019.

[23] Sörlin, "The Environment as Seen through the Life of a Journal." Sundin, "Environmental Protection and the National Parks."

in Villach, Austria, in October 1985 was an integrative moment in the scientific understanding of climate, assembling a wide range of experts to form a consensus on the dynamics and drivers of global warming. It was also a turning point in the politicization of climate change. Organized by the International Meteorological Institute (IMI) at Stockholm University and the Stockholm-based Beijer Institute, the conference was the culmination of a series of scientific meetings in the Alpine resort mandated by the 1979 World Climate Conference in Geneva and convened under the auspices of the World Climate Programme. The previous Villach workshops, one in 1980 chaired by Bolin and another in 1983, had been limited to the scientific aspects of climate and CO_2 Climate change had thus remained mostly apolitical through the first half of the 1980s, a status that would change dramatically after 1985.[24]

The 1985 Villach Conference brought together the heads of the sponsoring organizations UNEP, the World Meteorological Organization (WMO) and the International Council of Scientific Unions (ICSU) – the same trio that supported the Villach meetings in 1980 and 1983 – and scientists from twenty nine developed and developing countries. For the first time, a scientific assembly of such significance managed to reach a general consensus – a politically powerful concept in climate discourse – on the near-term threat posed by anthropogenic climate change. Mostafa Tolba (Figure 6.2), the Executive Director of UNEP, evoked climate-induced disasters in an opening statement that signaled the emerging politicization of climate change. In contrast to the previous Villach meetings, the 1985 conference foreshadowed calls for climate mitigation and adaptation policies by emphasizing the importance of government action to limit greenhouse gas emissions. It also brought up the need for societies to take a changing climate into account when making longer-term decisions; the climatic past was no longer a reliable baseline for future conditions. The conference statement put forward a range of recommended actions for policy and, importantly, tied climate change to the other politically salient atmospheric issues of the day: acid rain and ozone depletion.[25]

24 Wendy E. Franz, *The Development of an International Agenda for Climate Change: Connecting Science to Policy. ENRP Discussion Paper E-97-07* (Cambridge, MA: Harvard University Kennedy School of Government, 1997).

25 World Meteorological Organization, International Council of Scientific Unions, and United Nations Environment Programme, *Report of the International Conference on the Assessment of the Role of Carbon Dioxide and of Other Greenhouse Gases in Climate Variations and Associated Impacts, Villach, Austria, October 9–15, 1985* (Geneva: World Meteorological Organization, 1986).

FIGURE 6.2 As Executive Director of the United Nations Environment Programme from 1975 to 1992, the Egyptian scientist Mostafa Tolba played a key role in the promotion of the science and politics of ozone and climate, including the 1985 Vienna Convention for the Protection of the Ozone Layer and the establishment of the Intergovernmental Panel on Climate Change in 1988. Photo: United News/Popperfoto via Getty Images.

The signing of the Vienna Convention for the Protection of the Ozone Layer just six months before (and, incidentally, only a few hundred kilometers from) the 1985 Villach Conference had a profound influence on the trajectory of climate governance. The participants at Villach 1985, some of whom had also been involved with the ozone issue, were convinced that the decisive role of science in catalyzing government action on ozone-depleting chlorofluorocarbons (CFCs) could and should serve as a model for political engagement on greenhouse gases. Further, the success of the Vienna Convention – a watershed agreement in global environmental politics – encouraged UNEP and Tolba, who had since the late 1970s effectively leveraged scientific expertise in advancing the ozone agenda, to take a similar, scientized approach in fostering political engagement on climate change.[26] The 1985 Villach Conference was, in

[26] David George Hirst, *Negotiating Climates: The Politics of Climate Change and the Formation of the Intergovernmental Panel on Climate Change (IPCC), 1979–1992*, PhD diss. (University of Manchester, 2015).

effect, orchestrated to become the manifestation and coming-out party for the convergence of climate science and international politics.

The importance of Villach 1985 has not only been realized in retrospect; conference participants were aware that they were taking part in something momentous.[27] For some, including many of the scientists that animate this chapter, the conference marked their emergence as de facto political actors on the global stage. Although a scientific conference with all the trappings of such – keynote addresses, research papers, thematic working groups, and an extensive conference report – the enduring historical significance of Villach 1985 would prove to be this pivot toward the policy implications of climate change, in which scientists and their institutions would come to play a decisive role. The enhanced political dimension of the 1985 meeting was facilitated by the fact that the scientists were, by design of the conference organizers, attending in their capacity as individuals, rather than representing a particular country. They were thus at liberty to make statements based solely on their professional conclusions (and, perhaps, personal convictions) without inhibition over the potential political implications of their scientific claims.[28]

Beyond the synthesis of diverse climate knowledge that took place during the seven days in Villach, including the forcing effect of trace gases other than CO_2 and the potential regional impacts of climate change, the 1985 Conference and its scientific outputs would prove to be a rallying point. It also served as a venue of political leverage for a cadre of activist scientists who had become convinced that climate change posed a clear and present danger to humanity. As explored later in this chapter, several important outcomes followed in the wake of the conference – not only scientific but also political, institutional, and social, in terms of the emergence and strengthening of networks dedicated to bringing about government action on climate change. The ability to speak on behalf of science, however, was the primary source of influence that the networks constituted. More than just closed-door meetings of scientific elites, this required documentation grounded in research – and research, as has always been the case, required funding.

The scientific results of Villach 1985 were summarized in a seventy-eight-page official conference report written by Bolin and colleagues at Stockholm University, and a landmark, 500-page edited

[27] BBC World Service, "Climate Change: The Early Years," *Witness History*, October 9 (2014). Available at: www.bbc.co.uk/programmes/p027rh9c (accessed November 26, 2021).
[28] Franz, *The Development*; Hirst, *Negotiating Climates*.

volume in the SCOPE series. SCOPE – Scientific Committee on Problems of the Environment – is an international scientific body under ICSU that has since 1971 published influential reviews of contemporary environmental issues with human dimensions. The climate report, number 29 in the SCOPE series, was based on research specifically commissioned by UNEP, with additional support from WMO and ICSU, that was explicitly intended to serve as the scientific basis for the deliberations at the 1985 Villach conference.

The research project originated in a pivotal June 1982 meeting in Stockholm, arranged by the Swedish diplomat Göte Svensson, between Bert Bolin and Mostafa Tolba.[29] The two were primary protagonists in the formation of climate governance institutions later in the decade. Bolin would in 1988 become founding chairman of the IPCC, the organization that Tolba would as Executive Director of UNEP be instrumental in establishing.[30] The 1982 meeting in Stockholm was thus an early signal of the emerging alliance of scientists and international organizations, particularly UNEP under Tolba, that would, as with ozone, underpin the politicization of climate change. Not surprisingly, the project was placed at the Department of Meteorology at Stockholm University. Starting in 1983, the research there was carried out by Bolin, Bo Döös, and Jill Jaeger who, together with climate scientist Richard Warrick, co-authored and edited the SCOPE 29 report *The Greenhouse Effect, Climatic Change, and Ecosystems*.[31]

The scientific report also included an ominous message that was clearly legible even to the layman: "In the first half of the next century a rise of global mean temperature could occur which is greater than any in man's history."[32] Published in 1986, the report became a milestone in climate science and in understanding the carbon cycle. Some even considered it the "IPCC bible" in the early days of the Panel.[33] Yet as influential as the

[29] Bert Bolin, *A History of the Science and Politics of Climate Change. The Role of the Intergovernmental Panel on Climate Change* (Cambridge: Cambridge University Press, 2007), 35–36.

[30] Shardul Agrawala, "Context and Early Origins of the Intergovernmental Panel on Climate Change," *Climatic Change* 39(1998):4, 605–620.

[31] Bert Bolin, Bo Döös, Jill Jäger & Richard A. Warrick, *The Greenhouse Effect, Climatic Change, and Ecosystems*. SCOPE 29 (Chichester: John Wiley & Sons, 1986).

[32] Bolin et al., *The Greenhouse Effect;* Weart, *The Discovery of Global Warming*.

[33] Sonja Boehmer-Christiansen, "The International Research Enterprise and Global Environmental Change: Climate-Change Policy as a Research Process," In: Mark Imber & John Vogler, eds., *The Environment and International Relations* (New York: Routledge, 1995), 183–209.

SCOPE 29 report was, the Villach findings would receive their widest dissemination, and greatest political impact, not in a specialized publication for a narrow scientific audience but in the landmark report of the World Commission on Environment and Development (WCED), also known as the Brundtland Commission, chaired by former Norwegian prime minister Gro Harlem Brundtland.

GORDON GOODMAN, THE BEIJER INSTITUTE, AND THE BRUNDTLAND REPORT

Upon its release in October 1987, the WCED report *Our Common Future* attracted tremendous international attention and greatly increased public and political interest in issues of the environment and "sustainable development" – the term the Brundtland Commission is usually credited with coining, but which was of course rooted in ideas with a much longer lineage.[34] It also generated a serious discussion in the UN General Assembly specifically on climate change, which up to that point had not been an issue of significant international concern. The sections on climate change in *Our Common Future* drew heavily, indeed exclusively, on Villach, citing and extracting passages directly from the conference's report.[35] This was not a coincidence, nor done simply out of convenience. The use of the Brundtland Report as a high-profile platform to leverage, communicate, and advance the Villach science-policy agenda was part of a longer-term plan involving scientists and other actors both inside and outside of the Brundtland Commission. Stockholm figured prominently in the formation and execution of this strategy that would strongly influence the events that led to the establishment of an intergovernmental organization for climate science and help lay the foundation for a framework convention on climate change.

The energy and climate chapter of the Brundtland Report was written by Gordon Goodman (Figure 6.3), the director of the Stockholm-based

[34] G. H. Brundtland et al., *World Commission on Environment and Development. Our Common Future* (Oxford: Oxford University Press, 1987); Paul Warde, *The Invention of Sustainability: Nature and Destiny c. 1500–1870* (Cambridge: Cambridge University Press, 2018).

[35] Wendy Torrance, "Science or Salience: Building an Agenda for Climate Change," In: Ronald B. Mitchell, William C. Clark, David W. Cash & Nancy M. Dickson, eds., *Global Environmental Assessments: Information and Influence* (Cambridge, MA: The MIT Press, 2006), 29–56. Sonja Boehmer-Christiansen, "Scientific Consensus and Climate Change: The Codification of a Global Research Agenda," *Energy & Environment* 4(1993):4, 362–407.

FIGURE 6.3 Gordon Goodman (right) and Lars Kristoferson (left) were director and deputy director, respectively, for the entire duration of the Beijer Institute's original incarnation from 1977 to 1989, and remained in those positions for several years after Beijer's transformation into the Stockholm Environment Institute. While director of Beijer, Goodman – a Welsh biologist and energy expert who was well connected in international scientific networks – contributed to an array of seminal processes, including the Brundtland Commission and the climate meetings in Villach and Bellagio that facilitated the establishment of the AGGG and IPCC. Photo: Courtesy of Lars Kristoferson.

Beijer Institute.[36] A biologist by training, Goodman was also an energy specialist with exceptional communication, organizational, and social networking skills. He was moreover a committed environmentalist, known in his native Wales for applying his scientific research to help ameliorate real-world problems, particularly related to energy and air pollution, caused by human activity. His scientific activism encompassed a variety of remediation projects and environmental conservation initiatives that engaged local stakeholders in the degraded postindustrial landscape of the Lower Swansea Valley coal district.[37] Throughout his scientific career, which included a position at the University College of

[36] Torrance, "Science or Salience," 2006, p. 44.
[37] Miles Chadwick, "Gordon Goodman: Ecologist with an Early Interest in Environmental Issues," *The Guardian*, June 26, 2008.

Swansea and a professorship at the University of London, Goodman cultivated close relationships with a wide range of international organizations. Among these were UNEP, WMO, and ICSU, including the latter's SCOPE initiative,[38] through which he had established the Monitoring and Assessment Research Centre at Chelsea College in 1974.

Goodman's strong ties to Sweden stretched back to the early 1970s. Through ICSU and a Winston Churchill Memorial Fellowship, Goodman spent a great deal of time in Stockholm, receiving training on the detection and measurement of organic mercury in marine environments.[39] He spent the summer of 1971 in Stockholm working together with Swedish scientists Bengt Lundholm and Sören Svensson to prepare the ICSU-SCOPE 1 report, *Global Environmental Monitoring*. As discussed in Chapter 4, the report was written specifically, even strategically, as a contribution to the 1972 Stockholm Conference, laying the scientific foundation for the establishment of the UNEP Earthwatch program and its Global Environmental Monitoring System.[40]

As part of the British delegation to the Stockholm Conference, Goodman met Carl Gustaf Bernhard, secretary general of the Royal Swedish Academy of Sciences and at the time chairman of the Swedish SCOPE Committee.[41] Several years later, Bernhard would oversee the establishment of the Beijer Institute – the International Institute for Energy Resources and the Human Environment. It was on the basis of these scientific and organizational capabilities and his long background in environmental research, as well as his strong associations with Sweden and long-standing relationships within the international scientific community, that the Academy appointed Goodman to become the founding executive director of the Beijer Institute. He would hold that position for fourteen years, encompassing the entire duration of the Beijer Institute's original organizational form, as well as the first

[38] Goodman had moreover written the terms of reference for SCOPE, Chadwick "Gordon Goodman."

[39] Chadwick, "Gordon Goodman."

[40] Commission on Monitoring of the Scientific Committee on Problems of the Environment (SCOPE) of the International Council of Scientific Unions (ICSU), *Global Environmental Monitoring* (Stockholm: Scientific Committee on Problems of the Environment, 1971); Elena Aronova, "Environmental Monitoring in the Making: From Surveying Nature's Resources to Monitoring Nature's Change," *Historical Social Research* 40(2015):2, 222–245.

[41] Carl Gustaf Bernhard, *The Beijer Institute: The International Institute for Energy Resources and the Human Environment* (Stockholm: Royal Swedish Academy of Sciences, 1991).

two years of its rebranded incarnation as the Stockholm Environment Institute (SEI). The Beijer Institute would enjoy great success under Goodman's leadership, attracting hundreds of millions of Swedish crowns in project funding and initiating close to a hundred projects, mostly in the developing world.[42]

In his speech at the opening ceremony of the Beijer Institute in 1977, Goodman stated that the Institute would welcome researchers from all over the world, "from east and west, from north and south." The King of Sweden, H. M. Carl XVI Gustaf, ever the patron of the environment, expressed a similar sentiment in his official opening of the Beijer Institute, proclaiming, "I am proud that Sweden and its people, by making this Institute possible, have shown that they are also citizens of the world."[43] While such statements might seem like ceremonial boilerplate, this dovetailing of Sweden's perception of its place in the world, and the Beijer Institute's outlook and status as a bridge spanning North-South and East-West divides, gave Goodman and the Institute certain operational advantages in achieving their energy and environmental goals in an often-fraught geopolitical context.

The Institute would be based at the Royal Swedish Academy of Sciences, located in the Frescati district adjacent to the campus of Stockholm University. A nexus of expertise as well as institutional power and prestige, both the Academy and the University are situated within an extensive green wedge on the northern edge of downtown Stockholm that is today a protected area on environmental and cultural heritage grounds as part of the Royal National City Park.[44] The Institute's offices were on the top floor of a newly built Beijer wing that greatly expanded the Academy's capacity for holding larger events. The small staff steadily grew as Goodman managed to secure substantial project grants from various Swedish and international sources to supplement the base funding from the Beijer Foundation. The Beijer Institute even expanded internationally, opening offices in York, England, and Boston in the United States, under the leadership of Michael Chadwick

[42] Kristoferson, "Gordon and the Early Days at Beijer and SEI."

[43] Bernhard, *The Beijer Institute.*

[44] Peter Schantz, "The National Urban Park in Greater Stockholm: Background, Legislation and Implementation," *Garden History* 32(2004):2, 279–280; Ernstson & Sörlin, "Weaving Protective Stories"; Henrik Ernstson, Sverker Sörlin & Thomas Elmqvist, "Social Movements and Ecosystem Services: The Role of Social Network Structure in Protecting and Managing Urban Green Areas in Stockholm," *Ecology and Society* 13(2008):2, 39.

and Paul Raskin, respectively. Those Beijer satellites still exist today as two of the seven international centers of the SEI.

The composition of Beijer's advisory board signaled the strong international orientation of the Institute. The chairman for the entirety of the original Beijer Institute's twelve-year existence was the nuclear chemist and energy expert Jack Hollander of the Lawrence Berkeley Laboratory. Hollander had previously spent extended periods on sabbatical at Uppsala University and at the Nobel Institute for Physics in Stockholm. He therefore had strong ties at the Academy and within the Swedish research community, as well as a deep personal affinity for Sweden.[45] Sweden's special geopolitical position as a nonaligned state was reflected in the institute's international advisory board, which along with Americans such as Hollander and the economist Robert Solow also included biochemist Yuri Ovchinnikov of the Soviet Academy of Sciences. The board further encompassed Swedish science and technology elites and, particularly from the mid-1980s, an array of international dignitaries such as Aga Khan, Maurice Strong, Prince Claus of the Netherlands, and Y. Hayashi of the Toyota Foundation.[46] Interestingly, Bert Bolin and Gilbert White – a renowned American geographer closely involved with the climate question – briefly served on the advisory board at precisely the time, in 1984–1985, when the Beijer Institute was becoming fully engaged in the science and nascent politics of climate change.[47]

The Beijer Institute was also strongly engaged with the Global South, reflecting another hallmark of Swedish foreign policy since the 1960s. Under Goodman's leadership, the Institute launched a range of energy-related projects in the developing world, primarily in East Africa. The projects were largely focused on fuelwood – one of the ten most important environmental management priorities for the 1980s according to the experts at the Rättvik Conference, which included Goodman (who likely lobbied for the inclusion of the so-called fuelwood crisis on the list). A major, multi-year initiative on fuelwood in Kenya was inaugurated with a series of symposiums and meetings in Stockholm and Nairobi in the late 1970s and early 1980s. Beijer's

[45] Jack Hollander, *My Lunch with Shostakovich: Essays from the Serendipitous World of a Nuclear Scientist* (Morrisville, NC: Lulu, 2009).

[46] Bernhard, *The Beijer Institute*.

[47] In 1986, White would be selected as one of six experts, together with Bolin and Goodman, to become part of the blue-ribbon Advisory Group on Greenhouse Gases. He had also served as president of SCOPE from 1976 to 1982.

collaboration with local partners and government agencies, including the Kenya Academy of Science and the Nairobi-based UNEP, led to the establishment of the Kenya Ministry of Energy. A local office in Nairobi, the Beijer Institute Centre for Energy and Development in Africa, was also active during this period in order to directly involve stakeholders in the Institute's projects.

Goodman's political skills were exemplified at an international energy seminar in 1981 in Ethiopia. Presiding over a room full of energy experts, World Bank officials, and representatives of the Ethiopian government that he was advising at the time, Goodman, according to an observer from the Swedish International Development Agency, was able to adeptly navigate the political complexities and sensitivities of the situation and win over skeptics in the audience with actionable knowledge.[48] His scientific expertise and extensive experience with real-world matters of environment and development, including their political aspects, made Goodman singularly qualified to contribute to the commission that popularized the concept of sustainable development. His work with the Brundtland Commission would also provide him with another high-profile platform for advancing the cause of climate change in the arena of international politics.

THE ROLE OF ROCKEFELLER FUNDING IN FURTHERING THE CLIMATE MOVEMENT

The final five years of Goodman's career came to be dominated by climate change. Those years proved to be some of the most eventful in modern environmental history. With Stockholm and the Beijer Institute as his base, Goodman's prowess at operating at the intersection of science, policy, and transnational networks – a skillset similar to Bert Bolin's – enabled him to play a leading role in the establishment of climate change as a significant international issue by the end of the 1980s. The Beijer Institute was located only a few hundred meters from Bolin's world-leading center of atmospheric research at Stockholm University. In tandem with Beijer's own in-house research on greenhouse gases that grew out of its work on air pollution issues, this institutional proximity facilitated collaboration on climate as well as Goodman's previous areas of engagement, including energy, ecology, and development.

[48] Karin Lange, *SIDA från insidan: Minnen och tankar kring bistånd 1965–1995* (2004). Available at: https://utvecklingsarkivet.se/wp-content/uploads/2014/04/Sida-fr%C3%A5n-insidan-2013-fin.pdf (accessed June 10, 2022).

This was certainly reflected in Goodman's involvement with the Brundtland Commission, which became a key aspect of his influence at a critical juncture in the political history of climate change. Goodman served as part of a Group of Special Advisers on Energy for the WCED, where he developed a highly productive collaboration with the Commission's secretary general, Jim MacNeill, the Canadian former environment director of the OECD.[49] In this context, Goodman also came into contact with Tom Wahman, an executive working with environment and climate issues at the Rockefeller Brothers Fund (RBF) in New York.[50] Wahman had previously been a prominent figure in the civil rights movement, as well as an All-American ice hockey goaltender at Dartmouth College. Starting in 1985, this North Atlantic trio – which largely reflected the leading roles played by a cohort of Sweden, Canada, and United States-based climate experts and advocates during this period – closely coordinated their activities while also involving a small circle of other actors that would make key contributions to advancing the scientific and political agenda on climate.

The Rockefeller Brothers Fund saw the ongoing work of the WCED as an opportunity to enhance the urgency and political status of climate change, which they perceived as an emerging issue that fit well within the Fund's *One World* program that aimed to foster greater global interdependence.[51] The RBF considered it important to involve, and thus fund, organizations based in Europe, which they saw as the locus of climate leadership in comparison to the United States.[52] Global warming had been brought to the attention of the RBF by James Gustave Speth of the World Resources Institute, which had been collaborating with the Beijer Institute on a UNEP/WMO/ICSU-sanctioned project called Energy, CO_2 and Climate Change. Like Goodman and his scientific circle, Speth's institute was exploring ways to raise political interest in climate change. Receptive to Rockefeller's involvement in their climate agenda setting, and sharing Wahman and the Fund's conclusion that science could be effectively leveraged to precipitate government intervention against greenhouse gas emissions – perhaps even leading to a

49 Bolin, *A History of the Science and Politics of Climate Change*, 47.

50 Laura Schwarz, *A History of Climate Action through Foundations' Archives* (New York: Rockefeller Archive Centre, 2018).

51 Schwarz, *A History of Climate Action*, 3.

52 Edouard Morena, *Lost in Translation?: US Foundations as Mediators between US Interests and the International Climate Policy Space* (New York: Rockefeller Archive Center, 2020).

global convention – MacNeill and Goodman were "favorably disposed if not enthusiastic about the idea of using the Commission to issue a climatic change action agenda."[53]

Goodman thus became the critical link between WCED and the part of the climate community that advocated for political action. Like the Beijer Institute itself, funded by a wealthy Swedish businessman, the efforts of the network of climate advocates were facilitated by financial support from philanthropic foundations built upon the fruits of industrial capitalism; in this case, the Rockefeller family fortune. The correspondence between Goodman, Wahman, and MacNeill resulted in a $100,000 grant from the foundation in support of an emerging strategy to enhance the political salience of climate change. This objective was to be pursued primarily through a series of conferences meant to catalyze political action by consolidating the growing scientific consensus. Approved in December 1985, much of the grant went to Beijer, with the Institute receiving additional RBF funding in 1987. In internal RBF communications, the initiative came to be called "the Beijer project," with the two-part workshop that was in the works for late 1987 referred to as "the Beijer meeting."[54] Although it was largely orchestrated in Stockholm, the first part of the invitation-based workshop was to be held in the familiar confines of Villach, followed by an even more exclusive gathering at the Rockefeller Foundation's resort in Bellagio, Italy.

VILLACH-BELLAGIO 1987: ADVANCING CLIMATE SCIENCE AND POLITICS THROUGH SCIENTIFIC ACTIVISM

A core group of like-minded scientists who attended Villach 1985 were intent on building upon the breakthrough of that conference to shape a longer-term strategy for subsequent interventions.[55] The Villach Conference statement had called for the establishment of a small task force to provide scientific advice and assessments on greenhouse gases and even to "initiate, if deemed necessary, consideration of a global convention." This led to the formation of the Advisory Group on Greenhouse Gases (AGGG), a short-lived yet influential body often described as the immediate forerunner of the IPCC. Two members were appointed by

[53] Schwarz, *A History of Climate Action.*
[54] Ibid., p. 7.
[55] Michael Oppenheimer, "Developing Policies for Responding to Climate Change," *Climatic Change* 15(1989): 1–4.

each of the organizations that together convened the AGGG, the by-then usual suspects of UNEP, WMO, and ICSU. Although it is difficult to trace a direct political or institutional outcome to the work of the AGGG, it served an important transitional role in anchoring the network of scientific activists and in advancing the climate agenda during this formative moment of climate governance.

While the AGGG held a formal status as an advisory body answerable to its organizational sponsors, a less formal constellation of Villach veterans became the steering committee for the meetings that would be held in Villach and Bellagio in the autumn of 1987. Bert Bolin, Gordon Goodman, and Ken Hare of the University of Toronto sat on the steering committee while also serving on the AGGG. The other committee members included Michael Oppenheimer of the Environmental Defense Fund; ecologist George Woodwell, founder of the Woods Hole Research Center (later renamed Woodwell Climate Research Center) and, like Oppenheimer, a committed environmental activist with an extensive NGO background; Worknesh Degefa of Ethiopia's National Meteorological Service; William Clark of IIASA and Harvard University; Harold Ferguson of the Atmospheric Environment Service at Environment Canada; C. C. Wallén, a Stockholm climate scientist associated with UNEP and WMO; and Jill Jaeger, a scientific consultant affiliated with IMI and the Beijer Institute.

The Villach-Bellagio steering committee was somewhat top-heavy, composed mostly of senior scientists such as Bolin, Goodman, Woodwell, and Ferguson who had decades of experience operating within international scientific networks. The committee, however, also encompassed the energy and expertise of younger collaborators such as Clark and Oppenheimer, who held something of an activist outlook on climate change. The extended network of "the Villach Group" also included influential science administrators with convening power and the capacity to communicate climate knowledge in political and intergovernmental venues.[56] Network participants and patrons included UNEP executive director Mostafa Tolba and Jim Bruce, a Canadian government and WMO official who chaired the 1985 Villach meeting as well as the AGGG. While perhaps not demographically representative of the

[56] Peter M. Haas & David McCabe, "Amplifiers or Dampeners: International Institutions and Social Learning in the Management of Global Environmental Risks," In: The Social Learning Group, ed., *Learning to Manage Global Environmental Risks – Volume 1: A Comparative History of Social Responses to Climate Change, Ozone Depletion, and Acid Rain* (Cambridge, MA: MIT Press, 2001), 323–348.

FIGURE 6.4 Jill Jaeger at the IIASA Systems Analysis Conference in 2015. Jaeger played a critical role in the advancement and organization of climate science from the late 1970s onward, and became a leading member of the global change community. In association with Stockholm-based scientists Bert Bolin and Gordon Goodman, she co-authored the seminal SCOPE 29 report, contributed to the work of the AGGG, and was a central figure in the steering committee that arranged the 1987 climate meetings in Villach and Bellagio, for which she wrote the final report on behalf of the Beijer Institute. Photo: Matthias Silveri/IIASA.

global stakeholders at risk from climate change, the competencies, connections, and functional diversity of the extended network were major sources of strength in advancing the political objectives of the Villach Group and its supporters.

Jill Jaeger (Figure 6.4) was the only woman on the steering committee. A British-born climate researcher based in West Germany, Jaeger was also one of the youngest members of the climate network. She had initially come into contact with Bert Bolin in 1979 at the World Climate Conference in Geneva. Working as a research scholar for the International Institute of Applied Systems Analysis in Vienna, where she was leading the Institute's research on energy and climate, hers was the only scientific presentation made by a woman at the WCCC.[57] In a professional sphere dominated at the time by men, Jaeger – who would later hold senior positions in organizations operating at the intersection of global change and sustainable development, including as Deputy Director of

[57] Jill Jaeger, interview with Eric Paglia, June 15, 2021.

IIASA – established herself as a key contributor and central node in the network taking shape around IMI and the Beijer Institute in Stockholm.

Several years after the WCCC, she was recruited by Bolin, who had also attended the "Carbon Dioxide, Climate and Society" workshop that Jaeger had organized at IIASA in 1978,[58] to take part in the IMI research project commissioned by UNEP for the 1985 Villach conference. As a parent of young children at the time, Jaeger traveled to Sweden on weekends to carry out research at the Department of Meteorology at Stockholm University and assist with the Villach preparations.[59] Her contributions surrounding Villach 1985 were some of the most significant, including writing parts of the conference report, as well as co-authoring and editing the SCOPE 29 report that was based on the research project at Stockholm University. As it turned out, her association with Stockholm further deepened via Villach. There she met Gordon Goodman, who would later invite her to assist with the work of the AGGG and the steering committee for the 1987 Villach-Bellagio workshop (Figure 6.5),[60] for which she would write the final report on behalf of the Beijer Institute.

With Goodman acting as chairman and Jaeger playing a key coordinating role, the Villach-Bellagio steering committee operated in concert with the AGGG to organize activities and formulate a scientific and policy-oriented agenda. Committee members kept in close contact over the two years between the 1985 and 1987 Villach meetings, holding planning sessions in Toronto, Canada, and Mainz, West Germany.[61] The latter meeting was hosted by atmospheric chemist Paul Crutzen, the director of the Max Planck Institute, who had received his PhD training under Bert Bolin at Stockholm University and would later be awarded a Nobel Prize in Chemistry for his groundbreaking research on ozone conducted in Stockholm in the late 1960s and early 1970s (see Chapter 3).

Members of the steering committee were adept at finding financial support for their activities. Gordon Goodman, for one, had over the years

[58] Jaeger, whose surname was Williams at the time, edited the workshop's proceedings while also contributing a chapter: Jill Williams, ed., *Carbon Dioxide, Climate and Society: Proceedings of an IIASA Workshop, February 21–24, 1978* (Oxford: Pergamon Press, 1978).

[59] Jaeger interview, June 15, 2021.

[60] Ibid; Hirst, *Negotiating Climates*.

[61] Jill Jaeger, *Developing Policies for Responding to Climatic Change: A Summary of the Discussions and Recommendations of the Workshops held in Villach (September 28–October 2, 1987) and Bellagio (November 9–13, 1987), under the Auspices of the Beijer Institute, Stockholm* (Geneva: World Meteorological Organization, 1988).

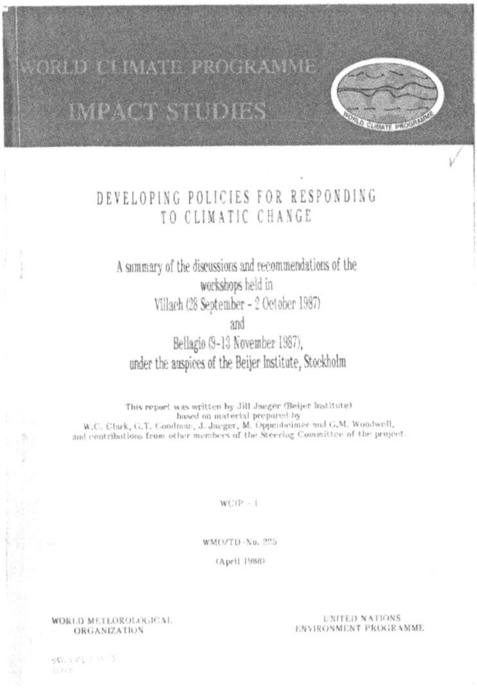
WORLD CLIMATE PROGRAMME
IMPACT STUDIES

DEVELOPING POLICIES FOR RESPONDING
TO CLIMATIC CHANGE

A summary of the discussions and recommendations of the
workshops held in
Villach (28 September - 2 October 1987)
and
Bellagio (9-13 November 1987),
under the auspices of the Beijer Institute, Stockholm

This report was written by Jill Jaeger (Beijer Institute)
based on material prepared by
W.C. Clark, G.T. Goodman, J. Jaeger, M. Oppenheimer and G.M. Woodwell,
and contributions from other members of the Steering Committee of the project.

WCIP - 1

WMO/TD-No. 225

(April 1988)

WORLD METEOROLOGICAL
ORGANIZATION

UNITED NATIONS
ENVIRONMENT PROGRAMME

FIGURE 6.5 Some of the scientists and institutions that played central roles in the landmark climate meeting in Villach, Austria, in 1985 were also involved in the smaller and more policy-oriented two-stage workshop in Villach and Bellagio, Italy, in autumn 1987. One of the science-policy contributions of the workshop's report, prepared under the auspices of the Beijer Institute, was the introduction of the idea of the 2°C temperature target. Photo: Courtesy of the World Meteorological Organization.

secured millions of dollars in project financing for the Beijer Institute from an array of international sources, including, in the case of climate, the Rockefeller Brothers Foundation. The RBF, as well as the Rockefeller Foundation, was not, however, the committee's sole sponsor. Additional support for the Villach-Bellagio workshop came from the governments of Austria and Sweden, UNEP, and the Beijer Institute. Two other American philanthropies also contributed to the efforts of the steering committee. One was the W. Alton Jones Foundation, a Charlottesville, Virginia-based environmental charity that, like the Rockefeller organizations, was built upon the oil fortune of its founder. The other was the German Marshall Fund of the United States, which not only provided funding for the workshop but also played an active role in supporting the network through one of its executives, Marianne Ginsburg, who collaborated with Goodman and Tom Wahman at the RBF in the planning phase of the workshop.[62]

Indicative of the steering committee's longer-term strategic vision, their trajectory of action extended beyond the 1987 Villach-Bellagio workshop to include the 1988 World Conference on the Changing Atmosphere in Toronto,[63] and the Second World Climate Conference in Geneva in 1990.[64] Most of the members of the Villach-Bellagio steering committee and extended network would in fact serve in similar capacities for those events, from conference planning and agenda setting to drafting the final statements and reports. Leading up to Toronto and Geneva, the much smaller Villach-Bellagio workshop served to maintain momentum after Villach 1985, building upon and further consolidating the scientific consensus and advances in climate knowledge and continuing the development of – particularly at Bellagio – the political dimension of the network's activities and ambitions. Senior Canadian scientific officials Jim Bruce and Howard Ferguson, who were leading the organizing efforts for the upcoming Toronto Conference – intentionally timed to follow the release of the Brundtland Report, and where Gro Harlem Brundtland herself would deliver the keynote address – were among those in attendance at Bellagio, demonstrating the direct connection between the scientific and political processes surrounding climate.[65]

Much had transpired on the environmental front in the two years between Villach 1985 and the 1987 Villach-Bellagio workshop in September–October and November that year. The autumn of 1987 in particular marked a turning point in global environmental and sustainability governance. Ozone negotiations culminated in the signing of the Vienna Convention's Montreal Protocol in September, and in October, the United Nations released the Brundtland Report after four years of work by the WCED. Meanwhile, behind the scenes, Mostafa Tolba – emboldened by the unexpected success of the ozone treaties – had been promoting the idea of an international convention on climate change, including a lobbying effort directed toward the US State Department.[66] In the realm of international scientific organizations, ICSU was also at this time preparing to launch a major initiative that would institutionalize the emerging field of Earth system science (below Chapters 6 and 7).

62 Schwarz, *A History of Climate Action.*
63 Franz, *The Development.*
64 Michael Oppenheimer, interview with Eric Paglia, June 7, 2021.
65 Franz, *The Development.*
66 Agrawala, "Context."

Taking place during precisely the same period as the pair of landmark events in global environmental history, the Villach-Bellagio workshop is, rightly, far less celebrated. The workshop is also eclipsed in terms of its historical significance by Villach 1985, which was indeed a turning point in the politicization of climate change. Villach-Bellagio 1987 and the almost two years of planning leading up to the workshop were, however, important exercises in organizing, expanding, and mobilizing the network of activists that carried climate change from scientific circles into the realm of international politics. Moreover, the workshop also generated the idea of global temperature targets, including the 2°C benchmark, which two decades later became a pillar of climate governance.[67]

The Villach-Bellagio process thus represented a key aspect of the 1985–1988 agenda-setting phase for climate change,[68] a period when greenhouse gases and global warming went from being a topic of scientific interest to an issue of high-level political concern. Furthermore, together with the watershed 1985 Villach conference, which served as the point of departure for the meetings that followed, the 1987 Villach-Bellagio workshop has become part of the narrative and lore of climate governance, with many of the main protagonists represented at the meetings in the Alpine retreats. Villach and Bellagio have since become almost synonymous with the scientific activism that sparked the politicization of climate change. Yet much of the impetus and organization surrounding the meetings took place well to the north.

VILLACH AND BELLAGIO VIA STOCKHOLM: CONVENING POWER AND SCIENTIFIC NETWORKS

The convening power of Stockholm extended to Villach and Bellagio despite the fact that the two workshops were held some 2,000 kilometers from the Swedish capital. While Bert Bolin and IMI had been main actors in the run-up to Villach 1985, it was Gordon Goodman, together with the rest of the AGGG and steering committee that took the lead in initiating and organizing the 1987 Villach-Bellagio workshop. Held under the auspices of the Beijer Institute, its workshop report, like the previous one, carried the imprimatur of the World Climate Programme and its

[67] Piero Morseletto, Frank Biermann, & Philipp Pattberg, "Governing by Targets: Reductio ad unum and Evolution of the Two-degree Climate Target," *International Environmental Agreements: Politics, Law and Economics* 17(2017): 655–676.

[68] Bodansky, "History of the Global Climate Change Regime."

sponsors, UNEP and WMO. However, this time ICSU was not directly involved, other than through its association with the AGGG, as the 1987 Villach-Bellagio process took on a more explicit policy orientation.[69]

The two-stage Villach-Bellagio workshop resulted in a single combined report.[70] Building on the scientific consensus of Villach 1985 and framed as a direct response to the recommendations of that conference, the 1987 workshop report projected accelerated sea level rise, described potential impacts of climate change on a regional basis, and outlined three warming scenarios based upon climate sensitivity estimates and different levels of future greenhouse gas emissions. It called for general reductions in deforestation and fossil fuel use and recommended the development of both societal adaptation and greenhouse gas limitation strategies. The report stated that "a coordinated international response will become inevitable," and that international organizations should examine the possibility of instituting a "law of the atmosphere as a global commons or the need to move towards a convention along the lines of that developed for ozone." The AGGG approved a draft of the report at a meeting in Paris in December 1987, changes were discussed by the steering committee in January 1988, and after further review and input from a small group of experts, Jill Jaeger of the Beijer Institute wrote the final version of the report, entitled *Developing Policies for Responding to Climate Change*.[71]

The workshop and its report have come to be seen in retrospect as a seminal moment in climate governance. They marked the beginning of the political history of the 2°C temperature target,[72] which has since become the primary benchmark of international climate policy under the UNFCCC.[73] Two of the scientists that attended the Villach part of the workshop, Pier Vellinga and Peter Gleick, would be appointed by the AGGG to chair a working group that produced the report *Targets and Indicators of Climate Change*.[74] The report was published under the auspices of the AGGG in 1990 by the SEI, where Gordon Goodman

69 Agrawala, "Context."
70 Jill Jaeger, *Developing Policies for Responding to Climatic Change.*
71 Jaeger, *Developing Policies.*
72 Morseletto et al., "Governing by Targets."
73 Eric Paglia & Erik Isberg, "On Record: Political Temperature and the Temporalities of Climate Change," In: Anders Ekström & Staffan Bergwik, eds., *Times of History, Times of Nature: Temporalization and the Limits of Modern Knowledge* (New York: Berghahn, 2022), 259–283.
74 F. R. Rijsberman & R. I. Swart, eds., *Targets and Indicators of Climatic Change* (Stockholm: The Stockholm Environment Institute, 1990).

had become the founding executive director, as part of a collection of four reports on climate change edited by Jill Jaeger. The *Targets* report explored both a 0.1°C increase per decade and a 1°C or 2°C total increase in global mean temperature as possible benchmarks that climate policy could be based upon. Vellinga would also co-author a key 1991 article that further developed the concept of the 2°C temperature target,[75] which was adopted five years later by the European Union as the basis of EU climate policy. In 2009, 2°C was endorsed by the UNFCCC in the Copenhagen Accord.

Only nine members of the extended climate network attended both parts of the 1987 workshop.[76] The guest list of invited experts participating in one or the other stage of the workshop indicated the functional differentiation between the Villach and Bellagio meetings. The forty-eight people in attendance at Villach 1987, including four from the Beijer Institute,[77] were for the most part technical experts and scientists from a range of disciplines. The discussions at Villach 1987 thus centered on scientific aspects of climate change, as well as potential societal responses for limiting and adapting to the changes in climatic conditions that the workshop participants anticipated.

Bellagio, by contrast, with a handful of government officials and policy experts participating alongside the scientists, took the findings of the first part of the workshop as a baseline in considering possible policy options for dealing with climate change. The meeting in Bellagio was half the size of Villach 1987 and far more posh, taking place at the Rockefeller Foundation's opulent resort on the shores of Lake Como. Among the twenty-four participants receiving white glove service at the Bellagio Center – signaling, perhaps, the rising interest in climate change beyond scientific circles – a quarter had close ties to Stockholm. In addition to Goodman, Bolin, and Jaeger, in attendance were Paul Crutzen, director of the Max Planck Institute; Måns Lönnroth, a senior Swedish government official who had been involved in legislation on climate already in the 1970s; and the Swedish climate scientist C. C. Wallén. Other notable participants included Jim MacNeill, Secretary General of the Brundtland Commission, and Jessica Mathews, the founding vice president of the

[75] P. Vellinga & R. J. Swart, "The Greenhouse Marathon: A Proposal for a Global Strategy," *Climatic Change* 18(1991): 7–12.

[76] These were: J. Bardach, W. Clark, G. Goodman, J. Jaeger, G. P. Hekstra, W. H. Mansfield, M. Oppenheimer, C. C. Wallén, and G. Woodwell.

[77] Gordon Goodman, Jill Jaeger, Beijer Institute deputy director Lars Kristoferson, and Michael Chadwick, director the Beijer Institute's center in York, England.

World Resources Institute, who would later become president of the Carnegie Endowment for International Peace.[78]

The 1987 Villach-Bellagio workshop marked the peak of AGGG influence. Several months after the Villach-Bellagio workshop, Bert Bolin would resign from the advisory group after accepting Mostafa Tolba's invitation to chair the new scientific body being set up by UNEP and WMO: the Intergovernmental Panel on Climate Change.[79] As founding chairman of the IPCC, a position he would hold from 1988 to 1997, Bolin was much better positioned, as it turned out, to shape climate governance during the establishment phase of the Framework Convention on Climate Change under the United Nations. The involvement of governments in the climate assessment process, a fundamental characteristic of the IPCC, provided the Panel with a level of legitimacy and political potency that the AGGG could not have hoped to achieve. Bolin had in fact, from the beginning, been somewhat skeptical of the AGGG, seeing the six-member circle of scientific elites as not sufficiently representative of the climate science community and as lacking the resources and influence to move the needle on climate change in the realm of politics.[80]

Others were however initially concerned that government involvement could compromise the scientific integrity of climate assessments,[81] and continued to prefer the AGGG format for providing policy-relevant expertise on climate change. Two parallel scientific assessment processes emerged. One under Bolin and the IPCC, which was hurrying to produce a report that could inform international negotiations leading up to an eventual climate convention. The other, under AGGG auspices,

[78] Mathews later also became a member of the Stockholm International Peace Research Institute's governing board.

[79] Bolin, *A History of the Science and Politics of Climate Change*, 49. Bolin accepted Tolba's invitation after consulting with Sweden's environment minister Birgitta Dahl. Bolin had also been lobbied to take the position by Jim Bruce and WMO Secretary-General Godwin O. P. Obasi during the 1988 Toronto Conference. According to Hirst (2014): "Bruce and Obasi believed Bolin was the perfect fit for the job of IPCC chairman, as did the scientific research community, the WMO, UNEP, and national policymakers. These three groups had differing but complimentary aspirations for the chairman of the IPCC to appear to be politically neutral and scientifically literate. Bolin emphatically embodied these criteria as a citizen of a nonaligned country, an internationally respected scientist (in meteorology), and someone well versed in the field of international scientific co-operation. The decision to appoint Bolin, a seemingly apolitical scientifically credible chairman, is representative of a wider ambition to create a broadly inclusive, scientifically credible and politically legitimate Panel." David G. Hirst, "Balancing Scientific Credibility and Political Legitimacy: The IPCC's First Assessment Cycle, 1988–1990," *History of Meteorology* 6(2014):6, 79–94.

[80] Bolin, *A History of the Science and Politics of Climate Change*.

[81] Michael Oppenheimer, interview with Eric Paglia, June 7, 2021.

centered on the SEI, with Goodman and Jaeger leading a team of experts in producing a comprehensive climate assessment based on the discussions in Villach and Bellagio. Funding for the latter was provided by Villach-Bellagio backers, the RBF, and the W. Alton Jones Foundation, as well as SEI.[82] Both the IPCC and the AGGG assessment processes were structured around three working groups, and each constellation delivered its final sets of reports within weeks of each other in 1990. Although the concurrent assessment processes were not considered to be in competition with one another, and the almost simultaneous arrival of their respective reports was apparently coincidental, only one of the two seminal climate science institutions would continue to exist after 1990.[83]

The collection of AGGG reports, including the ad hoc Working Group II report that introduced the idea of the 2°C target, was published by the SEI in October 1990. They received extensive distribution and discussion that same month at the Second World Climate Conference in Geneva – a hybrid science and policy event that included a ministerial meeting as well as the main scientific/technical component.[84] The SWCC in fact rested upon an exceptionally strong scientific foundation, as it had been scheduled to take place after the publication of the first IPCC assessment report, which had been finalized in August at a major meeting in Sundsvall, Sweden. The rapid rise of the Panel, producing its assessment in some eighteen months, had in effect rendered the AGGG redundant. Although the AGGG never formally disbanded, the Second World Climate Conference turned out to be the final occasion when the advisory group would meet in person. The Panel's intergovernmental mechanism, while a complicating factor in terms of reaching consensus, proved to be an effective means for producing high-profile scientific reports that governments had a direct stake in, thus enhancing the political salience of climate change.

SCIENCE DIPLOMACY IN SUNDSVALL: SECURING THE FIRST IPCC ASSESSMENT REPORT

The contentious IPCC meeting in Sundsvall, Sweden, in the end of August 1990 has been described as trench warfare by one of the over 400 participants.[85] The meeting's purpose was to adopt the first IPCC

82 Peter M. Haas, *Epistemic Communities, Constructivism, and International Environmental Politics* (London & New York: Routledge, 2015).

83 Jäger, "From Conference to Conference."

84 Ibid.

85 Peter Aldhous, "Trench Warfare at Sundsvall," *Nature* 347(1990):9, 9.

assessment report. The synthesis of that report – as finalized in Sundsvall – summarized for policymakers and other stakeholders the conclusions of the Panel's three working groups on the science, societal impacts, and possible policy responses to climate change. The first IPCC report would in turn serve as the scientific point of departure for the upcoming negotiations toward an international climate convention. Politics were thus not far under the surface in the discussions surrounding the results of the scientific assessments that had been prepared by an array of experts in advance of Sundsvall. Every line of text in the summary report was subjected to extreme scrutiny from delegates representing a wide range of scientific, political, and economic interests. An multitude of environmental NGOs, such as Friends of the Earth, were also on hand to observe the meeting and expose attempts to water down the report's wording by, for example, overemphasizing scientific uncertainties.[86]

Somewhat ironically, Sundsvall is one of the highest carbon-emitting cities in Sweden due to the presence of the country's only aluminum smelter, the massive Kubal facility with its five towering smokestacks, prominently located close to the downtown area. The port city of some 100,000 inhabitants, situated along the Baltic coast approximately 400 km north of Stockholm, is also a major center of Northern Sweden's large and influential forest industry. Scientists and government representatives from seventy-four countries gathered in Sundsvall rather than Stockholm because Sweden's environmental minister, Birgitta Dahl, was a strong advocate of a regional policy that promoted activity away from the capital. Even outside Stockholm, the high costs of Sweden turned out to be prohibitive for many of the delegates from developing countries, whose per diem was not sufficient to cover the cost of meals at local restaurants. Funds from the ministry of finance were therefore secured by the Swedish government official responsible for organizing the conference, Svante Bodin, to ensure that all delegates were adequately fed while in Sundsvall.[87]

With the consummate diplomat-scientist Bert Bolin chairing the proceedings in Sundsvall, the four days of deliberations came down to a final all-night session – the start of a pattern repeated through the six IPCC assessment reports so far – that was reminiscent of a high-stakes

86 Aldhous, "Trench Warfare at Sundsvall."

87 Magnus Jakobsson, "Mötet alla glömt bort – FN:s första klimatprotokoll skrevs i Sundsvall," *Sundsvalls Tidning*, November 30, 2015. Available at: www.st.nu/2015-11-30/motet-alla-glomt-bort--fns-forsta-klimatprotokoll-skrevs-i-sundsvall (accessed June 12, 2022).

political negotiation rather than a collegial scientific debate among experts. Despite the stark differences between countries and regional groupings on how to frame the level of scientific certainty on climate change, and what mitigation actions should be considered by governments, the meeting succeeded in adopting what is today referred to as the IPCC First Assessment Report.[88] The report, together with guidance provided by the scientists and government ministers at the Second World Climate Conference, would be a decisive factor in the December 1990 adoption by the UN General Assembly of resolution 45/212, *Protection of global climate for present and future generations of mankind*. The report also undergirded the work of the Intergovernmental Negotiating Committee that was established to prepare a Framework Convention on Climate Change.[89] By educating government officials on climate change, INC chairman Jean Ripert credits the IPCC with making the 1992 signing of the framework convention possible.[90]

Bolin's ten-year tenure as IPCC chairman was the pinnacle of his many years of work at the interface of science and politics. During his chairmanship, Bolin also oversaw the 1992 supplement to the first IPCC assessment report, which updated the state of climate knowledge ahead of the signing of the Framework Convention on Climate Change at the Rio Earth Summit, and the 1995 Second Assessment Report that informed international climate negotiations leading up to the landmark Kyoto Protocol in 1997. His contributions as a scientific statesman, particularly from the mid-1980s onward,[91] took place in parallel to two other major initiatives that drew upon Bolin's decades of experience as a prolific science organizer and institution builder. Like many of Bolin's previous projects, these required navigating networks of science, politics, and international organizations to lay the foundation for institutions that have played prominent roles in the evolution of global environmental governance. One of the initiatives led to the establishment of a major international research program headquartered in Stockholm, and the other to a Swedish

[88] Tony Brenton, *The Greening of Machiavelli: The Evolution of International Environmental Politics* (London & New York: Routledge, 1994).

[89] John Zillman, "A History of Climate Activities," *WMO Bulletin*, 58(2009):3, 141–150. https://public.wmo.int/en/bulletin/history-climate-activities

[90] Agrawala, "Context"; Eric Paglia & Charles Parker, "The Intergovernmental Panel on Climate Change: Guardian of Climate Science," In: Arjen Boin, Lauren A. Fahy & Paul 't Hart, eds., *Guardians of Public Value* (Cham: Palgrave Macmillan, 2021), 295–321.

[91] Thomas F. Malone, "Global Change," *Tellus B: Chemical and Physical Meteorology* 43(1991):4, 182–187.

government-funded institute with broad global engagement on environmental and sustainable development issues.

GLOBAL CHANGE AND THE INSTITUTIONALIZATION OF EARTH SYSTEM SCIENCE

As climate science during the 1980s became increasingly institutionalized through the World Climate Research Program and the IPCC, as well as new research institutes and informal networks like the Villach group, there was also a growing awareness among many scientists, Bert Bolin included, that the planetary-scale changes taking place were not limited to the climate system. The coalescing of many research strands and disciplinary fields engaged in systems approaches to the planet and its colossal components, such as oceans, atmosphere, and life itself, was during this period brought into the emerging integrative field of Earth system science. Named by NASA with its establishment of an Earth System Science Committee in 1983, this "new scientific endeavor" provided a powerful paradigm and conceptual framework for integrating the study of the entire spectrum of planetary processes and systems, and for understanding these as part of an interactive and cohesive whole – a single *Earth system*.[92] NASAs contribution to the new scientific paradigm also included the first graphic depiction of the biogeochemical functioning of the Earth system in the form of the 1986 Bretherton Diagram, which, in retrospect, can be seen as the first of a number of iconic images to emerge from the Earth system science community.

Until the 1980s, the component parts of the Earth system, including the biosphere, hydrosphere, atmosphere, and lithosphere, were associated with particular scientific disciplines and separate research programs that studied the various spheres independently of the others. The establishment, several years after NASA's 1983 innovation, of an entirely new international institution based on the integrative concept of Earth system science would mark a turning point in the scientific understanding of global change. The latter was not, however, solely an environmental phenomenon to be studied by a new category of experts that came to consider themselves Earth system scientists. In the Earth system paradigm, particularly as it evolved after the turn of the millennium, humans were

[92] Will Steffen, Katherine Richardson, Johan Rockström, et al., "The Emergence and Evolution of Earth System Science," *Nature Reviews Earth & Environment* 1(2020):1, 54–63.

fully immersed and implicated in global change, both cause and casualty of profound changes in the planetary environment. The Earth's habitability for humans over the coming decades and centuries was, for instance, a critical concern for adherents of Earth system science, who saw global change as a governance problem of enormous magnitude as well as an object of research for cutting-edge science that was increasingly empowered by technologies like satellites and supercomputers. Global change research was thus, from the outset, explicitly seen by its practitioners as a policy-relevant undertaking. The emergence of Earth system science in the 1980s became intertwined with expanding scientific and political interest in environmental issues like acid rain, ozone depletion, and climate change, as well as the promotion of sustainable development in the wake of the Brundtland Report.[93]

Neither the ambitions of global change research nor the auto-histories of the movement's leaders were modest. In the foundational works of the movement, Earth system scientists situated their revolutionary endeavor within a centuries-long history of science narrative that dated back to Newton.[94] As visionary agents of change, they saw themselves as engaged in bringing forth a paradigm shift in how the functioning of the planet should be understood. They also saw their institution-building project, following in the path of earlier large-scale initiatives like the International Geophysical Year and the International Biological Program, as part of a trajectory of "international science collaboration, devoid of political interest, with science bridging the geopolitical divisions of the Cold War."[95] The historically minded adherents of the Earth system science paradigm would in the ensuing decades put forward ideas that drew upon both Earth and science history. These included the Anthropocene and "a second Copernican revolution," as well as evocative concepts like "Hothouse Earth" and "A Safe Operating Space for Humanity" that implicated humans as drivers of global change in time and space.[96]

[93] Steffen et al. "Earth System Science."

[94] T. F. Malone & J. G. Roederer, eds., *Global Change: The Proceedings of a Symposium Sponsored by the International Council of Scientific Unions (ICSU) during Its 20th General Assembly in Ottawa, Canada on September 25, 1984* (Cambridge: ICSU press, 1985); Ola Uhrqvist & Björn-Ola Linnér. "Narratives of the Past for Future Earth: The Historiography of Global Environmental Change Research," *The Anthropocene Review* 2(2015):2, 159–173.

[95] Uhrqvist and Linnér, "Narratives."

[96] Paul Crutzen & Eugene Stoermer, "The Anthropocene," *Global Change Newsletter* 41(2000): 17; Hans J. Schellnhuber, "'Earth System' Analysis and the Second

IGBP: ESTABLISHING AN INTERNATIONAL HUB FOR EARTH SYSTEM SCIENCE IN STOCKHOLM

The institutionalization of Earth system science and global change research can be traced back to a meeting of the ISCU executive board in Stockholm in early 1983,[97] the same year NASA created its Earth System Science Committee. The American geophysicist Thomas Malone, who together with others had been advocating a large-scale study on global change, proposed to the executive board that ICSU establish an interdisciplinary research program on the emerging area of scientific interest. Malone's proposal led to a symposium on global change the following year in Ottawa, a milestone event where the ICSU General Assembly decided to assemble an ad hoc planning group to scope out an international research initiative centered on the "global dimension of chemical and biological processes."[98] With its emphasis on interactions between components of the Earth system, biogeochemistry – a field of science that studies interactions between planetary spheres – represented a foundational discipline for the new research program.

Bert Bolin was appointed chairman of the planning group in 1985. He had been an expert on biogeochemical cycles since the 1950s, after Carl-Gustaf Rossby recommended that he pursue research on the carbon cycle.[99] Throughout the course of his career, Bolin would play an instrumental role in bringing biogeochemistry into climate science.[100] Following his mentor's advice on scientific specialization, Bolin and

Copernican Revolution," *Nature* 402(1999): C19–C23; Will Steffen, Johan Rockström, Katharine Richardson, et al., "Trajectories of the Earth System in the Anthropocene," *Proceedings of the National Academy of Sciences of the United States of America* 115(2018):83, 8252–8259. Johan Rockström, Will Steffen, Kevin Noone, et al., "A Safe Operating Space for Humanity," *Nature* 461 (2009):1, 472–475; Will Steffen, Katherine Richardson, Johan Rockström, et al., "Planetary Boundaries: Guiding Human Development on a Changing Planet," *Science* 347 (2015):6223, 736–747.

97 Malone, "Global Change."

98 International Geosphere-Biosphere Programme (IGBP), *Global Change Report 1 – The International Geosphere-Biosphere Programme: A Study of Global Change: Final Report of the Ad Hoc Planning Group. Prepared for the ICSU 21st General Assembly, Berne, September 14–19, 1986* (Stockholm: International Geosphere-Biosphere Programme, 1986); Sybil P. Seitzinger et al., "International Geosphere–Biosphere Programme and Earth System Science: Three Decades of Co-evolution," *Anthropocene* 12(2015): 3–15.

99 Henning Rodhe, "Bert Bolin and His Scientific Career," *Tellus B: Chemical and Physical Meteorology* 43(1991):4, 3–7.

100 T. S. Bianchi, "The Evolution of Biogeochemistry: Revisited," *Biogeochemistry* 154(2021): 141–181.

his Stockholm University colleague Erik Eriksson co-authored a seminal analysis, published in a 1959 memorial volume to Rossby, on the exchange of carbon dioxide between the atmosphere and ocean.[101] Bolin fortified his status as a leading authority on biogeochemistry by editing four SCOPE reports in the 1970s and 1980s that further developed the discipline.[102] He had moreover collaborated with Malone, also a member of the ICSU planning group, for decades on climate-related research as well as in organizing major scientific events such as the 1967 ICSU-sponsored meeting in Stockholm that gave rise to the GARP initiative (see Chapter 4).[103] Bolin had even been involved with the 1957–1958 International Geophysical Year as well as the World Climate Research Program, which were considered forerunners of the new initiative. The selection of Bolin to lead the ICSU planning group for a major global change research program largely based on biogeochemistry was thus anything but coincidental.

The ad hoc planning group delivered its final report to the ICSU General Assembly in September 1986.[104] The report provided a detailed plan and scientific rationale for establishing a new institution, the International Geosphere Biosphere Program: a Study of Global Change. Over the next thirty years, the IGBP would coordinate research within an international network consisting of hundreds of natural scientists working in disciplines related to global change. Anchored in the science of biogeochemical cycles, the IGBP was also from the outset conceived as an institution that would inform environmental and societal policymaking, "providing the information we need to assess the future of the Earth in the next 100 years, with an emphasis on processes that change on time scales of decades to centuries. It will be a program of basic research with almost immediate practical applications in the management of resources at national and international levels and as a means of improving the reliability of warnings of global change of significance to our environment and to humankind."[105] Despite its intention to incorporate anthropogenic forces and impacts on human societies in providing policy-relevant

[101] Bert Bolin & Erik Eriksson, "Changes in the Carbon Dioxide Content of the Atmosphere and Sea Due to Fossil Fuel Combustion," In: Bert Bolin, ed., *The Atmosphere and the Sea in Motion: Scientific Contributions to the Rossby Memorial Volume* (New York: The Rockefeller Insitute Press, 1958), 130–142.

[102] Rodhe, "Bert Bolin."

[103] Malone, "Global Change."

[104] International Geosphere-Biosphere Programme (IGBP), *Global Change Report 1*, p. V.

[105] International Geosphere-Biosphere Programme (IGBP), *Global Change Report 1*.

knowledge to decision-makers, the IGBP initially excluded the social sciences from its study of global change.[106]

With financial support from the Swedish government, Bolin facilitated the placement of the IGBP secretariat in Stockholm. He had for many years participated in various advisory councils and scientific investigations in Sweden and had close ties to the highest levels of Swedish politics, including as a personal science advisor to Prime Minister Ingvar Carlsson.[107] This, together with Bolin's prominent position in the Swedish scientific community, led to an invitation to establish the IGBP secretariat at the Royal Swedish Academy of Sciences. Thomas Rosswall, a Swedish soil ecologist and expert on carbon and nitrogen cycling, was appointed as the first executive director of the IGBP. He personified the kind of Swedish scientist that, as their career evolved, would find their calling in scientific leadership and international engagement. Rosswall, who had earlier in his career served as secretary of Sweden's International Biological Program committee and would later become executive director of ICSU, had published extensively on biogeochemistry, including in SCOPE volumes edited by Bolin. A decisive moment in Rosswall's engagement with biogeochemistry, and in the evolution of the field itself, was a project he was in 1975 asked by SCOPE chairman J. W. M. la Rivière to establish at the Royal Swedish Academy of Sciences. Bringing in Bert Bolin and Erik Eriksson, the three Swedes and the Swedish National SCOPE Committee organized an international workshop at Friibergh Manor outside Stockholm that included many of the leading scientists in the disciplines, such as ecology, hydrology, and meteorology, that underpin the study of biogeochemical cycles.[108] The workshop resulted in the SCOPE 7 report, *Nitrogen, Phosphorus and Sulphur: Global Cycles*, which was published under the auspices of the Swedish Natural Science Research Council's series *Ecological Bulletins*,[109] the same outlet where a seminal 1968 article on acid rain by Svante Odén had appeared.[110] SCOPE would further support the development of the scientific basis of the IGBP

[106] Harold A. Mooney, Anantha Duraiappah & Anne Larigauderie, "Evolution of Natural and Social Science Interactions in Global Change Research Programs," *Proceedings of the National Academy of Sciences of the United States of America* 110(2013):Supplement 1, 3665–3672.

[107] Bolin, *A History of the Science and Politics of Climate Change.*

[108] Thomas Rosswall, personal communication, August 16, 2022.

[109] B. H. Svensson & R. Söderlund, eds., *Nitrogen, Phosphorus and Sulphur: Global Cycles.* SCOPE Report 7, *Ecological Bulletins* 22(1976), 1–192.

[110] Odén, "The Acidification of Air Precipitation."

with a 1988 report, *Scales and Global Change: Spatial and Temporal Variability in Biospheric and Geospheric Processes* (SCOPE 35), that included papers on aspects of biogeochemical and Earth system processes by Bolin, Rosswall, and Paul Crutzen.[111]

When the IGBP secretariat opened at the Academy in 1987, it joined the Beijer Institute and the journal *Ambio*, which thereafter became an important outlet for global change research.[112] Further, *Global Change* magazine,[113] with eighty-four issues published by the IGBP secretariat between 1989 and 2015, provided a focal point and platform for the development of scientific concepts – including the first article on the Anthropocene – for the emerging epistemic community of Earth system scientists and others engaged in the study of global change. The secretariat's location at the Academy and in the heart of Stockholm's main research district afforded the IGBP strong operational support and an array of scientific synergies.[114] Similar to the leading role played by its neighboring institutions, IMI and the Beijer Institute, in organizing pivotal events in locations other than Stockholm, the 2001 open science conference in Amsterdam "Challenges of a Changing Earth: Global Change," was largely orchestrated in Stockholm by the IGBP secretariat.[115] The landmark Amsterdam Declaration on Earth System Science definitively situated humankind in the center of global change, equating human activities with "some of the great forces of nature in their extent and impact" and called for "an ethical framework for global stewardship and strategies for Earth system management."[116]

Attended by some 1,400 experts of different scientific backgrounds, the conference also demonstrated the expansion of the global change research agenda since the 1980s to also include the human sciences. The Amsterdam Declaration brought the IGBP together with the International Human Dimensions Programme on Global Environmental Change, as well as the World Climate Research Program and the international biodiversity program DIVERSITAS, to create the Earth System Science

[111] SCOPE, *Scales and Global Change: Spatial and Temporal Variability in Biospheric and Geospheric Processes*, eds. Thomas Rosswall, Robert G. Woodmansee & Paul G. Risser (Hoboken: Wiley, 1988).

[112] Sverker Sörlin, "The Environment as Seen through the Life of a Journal: *Ambio* 1972–2022," *Ambio* 50(2021):1, 10–30.

[113] Originally called *IGBP Global Change Newsletter*.

[114] IGBP executive director (1998–2004) Will Steffen, personal communication, December 6, 2018.

[115] Ibid.

[116] Sybil P. Seitzinger et al., "International Geosphere–Biosphere Programme and Earth System Science."

Partnership (ESSP). The declaration and partnership were a recognition of the need for a new knowledge base that integrated a wide range of expertise that could collectively inform governance efforts in the face of global change.[117] Continuing in the vein of producing policy-relevant integrated knowledge, the IGBP secretariat a decade later co-arranged with its ESSP partners the 2012 Planet Under Pressure Conference in London. The conference provided a scientific foundation for the Rio+20 summit, which in turn gave rise to the Sustainable Development Goals, and initiated preparations for the Future Earth research program that would three years later succeed the ESSP and its component institutions.

Taking over the offices of the IGBP, which was dissolved at the end of 2015, one of Future Earth's five global hubs was established at the Royal Swedish Academy of Sciences. There at the Academy, the Future Earth Secretariat Stockholm joined the Beijer Institute of Ecological Economics. Although the latter bore the name Beijer, its organizational history was distinct from that of its predecessor, the first incarnation of the Beijer Institute. Beijer II, as it were, had been founded, with core funding from the Kjell and Marta Beijer Foundation, by the Swedish economist Karl-Göran Mäler together with the Cambridge-based Indian-British economist Sir Partha Dasgupta in 1991.[118] By that time, the operations and senior staff of the original Beijer Institute had been transplanted into a new organization that had been created on the initiative of the Swedish government to signal its commitment to the environment and the new concept of sustainable development.

BEIJER REBRANDED: THE STOCKHOLM ENVIRONMENT INSTITUTE

On the eve of the election year of 1988, during which the environment became the dominant electoral issue of the day,[119] Sweden's environment minister Birgitta Dahl assembled an expert commission to investigate

[117] Ola Uhrqvist and Eva Lövbrand, "Seeing and Knowing the Earth as a System – Tracing the History of the Earth System Science Partnership." Paper presented at the Amsterdam Conference on the Human Dimensions of Global Environmental Change – "Earth System Governance: People, Places, and the Planet," December 3, 2009.

[118] Beijer Institute, "Karl-Göran Mäler 1939–2020," https://beijer.kva.se/news-item/karl-goran-maler-1939-2020/ (retrieved November 24, 2021). Close friends and long-time collaborators, Mäler and Dasgupta, were co-recipients of the 2002 Volvo Environment Prize.

[119] Lars Nord & Elisabeth Stúr, *Från ödesfråga till övrig fråga: En studie av den politiska debatten om kärnavfallet i Sverige 1976–2009* (Stockholm: Svensk Kärnbränslehantering AB, 2010).

the potential orientation, scope, and organization for a new institute dedicated to the development of environmentally appropriate technology. The idea had emerged at a Social Democratic Party conference in 1987, and by the time of the commission's creation in December that year, the Swedish Parliament had already earmarked 25 million crowns per annum in government core funding for the proposed institute.[120] The ubiquitous Bert Bolin was appointed as one of the five members of the blue-ribbon commission chaired by Ambassador Göte Svensson – the facilitator of the consequential 1982 meeting in Stockholm between Bolin and UNEPs Mostafa Tolba.[121]

A generation after the original 1967–1968 environmental awakening in Sweden, which witnessed the activism inspired by Hans Palmstierna's *Plunder*, the establishment of the world's first EPA, and the diplomatic initiative that launched preparations for the Stockholm Conference, the idea of a state-funded environmental institute dovetailed with the domestic politics of 1987–1988. The Swedish Ministry of Environment and Energy had been established on January 1, 1987, and the environment became an important part of the Social Democratic Party's platform for the 1988 national elections. The latter took place in the wake of the Chernobyl disaster, the strong reception in Sweden of the Brundtland Report, and a mysterious mass-death event of seals on Sweden's west coast that was generally assumed to have been caused by some undetermined human activity.[122] An environmental *re*awakening was thus in progress. Launching an institute that complemented the Social Democrats' efforts on environment and sustainable development issues as well as the party's campaign pledges during the "environmental election" of 1988, in which the upstart Green Party complicated attempts to form a government, could thus be seen as good politics.[123]

[120] Mike Chadwick, "The Stockholm Environment Institute: The First Ten Years and Beyond," Unpublished manuscript, 1999. Chadwick served as director of SEI from 1991 to 1995, prior to which he had led the SEI office in York, England. The York office had originally been founded by Chadwick as an office of the Beijer Institute before the founding of SEI.

[121] Statens offentliga utredningar (SOU) 1988: 23. *SIIESTA – Ett internationellt institut för värdering av miljöriktig teknik: Betänkande av miljöinstitutsutredningen Stockholm 1988* (Stockholm: Allmänna Förlaget, 1988). The other members of the Commission were Kerstin Nibleus, Ulf Svensson, and Hans G. Forsberg.

[122] It was later determined that the cause of the mass-death event was a seal virus.

[123] Craig R. Whitney, "Environment Seen as Key Issue in Swedish Vote," *New York Times*, September 17, 1988; Jonas Anshelm, *Socialdemokraterna och miljöfrågan: En studie av framstegstankens paradoxer* (Stockholm: Brutus Östlings bokförlag Symposion, 1995); Henrik Ekman, *Naturen vi ärvde: En miljöresa från tyst vår till het sommar* (Stockholm: Norstedts, 2021).

The commission delivered its report in May 1988, proposing the creation of SIIESTA: The Stockholm International Institute of Environmentally Sound Technologies Assessment. The extensive analysis encompassed input from a range of external experts and included the report from the 1987 Villach-Bellagio workshop, reproduced in its entirety as an appendix attributed to its author, Jill Jaeger. Lars Kristoferson (Figure 6.1), deputy director of the Beijer Institute since its inception, also contributed a section to the SIIESTA report that strongly argued for the inclusion of climate change – a core research area at Beijer – in the remit of the proposed institute. The report in fact singled out Beijer as an existing Stockholm-based institute that was already engaged in many of the international missions SIIESTA would aspire to. Some 50 percent of the research programs proposed in the SIIESTA report duplicated roughly 90 percent of the ongoing activities at the Beijer Institute.[124] The commission, clearly in close contact with Beijer's leadership during the course of the investigation, went so far as to propose initiating negotiations, as soon as possible, with the Beijer board and other interested parties for a potential organizational merger between the Beijer Institute and the SIIESTA initiative. Integrating Beijer's in-house expertise and experience in the new institute would, as the report noted, facilitate a rapid launch of SIIESTA. A merger would also avoid the inefficient and somewhat awkward redundancy of two very similar environmental institutes being situated in Stockholm.

Negotiations between Birgitta Dahl, Gordon Goodman, and Beijer Foundation representatives Anders Wall and Jan-Erik Wikström paved the way for the establishment of the SEI in 1989 (Figure 6.6).[125] Besides jettisoning SIIESTA's sleepy acronym and technology-centric title, much of the SEI's activities were based on the broader range of research and sustainable development programs that the Beijer Institute had built up over the previous twelve years. Except now with five to six times Beijer's core funding. A great deal of the early effort at SEI went into supporting the upcoming 1992 Earth Summit in Rio,[126] as well as the climate research associated with the AGGG. Most of the Beijer staff shifted to SEI, including Goodman, who stayed on as the new institute's founding director until his retirement at the end of 1990, and Lars Kristoferson, who remained deputy director. In addition to its main office, which

[124] Bernhard, *The Beijer Institute*.
[125] Ibid.
[126] Chadwick, "The Stockholm Environment Institute."

FIGURE 6.6 First board meeting in September 1989 of the Stockholm Environment Institute, which has since its establishment consistently ranked as one of the world's top environmental think tanks. Most of the original staff and leadership of the Institute, as well as the network of SEI satellite centers outside Sweden, were inherited from the original Beijer Institute. Photo: Courtesy of Lars Kristoferson.

left the Academy for new premises in Stockholm's Old Town, SEI also absorbed the Beijer Institute's international network, including its centers in York and Boston. Since the establishment of SEI, new centers and satellite offices have been opened in Tallinn, Bangkok, Bogotá, Nairobi, Oxford, Seattle, and Davis, California.

BEIJER II: THE BEIJER INSTITUTE OF ECOLOGICAL ECONOMICS

The advent of the Stockholm Environment Institute did not, however, represent the end of the Beijer Institute as an international brand and institution centered on the environment and sustainable development under the Royal Swedish Academy of Sciences. Even before the departure of Beijer's staff for SEI, the Academy created a study group to assess possible focus areas for a reconstituted Beijer Institute that, like its forerunner, would receive core funding from the Beijer Foundation. Gordon Goodman was included in the study group, chaired by zoophysiologist Kerstin Lindahl-Kiessling, to ensure that the orientation of the new Beijer Institute would not be redundant with that of SEI. Lars Kristoferson, Bert Bolin, and Thomas Rosswall were among a handful of experts who provided additional advice.[127]

Fielding ideas from members of the Academy, the study group ultimately threw its support behind a proposal from ecologist Bengt-Owe Jansson, and the economist Karl-Göran Mäler – a longtime member of the committee that awarded the Nobel Prize in economic sciences. The transdisciplinary field of ecological economics, they argued, offered a powerful, integrated approach for modeling and understanding the coupled environmental and socio-economic systems at the heart of global change and sustainable development. The Academy's executive board's endorsement of their proposal led to the founding of the Beijer Institute of Ecological Economics in 1991. Mäler would serve as the Institute's executive director for the next fifteen years, with his close friend and colleague Sir Partha Dasgupta – lead author of the 2019 British government-commissioned report *The Economics of Biodiversity: The Dasgupta Review* – the first chairman of its scientific advisory board.[128] The board also included prominent experts from other disciplines, for example, biologist Paul Ehrlich, who the Academy had the previous year awarded the prestigious Crawfoord Prize for his work in the biosciences.[129]

The new incarnation of the Beijer Institute played an instrumental role in reconciling economics and ecology. Like the prevailing outlook on economic growth and environmental protection, the two academic disciplines had previously been considered incompatible by many professionals in those fields.[130] The gap had begun to be bridged at a seminal 1982 symposium in Saltsjöbaden outside Stockholm. Organized by the Stockholm University systems ecologist AnnMari Jansson with funding from the Wallenberg Foundation, the meeting brought together forty-eight international experts, many of whom would several years later be involved in establishing the journal *Ecological Economics* and the International

[127] Bernhard, *The Beijer Institute*.
[128] Ibid.
[129] Paul Ehrlich, *Life: A Journey through Science and Politics* (New Haven and London: Yale University Press, 2023).
[130] Carl Folke, "Ecologists and Economists Can Find Common Ground," *Bioscience* 45(1995): 283–284. Anastasios Xepapadeas & Aart de Zeeuw, "Obituary: Karl-Göran Mäler," *Environmental and Resource Economics* 76(2020): 195–200.

Society for Ecological Economics.[131] Another milestone in the convergence of ecology and economics was an international workshop in 1993 convened by the Beijer Institute on the island of Askö where Stockholm University manages a marine laboratory.[132] The first Askö meeting marked the beginning of a long-term effort in bridging disciplines and developing concepts and synthetic knowledge on interlinked social and ecological systems. Askö has since become the host of a much-appreciated – socially as well as scientifically – annual event that has enabled the evolution of an epistemic community of ecological economists. It has also produced a series of high-impact publications by constellations of ecologists and economists together with experts from other disciplines.[133] The roll call of Askö participants and co-authors of influential articles is long. Besides Mäler, Jansson, Dasgupta, and other ecological economists, the list includes the likes of the biologist Paul Ehrlich and climate scientists such as Stephen Schneider and Bert Bolin.[134] Another frequent participant in the Askö workshops was the charismatic Canadian ecologist C. S. Holling, whose 1973 paper "Resilience and the Stability of Ecological Systems" introduced the concept of ecological resilience.[135] Holling was also an early member of the global change movement, contributing two chapters, including one with William Clark, to Thomas Malone and Juan Gualterio Roederer's seminal 1985 volume *Global Change*.[136]

[131] Robert Costanza, "The Early History of Ecological Economics and the International Society for Ecological Economics (ISEE)" (2003). In: *International Society for Ecological Economics Internet Encyclopedia of Ecological Economics*. Available at: https://isecoeco.org/pdf/costanza.pdf (accessed July 9, 2023).

[132] Mooney, Duraiappah, & Larigauderie, "Evolution of Natural and Social Science." Anna Tunlid, "The Askö Laboratory: The Field Station as a Place for Fostering Scientific Collaboration and Development," In: Helena Ekerholm, Karl Grandin, Christer Nordlund & Patience A. Schell, eds., *Understanding Field Science Institutions* (Sagamore Beach, MA: Science History Publications, 2018), 315–342.

[133] Tore Söderqvist, Anna Sundbaum, Carl Folke & Karl-Göran Mäler, eds., *Bringing Ecologists and Economists Together: The Askö Meetings and Papers (Askö 1993–2002)* (New York: Springer, 2010).

[134] Examples include: K. Arrow, B. Bolin & R. Costanza, et al., "Economic Growth, Carrying Capacity, and the Environment," *Science* 268(1995):5210, 520–521; G. C. Daily, T. Söderqvist & S. Aniyar, et al., "The Value of Nature and the Nature of Value," *Science* 289(2000):5478, 395–396; A Kinzig, D. Starrett, & K. Arrow, "Coping with Uncertainty: A Call for a New Science-Policy Forum," *Ambio* 32(2003):5, 330–335.

[135] C. S Holling, "Resilience and the Stability of Ecological Systems," *Annual Review of Ecology and Systematics* 4(1973): 1–23.

[136] C. S. Holling, "Resilience of Ecosystems: Local Surprise and Global Change," In: J. G. Roederer & T. F. Malone, eds., *Global Change* (Cambridge: Cambridge University Press, 1985), 228–269; W. C. Clark & C. S. Holling, "Sustainable Development of the Biosphere: Human Activities and Global Change," In: *Global Change*, 474–490.

Years later, Holling's resilience concept gave rise to another major scientific movement that formed the basis of a new Stockholm-based institute. Some of the leading Askö actors would also be deeply involved in the new initiative. As it had earlier done with the emerging discipline of ecological economics, the Beijer Institute, together with its organizational cousin, the Stockholm Environment Institute, and Stockholm University, seized upon the resilience concept to help launch the Stockholm Resilience Centre in 2007. The new institution, like its predecessors, was dedicated to the study of socio-ecological systems and the advancement of sustainability. Soon after its establishment, the SRC would initiate and serve as the central node for a major international research project that would result in a comprehensive new framework for conceptualizing the Earth system, and for assessing the catastrophic risks humankind collectively faced if so-called planetary boundaries were transgressed. While the quantified limits for most of the nine planetary systems included in the framework had not yet been breached, one indicator – the concentration of carbon dioxide in the atmosphere – signaled clearly that the climate system was already in a state of overshoot.

7

Planetary Boundaries and Big Tent Science

"350. Remember that number for the rest of your life."

So read the headline of a full-page advertisement in the *New York Times*, the *Financial Times*, and the *International Herald Tribune* on June 23, 2008 – twenty years to the day after NASA scientist James Hansen testified before the US Congress on the rising risks of the greenhouse effect and global warming. Undersigned by over 150 well-known scientists, activists, public intellectuals, and political leaders, the advertisement called for reducing the atmospheric concentration of carbon dioxide to below 350 parts per million, the level at the time of the famous June 1988 Hansen hearings. The diverse signatories included Hansen, climate writer Mark Lynas, creator of the Gaia hypothesis James Lovelock, human rights advocate Bianca Jagger, former Swedish prime minister Göran Persson, and Bill McKibben, the environmental activist who employed 350 ppm as the basis for his new climate NGO, 350.org. (Figure 7.1).[1]

One notable name that was not included in the 350 advertisement was that of Bert Bolin, who had passed away in the end of December 2007 – two weeks after Al Gore and the Intergovernmental Panel on Climate Change (IPCC) were presented with the Nobel Peace Prize in Oslo, Norway. Whether Bolin would have signed such a statement even if he were still alive is however uncertain. In an op-ed published posthumously in the Swedish daily *Svenska Dagbladet* several days after

[1] "350. Remember that number for the rest of your life," *New York Times*, *Financial Times*, and *International Herald Tribune*, June 23, 2008.

<350

Remember this number for the rest of your life.

We've been there: atmospheric concentration of carbon dioxide (CO_2) of 350 parts per million (ppm).

On this day twenty years ago, when Dr. James Hansen of NASA's Goddard Institute of Space Studies warned the US Congress of global warming forced by greenhouse gases emitted through human activities, CO_2 stood at 350 ppm.

That was then. Today it is up to 385 ppm. Sixteen years after the UN Convention on Climate Change was agreed upon to stabilize greenhouse gases at a level preventing dangerous human interference with the climate system, global carbon emissions are still increasing – *more rapidly than ever.* We are well on our way to 450 ppm and beyond.

What does the best science tell us?

Until recently, scientific consensus set the safe zone to avoid the worst effects of climate change at 450 ppm. But today the latest science tells us the danger zone may already begin at *350 ppm.* Catastrophic effects cannot be ruled out if *levels above 350 ppm* are maintained for a long time.

What does this mean?

We've gone too far – in a dangerous direction. Now we know enough. To act now. To foresee and forestall any risk of massive and irreversible damage to the earth and all its inhabitants for generations to come, we must reduce atmospheric CO_2 to *levels below 350 ppm* Scientific insights bring political responsibilities. We need leadership that respects the planetary boundaries of life.

We, the signatories of this message from all continents, call upon all nations in the ongoing climate negotiations to adopt 350 as the target to be reached peacefully and deliberately, *with all possible speed.*

350 is one of our planet's boundary conditions. It should not have been transgressed. We must go back for a future:

<350

essential to maintain human and planetary well-being

Poonam Ahluwalia, President, Youth Employment Systems
Pindarous W.T. Allison West Africa Coordinator, Youth Employment Systems (YES)
Dennis Altman, Professor
Peter Anderberg
Ray C. Anderson, Interface
Thomas Andersson, President, Jönköping University
Gunilla Arhen, Ruler Dam
W. Brian Arthur, Economist, Santa Fe Institute
Alan AtKisson, President, The AtKisson Group
Brian Bacon, Founder & President, Oxford Leadership Academy
Graham Barnes, Ph.D., psychotherapist, author
Steve Bass
Héctor Castillo Berthier
Göran Bexell, Vice-Chancellor, Lund University
Mihir R. Bhatt, Director, All India Disaster Mitigation Institute
Peter Blom, CEO, Triodos Investment
David Blood, Generation Investment Management
Nicole-Anne Boyer, Managing Director, Adaptive Edge
Derek Brown, Peace Appeal Foundation
Karin Bruce
Dan Brändström
Göran Carstedt
Per Carstedt
Robert Charpentier, CEO, Kaupthing Bank Sverige AB
Nayan Chanda
David Chaytor, MP Bury North
Göran Collert
Napier Collyns
Audrey Codera, Youth Works Inc.
Robert Corell
Hilary Cottam, Participle
Alexander Crawford
Tom Cummings, Chairman, Executive Learning Partnership
Bei Dao, Poet, Professor of Humanities, Chinese University of Hong Kong
Avril Doyle, MEP
Gwynne Dyer
Bo Ekman
Göran Ekman
Sara Ekman
Kristin Engvig, Founder, Women's International Networking
Profs Alice and Bob Evans, Plowshares Institute
Tim Flannery
José María Figueres Olsen, Ex-President of Costa Rica
Ulla-Britt Fräjdin-Hellqvist
M.Shafik Gabr, Chairman & Managing Director, ARTOC Group for Investment & Development
Yubaraj Ghimire, Editor, Samay/Newsfront Kathmandu
Marilou van Golstein Brouwers, Managing Director, Triodos Investment Management
Lennart Grebelius
Hal Hamilton, Sustainability Institute
KG Hammar
Dongfang Han
James Hansen
Molly Harriss Olson CEO, President Clinton's Council on Sustainable Development
Satu Hassi, Member of the European Parliament
Martin Hedberg, Swedish Weather & Climate Centre
Pär Holmgren, Meteorologist
Thomas Homer-Dixon, Professor, University of Toronto
Will Hutton
Ross Jackson, Gaia Trust, Denmark
Bianca Jagger, Chair, World Future Council
Oliver James, Psychologist, author of Affluenza
Colin Vivian Jones
Bill Joy, Partner, KPCB
Koosum Kalyan
Madhu Kishwar
Niclas Kjellström-Matseke, CEO, The Swedish Post Code Lottery
Igor Kluin, Founder Current
Shulamith Koenig, Founding President- PDHRE, People's Movement for Human Rights Learning
Jules Kortenhorst
Anna Langenius
Charles Laroche, Laroche Conseil
Alejandro Litovsky, Volans Ventures
Christine Loh, CEO, Civic Exchange, Hong Kong
Prof. Jack Lohman, Director, The Museum of London
Augusto Lopez-Claros, Ph.D. Director, EFD - Global Consulting Network
James Lovelock
Caroline Lucas, Member of European Parliament, Green Party
Mark Lynas
Dirk Messner, Director, German Development Institute
Fred Matser, Founder/Chairman Fred Foundation
Anand Mahindra
Manfred A. Max-Neef, Professor of Ecological Economics, Universidad Austral de Chile.
Jacqueline McGlade, Professor
Bill McKibben
John Morelli, McCarthy Chair, Rochester Institute of Technology
Ieva Morica
Carl Mossfeldt
Herman Mulder
Geoff Mulgan
Robert Myall, Chairman, Nanyuki River Water Association
Carlos Nobre
Cecilia and Lars Nordstrom
Karl-Erik Norrman, Sec-Gen, European Cultural Parliament
Harald Norvik
Paul O'Hara
Rebecca Oliver
Hans-Olov Olsson
Marcello Palazzi
Anthony Patt, International Institute for Applied Systems Analysis
Göran Persson, former Prime Minister of Sweden
Iqbal Z. Quadir, Founder, MIT Legatum Center
Sigrid Rausing
Ralph Rayner, London School of Economics
Pierre Rehnlund, Vice President, Swedbank AB
Howard Rheingold
Mike Robinson, Chair of Stop Climate Chaos Scotland
Johan Rockström
Yasuhiro Sakakibara
Mohammad Sammak
Peter Schwartz
Michael Schragger, The Forum For Design & Sustainable Enterprise
Dr. Ismail Serageldin
Clare Short, MP
Anthony Simon
Jens Martin Skibsted, Creative Director of Skibsted Ideation/Biomega
Youba Sokona
Paolo Soleri, Arcosanti
Hans Sollerman CEO, e-man AB
Jaap Spier, Prof. Dr, Netherlands Supreme Court
Ingrid Srinath, CIVICUS
Grace K and John R. Stanek, Stanwood Associates
Caroline Stiernstedt Sahlborn
Alan Stoga, Zemi Communications
Ira Sundberg, Sustainable Sweden Association
Uno Svedin
Sverker Sörlin, Professor of Environmental History, author
Pedro Tarak, Fundacion Avina, Panama
Camilla Toulmin
Domien Van Gool, Founder, Leader Academy
Sasa Vucinic
Denny Vågerö, Professor, Centre for Health Equity Studies
Dr. Marc A. Weiss, Chairman and CEO, Global Urban Development, Washington, DC
Anders Wijkman
Lawrence Wilkinson, Chairman, Heminge & Condell
James and Eva Wine
The Wirth Family
Frank Wisner, Vice Chairman, External Affairs, American International Group
Xu Sitao, Chief Representative, China The Economist Group
Oran Young, Bren School, UC Santa Barbara
Niklas Zennström, Co-founder Atomico, Joost, Skype
Zou Xinqing, Professor
American Indian Institute
China Labour Bulletin
Ejendals AB
e-man
The Foresight Group AB
Generation Investment Management LLP
Good Energies
H & M Hennes & Mauritz AB
The Lemelson Foundation
Living Partnership, International NGO (Belarus)
Mahindra & Mahindra Limited
NTR plc
Schweisfurth Foundation, Munich
Sundaram Tagore Galleries
The Swedish Post Code Lottery
The David Suzuki Foundation
Triodos Bank N.V.
Vsemirnyi Fond Prirody
350.org
European Environment Agency
Stockholm Environment Institute
Tällberg Foundation

Individual signatures represent the personal and professional perspectives of the signatories, not necessarily the opinion of the institution associated with their names. Institutional signatories represent solely themselves.

We invite you to watch the live webcasts from the Tällberg Forum June 26-29 from the village of Tällberg in Sweden. Information and the schedule can be found online at www.tallbergfoundation.org. The opening session includes a panel on "Defining planetary boundaries and tipping points" presented in partnership with the Stockholm Environment Institute, featuring James Hansen, Diana Liverman, Johan Rockström, Will Steffen. For more information on the global movement, please visit www.350.org.

TÄLLBERG FOUNDATION
How on Earth can we live together?
Tällberg, Sweden

FIGURE 7.1 The 350 advertisement. *New York Times*, June 23, 2008. Courtesy of the Tällberg Foundation.

FIGURE 7.2 Senior Swedish science diplomats meeting in the 1990s. Ambassador Bo Kjellén (left), head of the Swedish Kyoto delegation, and Bert Bolin (right), founding Chairman of the United Nations Intergovernmental Panel on Climate Change (IPCC). Photo: Jessica Gow/Scanpix.

his death, Bolin and the article's co-author Ambassador Bo Kjellén – Sweden's former chief climate negotiator who had played an instrumental role in establishing the United Nations Framework Convention on Climate Change (UNFCCC) and Kyoto Protocol as well as the UN Convention to Combat Desertification – argued that it was scientifically impossible to identify an exact point beyond which humanity would not be able to manage climate change. Bolin and Kjellén (Figure 7.2), citing IPCC science, also took exception to the more alarmist statements of Al Gore and others on the melting of Arctic sea ice and Greenland's glaciers and emphasized the uncertainties surrounding weather-related catastrophes such as Hurricane Katrina that many had linked to climate change.[2]

As implied by the editorial's title, "Serious – but not Hopeless," Bolin and Kjellén – who had worked closely with one another on climate science and diplomacy throughout the 1990s – strongly preferred the

[2] Bert Bolin & Bo Kjellén, "Allvarligt – men inte hopplöst," *Svenska Dagbladet*, January 4, 2008, www.svd.se/a/3b8b9e79-74bb-30bc-a745-e41abdc6c7b5/allvarligt-men-inte-hopplost.

deliberate, scientifically informed diplomatic process that was moving toward an inflection point over what they saw as the counterproductive discourse of the climate crisis, based on fear, that had become prominent in 2007. The editorial was written in the wake of the December 2007 UNFCCC meeting in Bali, where the Bali Road Map was adopted to guide international climate negotiations over the next two years. With the Kyoto Protocol's first commitment period expiring in 2012, climate diplomats' sights were set on the COP15 meeting in Copenhagen at the end of 2009, where they hoped to reach a legally binding agreement that would extend and enhance commitments by signatory nations to limit greenhouse gas emissions.

Following the 2007 breakthrough of climate change in terms of public awareness – epitomized by Nobel Peace Prizes and Oscar awards, for Al Gore's movie *An Inconvenient Truth*, as well as landmark scientific reports – the interim years between Bali and Copenhagen would be animated by an almost overwhelming sense of urgency, opportunity, and excitement. It was shared among scientists, activists, and diplomats who had devoted much of their careers to the climate issue. It was in the context of this decisive moment of climate governance that a large gathering of global elites took place in a scenic Swedish village, and a new framework for reconceiving humanity's relationship with the Earth System emerged.

TÄLLBERG 2008 – FRAMING A FRAMEWORK

On the same day as its publication in the three major international broadsheets, the 350 advert also appeared in a fourth newspaper, the *Falu-Kuriren*, a local daily covering the largely rural province of Dalarna in Sweden. It was there, in the hillside village of Tällberg, overlooking Lake Siljan, that many of the signatories and hundreds of other experts, industry leaders, and international elites, including UN Secretary General Kofi Annan, would gather later that week to discuss issues related to sustainable development and global interdependence. The event was the fourth installment of the Tällberg Forum (Figures 7.3 and 7.4), a major annual gathering convened by the Stockholm-based Tällberg Foundation. Founded by former Volvo executive Bo Ekman, the Tällberg Foundation has since 1981 been arranging, at first small informal workshops and, after the turn of the Millennium, larger events, in Tällberg and other locations in Sweden and around the world. The meetings were dedicated to promoting globalism and intellectual

FIGURE 7.3 Tällberg, a small village in the province of Dalecarlia, Sweden, was the place of several years of Midsummer conferences where Earth System scientists and other scholars engaged and networked with politicians, activists, artists, businessmen, trade unions, sponsors, media, royalty, and members of the general public. Workshops took place in nineteenth-century cottages, talks were held, and music was played for hundreds of people in big tents. Collective nature walks took place in the surrounding landscape in an old Swedish out of doors tradition. Photo: Courtesy of the Tällberg Foundation.

exchange under the banner "How on Earth can we live together?." In a village of small wooden houses, Tällberg Forum sessions took place inside large circus-style tents that could be seen as symbolizing the "big tent" approach to global problems that the Tällberg Foundation had fostered for forty years.

The 2008 edition of the Tällberg Forum opened with a session called "Defining planetary boundaries and tipping points," the web-cast of which was promoted in the 350 advertisement that the Tällberg Foundation had taken out in the four newspapers. Its scientific panel consisted of former International Geosphere Biosphere Programme (IGBP) director Will Steffen, Oxford geographer Diana Liverman, Stockholm Environment Institute (SEI) director Johan Rockström, and James Hansen of NASA, who had in a recent scientific article argued that CO_2 levels beyond 350 ppm risked triggering dangerous tipping

FIGURE 7.4 Convening and connecting at Tällberg. Here is a panel discussion in 2008 with British geographer Diana Liverman, NASA scientist Jim Hansen, climate scientist Will Steffen (Australia), Bangladesh environmental scientist Tariq Banduri, facilitated by climate scientist Johan Rockström. Photo: courtesy of the Tällberg Foundation.

points in the climate system.[3] Hansen's 350 threshold would constitute one of the nine pillars of a new conceptual framework that was under development by an international network of scientists centered on the recently established Stockholm Resilience Centre (SRC). During the Tällberg Forum's opening session, the panelists provided a sneak preview of what would a year later become one of modern environmental history's most impactful, controversial, and all-encompassing conceptualizations of the human–environment relationship, with implications for both scientific understanding and governance of the global environment: the Planetary Boundaries framework.

The four panelists were among around twenty of the framework's intellectual architects that had arrived in Tällberg in advance of the Forum to further develop its conceptual and scientific basis. The ideas and activities behind Planetary Boundaries (PB for short) date back to the early 2000s, although, as is the case with so many conceptual developments, it is hard to pinpoint a particular moment in time and place when the idea was initially conceived. In a recent book based on the

[3] James Hansen, Makiko Sato & Pushker Kharecha, et al., "Target Atmospheric CO_2: Where Should Humanity Aim?," *Open Atmospheric Science Journal* 2(2008):1, 217–234.

framework, Rockström and his co-author Owen Gaffney describe the birth of Planetary Boundaries as an intellectual process that evolved over time. In a chapter centered on the Scottish amateur scientist James Croll, who discovered the astronomical reason for the ice ages, Rockström and Gaffney cite a proverbial quote attributed to Alexander von Humboldt, "There are three stages of scientific discovery: first people deny it is true; then they deny it is important; finally they credit the wrong person."[4]

The quote fits well in explaining not only Croll's unlikely nineteenth-century scientific achievement, which was of course denied and disregarded, but also for the Planetary Boundaries idea. The Planetary Boundaries framework was published in *Nature* in September 2009 under the title "A Safe Operating Space for Humanity," and has since become well known, and with time, even famous. But where does the boundaries idea actually originate? Rockström and Gaffney focus mostly on the boundaries' content and actual meaning. They say little about the process that led up to the article, although they do say that it took place in Sweden, a circumstance to which we will return in some detail. It should also be mentioned that the Planetary Boundaries article in *Nature* has achieved phenomenal visibility and made a huge impact, especially in scientific circles, approaching a stunning 20,000 citations on Google Scholar. The world of environmental policy and governance, especially in the Global South, has, however, been somewhat less impressed. The framework has also been controversial, drawing criticism from North and South, and from left, right, and center.[5]

Another way of telling the PB story is to start with a broader article published in 2015, with Will Steffen as the lead author together with several Stockholm-based scholars, in the then recently established journal *The Anthropocene Review*.[6] In that article, Steffen, a central co-author

[4] Johan Rockström & Owen Gaffney, *Breaking Boundaries: The Science Behind Our Planet* (London: Dorling Kindersley, 2021), ch 15. The original attribution to Humboldt for a quote that has multiple roots and attributions is actually about Louis Agassiz' theory, for a long time controversial, of the Ice Age. This theory was in fact previously suggested by other scholars, and in addition, it was a theory Humboldt first did not believe in. It is cited in Andrea Wulf, *The Invention of Nature: Alexander von Humboldt's New World* (New York: Knopf, 2015), and earlier by Bill Bryson, *A Short History of Nearly Everything* (2003), new ed. (London: Doubleday, 2016), 508.

[5] Frank Biermann & Rakhyun E. Kim, "The Boundaries of the Planetary Boundary Framework: A Critical Appraisal of Approaches to Define a 'Safe Operating Space' for Humanity," *Annual Review of Environment and Resources* 45(2020):1, 497–521.

[6] Will Steffen, Wendy Broadgate, Lisa Deutsch, Owen Gaffney & Cornelia Ludwig, "The Trajectory of the Anthropocene: The Great Acceleration," *The Anthropocene Review* 2(2015):1, 81–98.

of the *Nature* piece as well as the lead author of the second, follow-up PB article in *Science* in 2015, and his colleagues trace the longer intellectual trajectory of Anthropocene thought. They point in particular to the integrative work conducted at the IGBP, which Steffen directed from its secretariat in Stockholm in the 1990s up until 2004, and in similar international fora of science and governance.

Steffen and colleagues emphasize the importance of meetings, just as we do in this book. In fact, some of the major conceptual and institutional advances during this multi-decadal trajectory from the 1980s to the 2000s were made at conferences of various kinds. One of the more intellectually consequential was the IGBP meeting in February 2000 in Cuernavaca, Mexico, where Paul Crutzen first uttered the word "Anthropocene," conceptualizing the link between the Earth System, its science, and human agency. Another was a 2005 Conference in Dahlem, a leafy suburb in the academic quarters of Berlin, when the newly formed interdisciplinary IHOPE network – Integrated History and Future of People on Earth – organized a comprehensive meeting from which the idea of "The Great Acceleration" is said to have emerged.

Meetings of scientists and thinkers from fields wide apart make possible a kind of interaction that traditional seminars and congresses devoted to individual disciplines do not. They relieve the participants of the preset agendas and free them from the methodological routines and theoretical shackles that typically define disciplinary gatherings. They create the space of intellectual freedom that is necessary and the confidence that the presence of a wide range of expertise provides, so that even very large and complex problems can be investigated and discussed.

The 2015 *Anthropocene Review* article starts by noting that Earth System science objects under IGBP had been in the making for more than a decade from the late 1990s, when the Programme decided to embark upon a stock-taking and synthesizing project to get a better overview of the field. It was conducted between 1999 and 2003, building upon the growing evidence that demonstrated how rapid change was becoming increasingly legible along a wide set of indicators, including ocean acidification, atmospheric CO_2, phosphate circulation, biodiversity loss, freshwater use, and multiple others. Little of this was entirely new information. However, what was unprecedented was the massive effect that these pressures were imposing on the Earth System if they were looked at *together* and not divided up along the lines of existing disciplines and deferred to different seminar rooms in separate faculties. This *integrative approach*, taking into account the combined and summative effect of the multiplicity of Earth System indicators, became the central issue of the analysis and provided, in all its simplicity (one may say afterward), the added value of the enterprise. As the article states already in its first paragraph: "The increasing human pressure on the Earth System was a key component of the synthesis."[7]

FINDING A FRAMEWORK – THE USEFULNESS OF HISTORY

Also new was the interpretive framework. Scientists knew the facts, but they were not at the outset of the synthesis project so well aware of how formidable the rates of change in the trajectories of the Earth System dimensions were, especially when they were seen as an ensemble. This was, however, already discussed, not least by historians. In the year 2000, John R. McNeill, an environmental historian at Georgetown University in Washington, DC, published a book on the history of the twentieth century as seen from the angle of human relations with the natural world. It was entitled *Something New Under the Sun* to underscore the main point of the book, namely that, in contrast to the well-known biblical words, in *Ecclesiastes* 1:9, the twentieth century was precisely *not* something that wasn't new under the sun. On the contrary, almost everything was new. It was the century when humankind's impact on the Earth literally exploded. Not just the first atomic bomb, but population growth, the widespread application of chemicals and pesticides in agriculture, mass mobility and transportation, urbanization, the widespread application of fertilizers, the felling of rainforests, the extraction of minerals and fossil fuels, the exploitation of other natural resources, and much more. The twentieth century was the century when global GDP grew by almost 2,000 percent and the population of blue whales diminished by 98 percent.[8]

Was there a connection? Will Steffen and his co-authors suggest that it was a revelatory moment, when the insight of the rates of change of social, economic, and demographic indicators were set alongside those of the Earth System. And clearly, something had happened. In the IGBP synthesis volume, published in 2004, twelve "Socio-economic trends," including the global number of McDonald's restaurants (later replaced

[7] Steffen et al., "The Trajectory."

[8] John R. McNeill, *Something New under the Sun* (New York: W. W. Norton & Company, 2000). Global inflation-adjusted historical GDP data, retrieved May 20, 2024: ourworldindata.org/grapher/global-gdp-over-the-long-run.

by primary energy use), were presented as graphs demonstrating exponential growth from around the mid-twentieth century and juxtaposed with twelve similarly steep curves, the "Earth System trends."[9] It was to further investigate the historical trajectories reflected in what would come to be known as the "Great Acceleration curves" that the 2005 Dahlem meeting was organized. The meeting was expanded to include archaeologists as well as historians, such as John McNeill, who was on the meeting's organizing committee, to further integrate the Earth System and its societal dimensions.[10] It was at Dahlem that the concept "the Great Acceleration" was coined,[11] later to become the title of a book by McNeill and historian Peter Engelke.[12]

The Steffen et al. article not only sums up the trajectory of the Anthropocene, it also brings to light the ongoing changes in the composition of the multiple pressures on the Earth System. The research team presents the recent leveling off in the building of new dams and the exponential growth rate of the global population and notes the reduction of stratospheric ozone following the Montreal Protocol coming into effect in 1989. But, by and large, the trends prevail and are in some cases reinforced. The article also takes up criticisms of the Anthropocene concept – the most articulated versions coming from scholars in the humanities and social sciences, but also from politicians and activists in developing countries and among less privileged social groups in the West.[13] This

[9] Will Steffen, Angelina Sanderson, Peter Tyson, Jill Jäger, Pamela Matson, Berrien Moore III, Frank Oldfield, Katherine Richardson, John Schellnhuber, B. L. Turner II & Robert Wasson, *Global Change and the Earth System: A Planet under Pressure*, IGBP Book Series (Berlin, Heidelberg, New York: Springer, 2004).

[10] Robert Costanza, Lisa J. Graumlich & Will Steffen, eds., *Sustainability or Collapse? An Integrated History and Future of People on Earth, Dahlem Workshop Report 96* (Cambridge, MA: MIT Press, 2007).

[11] Steffen et al., "The Trajectory," 82. According to John McNeill, he, at Dahlem, first proposed the Great Acceleration framing of the post-war period as an adaptation of Karl Polanyi's concept and 1944 book *The Great Transformation*. John McNeill, personal communication (Paglia), July 11, 2014. See also Eric Paglia, *The Northward Course of the Anthropocene: Transformation, Temporality and Telecoupling in a Time of Environmental Crisis*, PhD diss. (Stockholm: KTH Royal Institute of Technology, 2016), 11.

[12] J. R. McNeill & Peter Engelke, *The Great Acceleration: An Environmental History of the Anthropocene since 1945* (Cambridge, MA: Harvard University Press, 2016).

[13] Among the more vocal and articulate humanities critics, Eileen Crist, "On the Poverty of Our Nomenclature," *Environmental Humanities* 3(2013):1, 129–147. See also, for a broad register of critical responses, Gísli Pálsson, Bronislaw Szerzynski, Sverker Sörlin, John Marks, Bernard Avril, Carole Crumley, Heide Hackmann, Poul Holm, John Ingram, Alan Kirman, Mercedes Pardo Buendía & Rifka Weehuizen,

critique, which has been intense, points to the fact that only small portions of the global population have in fact contributed to the pressures on the Earth System, while the Anthropocene literature – from the first articles in the early 2000s – has treated "humanity" as a single entity. The 2015 Steffen et al. article provided new data and trend curves, with the global population broken down into Organisation for Economic Co-operation and Development (OECD), BRICS (Brazil, Russia, India, China, and South Africa), and low-income countries in the rest of the world. Unsurprisingly, the data demonstrate gross historical differences but also clear signs that the BRICS and some middle-income countries had started taking up significant space in several indicators and that China had by 2012 surpassed the per capita CO_2 emissions of Europe.

In other respects, the article was a summation of state-of-the-art knowledge that had emerged over an extended period of time. It by and large reflected an understanding of the planetary human–environment relationship that had been the consensus view of the global change community since at least the 2001 Challenges of a Changing Earth Conference in Amsterdam. Largely planned at the IGBP secretariat in Stockholm,[14] the conference was a turning point in the evolution of Earth System science and resulted in the discipline-defining Amsterdam Declaration.[15] The general insight that the *Anthropocene Review* article put forward after the first fifteen years of Anthropocene science and debates may be summarized by this quote: "Hitherto human activities were insignificant compared with the biophysical Earth System, and the two could operate

"Reconceptualizing the 'Anthropos' in the Anthropocene: Integrating the Social Sciences and Humanities in Global Environmental Change Research," *Environmental Science and Policy* 28(2013):1, 3–13; Andreas Malm & Alf Hornborg, "The Geology of Mankind?: A Critique of the Anthropocene Narrative," *The Anthropocene Review* 1(2014):1, 62–69; J. J. Schmidt. "The Moral Geography of the Earth System," *Transactions of the Institute of British Geographers* 44(2019):4, 721–734. Equity issues were discussed in Kate Raworth, *A Safe and Just Space for Humanity: Can We Live within the Doughnut?* (Nairobi: Oxfam International, 2012), www.oxfam.org/en/research/safe-and-just-space-humanity, and in Will Steffen & Mark Stafford Smith, "Planetary Boundaries, Equity and Global Sustainability: Why Wealthy Countries Could Benefit from More Equity," *Current Opinion in Environmental Sustainability* 5(2013):2–3, 403–408. For an overview of the Anthropocene critique, see Biermann & Kim, "Boundaries."

14 Interview (Paglia) with Will Steffen, December 6, 2018.

15 Will Steffen, "The Evolution of Earth System Science." Available at: https://futureearth.org/2015/12/14/the-evolution-of-earth-system-science/ (accessed August 6, 2022); Jill Jaeger, personal communication, August 4, 2022. See also the declaration itself: www.igbp.net/about/history/2001amsterdamdeclarationonearthsystemscience.4.1b8ae20512db692f2a680001312.html (accessed January 1, 2023).

independently. However, it is now impossible to view one as separate from the other." It could be said that this, on one level, was a quite moderate articulation of a profound and entangled relationship, given that "the environment" as a social and political issue had been around for almost seventy years – about the same age as the Great Acceleration[16] – and that its advent as an idea was by and large the clarion call that instigated the 1972 Stockholm Conference. Certainly, "the Earth System" did not exist as a term in 1972. It emerged gradually in the decades that followed, not receiving significant application until the 1980s, with its evolution closely monitored by the Stockholm-based science and policy journal *Ambio* (Latin for surroundings, or environment) that had been launched in 1972 in conjunction with the Stockholm Conference.[17]

Even more important, therefore, was the understanding from around 2000 that the human enterprise, as a whole, in all its countless dimensions and copious and ever-expanding scale, is directly interacting, impacting, and interfering with the Earth itself, *seen as a system*. That is the Anthropocene idea. Sometimes it *is* important how things are actually articulated. Because, in the articulation, the deeper significance of the idea becomes evident. In 1972, Stockholm's year of global revelation, the understanding was predominantly dualistic. The societal realm was, as it had always been, kept quite separate from the natural, (geo)physical realm. True, they were connected; humans and societies impacted – or "destroyed" as it was often presented – "nature," or "the environment." Society was on one side of an invisible fence, and the environment was on the other. It was a *human* environment, still two separate spheres, but with that qualifying word to signal the link.

Thus, when the Planetary Boundaries article started taking shape in the minds of some of the scholars in the Anthropocene and Earth System science community, including a few from the human sciences, the framing of the environment had already undergone a profound change. It had added a historical dimension. Rates of change over time across a wide range of *linked* social and natural parameters had started to play a major role. The human–Earth relationship was dynamic, not a static war of attrition. Societies and the Earth were profoundly entangled into a dynamic whole. They co-evolved, just as another environmental historian, Alfred W. Crosby, had suggested in the likely most influential

[16] Warde, Robin & Sörlin, *The Environment*, ch 1.

[17] Sverker Sörlin, "The Environment as Seen through the Life of a Journal: *Ambio* 1972–2022," *Ambio: A Journal of Environment and Society* 50(2021):1, 10–30, on 24.

work ever of environmental history, *Ecological Imperialism*. In that 1986 book, Crosby showed that the expansion of Europe over the course of a millennium from 900 to 1900 – starting in Iceland and Greenland, continuing with the Canary Islands and Madeira, expanding into overseas continents that the Europeans baptized as America and Australia, and a range of Pacific islands – entangled the "new Europes" within a biological web of plants, animals, epidemic diseases, and the gene pool of the Old World. Similar transplantations had certainly happened before, not least when early humans made their exodus out of Africa and spread across the world in a process that took hundreds of thousands of years. Now it accelerated fantastically, spurred, Crosby argued, by the evolutionary and epidemiological benefits that the Old World populations and biota had gained from higher population densities, in addition to their technology, science, capitalist ruthlessness, Christian sense of exclusivity, and racial supremacy ideas.[18]

Crosby called this intercontinental trade with and travel of seeds, weeds, animals, and disease "the Columbian exchange." The phrase was also the title of an earlier book of his from 1972, referring to Columbus as precisely the kind of iconic, restless, and ruthless agent of history that the expanding Europeans had become at that historical moment.[19] Historical understanding of the deep entanglement of human societies with continent-sized, indeed global, biological realities was thus not new when the Earth System science community began to engage with such ideas. But intellectual developments at the turn of the millennium still represented a significant step forward in the understanding of the human–Earth relationship. First, they were interdisciplinary, allowing for a wider analysis. Second, the comparisons were systematic, across a number of planetary dimensions, with more researchers engaged and hence a greater scientific capacity. Third, they were quantified and presented in a fashion that allowed comparison. Fourth, it universalized and generalized the historical change, which, although limited to a few cases, had been largely detailed in previous versions, presented with the usual care and precision of history writing. The Earth System science community further exposed the deep human–Earth entanglement, as it were, and allowed for an act of intellectual engineering that could bring out and – fifth – visualize in graphs and curves the magnitude and rate of the changes taking place.

[18] Alfred W. Crosby, *Ecological Imperialism: The Biological Expansion of Europe, 900–1900* (Cambridge: Cambridge University Press, 1986).

[19] Alfred W. Crosby, *The Columbian Exchange: The Biological Consequences of 1492* (Westport, CT: Greenwood Press, 1972).

This process was further assisted by the increasing interest among small groups of historians to write history along much bigger timescales than before. Several books were published in the first years of the new century that located human history on a planetary, if not astronomical, timescale identifying connections across thousands and hundreds of thousands of years. One work often cited in the IHOPE community was David Christian's *Big History* (2004), soon to be followed by books on "deep history" with links to neuroscience. The interest coincided with new kinds of historiography inspired by Anthropocene thought and a search, which has since then expanded, for synchronization of temporal scales, both natural and historical.[20] These kinds of material histories of Earth System timescales circulated among the Earth System science community and were cited in the work of the latter. The discussions took place, for example, in the IHOPE community and at the Dahlem Conference in 2005. Their results were published in academic journals, typically less in history than in outlets oriented toward Earth System science and global change research, and in a handful of important books.[21]

[20] David Christian, *Maps of Time: An Introduction to Big History* (Berkeley: University of California Press, 2004). Daniel Lord Smail, *Deep History and the Brain* (Berkeley & London: University of California Press, 2008). Dipesh Chakrabarty, "The Climate of History: Four Theses," *Critical Inquiry* 35(2009):2, 197–222. Daniel Lord Smail & Andrew Shryock, eds., *Deep History: The Architecture of Past and Present* (Berkeley, CA: University of California Press, 2011). Bruno Latour, *Facing Gaia: Eight Lectures on the New Climatic Regime* (Cambridge: Polity, 2017). Marek Tamm & Laurent Olivier, eds., *Rethinking Historical Time: New Approaches to Presentism* (London: Bloomsbury Academic, 2019). Anders Ekström & Staffan Bergwik, eds., *Times of History, Times of Nature: Temporalization and the Limits of Modern Knowledge* (New York: Berghahn, 2022).

[21] Some examples: S. Cornell, R. Costanza, S. Sörlin, S. E. van der Leeuw, "Developing a Systematic 'Science of the Past' to Create Our Future," *Global Environmental Change* 20(2010):3, 426–427. S. van der Leeuw, R. Costanza, S. Aulenbach, S. Brewer, M. Burek, S. Cornell, C. Crumley, J. A. Dearing, C. Downy, L. J. Graumlich, S. Heckbert, M. Hegmon, K. Hibbard, S. T. Jackson, I. Kubiszewski, P. Sinclair, S. Sörlin & W. Steffen, "Toward an Integrated History to Guide the Future," *Ecology and Society* 16(2011):4, 2. Robert Costanza, et al., "Developing an Integrated History and Future of People on Earth (IHOPE)," *Current Opinion in Environmental Sustainability*, 4(2012):1, 106–114. This trend has developed into new strands of nuanced Anthropocene history writing, both *about* the rise of the Anthropocene and its implications, as in Andrea Westermann & Sabine Höhler, "Writing History in the Anthropocene: Scaling, Accountability, and Accumulation," *Geschichte und Gesellschaft: Zeitschrift für Historische Sozialwissenschaft/Journal for Historical Social Sciences*, 46(2020): 4, 579–605, introducing an entire special issue with multiple contributions on the topic; Dipesh Chakrabarty, *The Climate of History in a Planetary Age* (Chicago, IL & London: The University of Chicago Press, 2021); and *as part of* the Anthropocene research enterprise and aligned with its main assumptions and claims, as in, for example, Jürgen Renn,

PLANETARY BOUNDARIES – FROM THE GROUND UP

Such was the situation among the Earth System sciences in the early years of the 2010s: a rapidly progressing discussion reaching across all possible disciplinary domains to establish a truer and more insightful image of the scale and impacts of the human dimensions of the human–Earth relationship. This emerging discourse and body of work aimed to better describe this relationship, and ultimately to be able to more effectively transform it or at least mitigate the most negative consequences of a co-evolution that was clearly moving in the wrong direction.

Seen in this way, it was perhaps not far-fetched to think that the dimensions of impact that the previous evolution of Anthropocene knowledge suggested could be turned into a set of quantifiable Earth System indicators, which are essentially what the planetary boundaries are. With the important qualification that these also have the potential to become policy objects, the term used in the text of the PB article is "guard rails," while in the title, these are described, perhaps more evocatively, as the "safe operating space for humanity."[22] The quantified boundaries represented a "space" within which "humanity" could "operate" without risking to tip the "Earth" out of its Holocene state. The latter was presented as a state of stability, in which humans had been able to proliferate, innovate, and evolve into complex modern societies. So progress was here defined as being related to Earth System stability rather than to other factors usually presented by archaeologists, anthropologists, and historians. This in some sense linked back to theories of geographical and climatic explanations, sometimes called "determinist," that had become less common after World War II. In retrospect, it may seem easy to explain, but before it had taken place, it was not obvious. The choice of Holocene stability as a boundary criterion, with all its vagueness, rather than precise calculations of risk,

The Evolution of Knowledge: Rethinking Science for the Anthropocene (Princeton, NJ: Princeton University Press, 2020); Julia Adeney Thomas, Mark Williams & Jan Zalasiewicz, *The Anthropocene: A Multidisciplinary Approach* (Cambridge: Polity, 2020); and, Julia Adeney Thomas, ed., *Altered Earth: Getting the Anthropocene Right* (Cambridge & New York: Cambridge University Press, 2022).

22 J. Rockström, W. Steffen, K. Noone, Å. Persson, F. S. Chapin, III, E. Lambin, T. M. Lenton, M. Scheffer, C. Folke, H. Schellnhuber, B. Nykvist, C. A. De Wit, T. Hughes, S. van der Leeuw, H. Rodhe, S. Sörlin, P. K. Snyder, R. Costanza, U. Svedin, M. Falkenmark, L. Karlberg, R. W. Corell, V. J. Fabry, J. Hansen, B. Walker, D. Liverman, K. Richardson, P. Crutzen, & J. Foley, "Planetary Boundaries: Exploring the Safe Operating Space for Humanity," *Ecology and Society* 14(2009):2, 32. [online] URL: www.ecologyandsociety.org/vol14/iss2/art32/, on 7.

human lives lost, or other standard indicators, may in fact have made PB more palatable and more "scientific" than "human."[23]

So, how did it happen? How could the PB article emerge from this mass of circulating ideas and initiatives? In their above-cited book *Breaking Boundaries: The Science of Our Planet*, Johan Rockström and Owen Gaffney (who was not involved at the time; he came to Stockholm in 2009) talk about the origins of the article in a short paragraph in Chapter 6. "We" [Johan] "invited scientists to Sweden in 2007." "[W]e turned every scientific stone we could find," went through aspects of the Earth System, identified nine processes, and quantified them.[24] These are then presented in various ways in the book. Although that is a very short but correct summary of the origins of the planetary boundaries concept, a closer look shows that "Sweden" was in fact several venues across the country, and the entire genealogy of the article is quite a bit more complex and indeed revealing, not least of the convening and converging power that Swedish institutions, scholars, and civic society could offer.

One such venue was Tällberg, where we must first return. The year 2007 was an entire year before the *New York Times* advertisement and the Planetary Boundaries workshop. At that point in time, Planetary Boundaries as a concept did not yet exist. But the Stockholm Resilience Centre did, having been selected for funding in June 2006, with the news from the Mistra Foundation in fact breaking at the 2006 Tällberg Forum. Preparations to set up the Centre were made in the fall of that year, including debates over the name of the new institute, and on January 1, 2007, the SRC formally started. At the 2007 Tällberg meeting, a discussion took place over the idea of a "Tällberg consensus," which was intended to start as a collaboration between the Tällberg Foundation and the SEI, a constituent part of the SRC that was still headed by Rockström, who maintained directorships of both SEI and SRC during the first five years of the latter's existence.

In the fall of 2007, the idea of a Tällberg consensus was prepared by Åsa Persson of SEI in collaboration with consultant Mike Schragger.[25]

[23] This assumption, with which we concur, comes from Åsa Persson, personal communication, e-mail September 4, 2022.

[24] Rockström & Gaffney, *Breaking Boundaries*, ch 6.

[25] Some details and quotations in this and several following paragraphs stem, unless other attributions are indicated, from large numbers of e-mails and working documents that one of us (Sörlin, a PB co-author) received and sent from November 2006 through to the publication of the two PB papers in September 2009.

It was vague at first, with talk of changing the focus of "sustainability" from vision to "business." Funding was sought from the Swedish government, including ten dimensions for reinterpreting sustainability. Not much came of this effort at first, and no funding was granted. In a later stage, the focus shifted after discussions within the SRC, and by March 2008, a new twist upon the idea was presented. Along the way, some of the scholars had been in touch with Schragger, who assembled ideas and took notes. In February, Rockström invited Carl Folke, Kevin Noone, and Sverker Sörlin to participate in preparations for the 2008 Tällberg Forum where the consensus workshop was set to take place. By that point, the terminology had changed to "tipping points" and "boundary conditions"; the two were used almost interchangeably.

On March 17, after the planning session (where only Rockström, Sörlin, and Persson participated), Persson circulated notes from the meeting containing the following: "A concrete idea that Johan Rockström and colleagues is developing is to formulate around ten absolute social-ecological boundary conditions at the global level that address the various 'spheres' (atmosphere, biosphere, etc.), such as 350 ppm CO_2, pH >7 in the oceans, maximum rate of biodiversity loss, etc. We are convening an international workshop for this purpose in conjunction with this year's Tällberg Forum in June." This was a concrete outcome of the meeting. Something pointing in this direction had also been on the program of a consensus meeting at SRC on October 12, 2007, where Rockström addressed the following topic, "The tipping points that must not happen: The boundary conditions – the physical ecosystem limits we must not transgress."

FROM BREVIK TO TÄLLBERG – AND BEYOND

Nonetheless, it is clear that it was the meetings in the first months of 2008 that sparked the process and in earnest provided the momentum toward the eventual article. The meeting notes from March 17 were distinct in that regard. They already had references to ten "boundary conditions." By now, it was also clear that presenting these boundaries was a central aim of the process and that it could also be the foundation of a research effort conducted by SEI. Plans were drafted for a two-day workshop at the Tällberg Forum in late June 2008, with discussions of both the scientific foundation of the ten boundaries and the policy implications, as well as the "ethics" of the limits in the assumption that their introduction would imply ethical concerns. There was also mention of a link to "Regime shifts," a line of research at the SRC that at this time engaged among others Carl Folke and Sverker Sörlin. The first decision bullet point read:

> The workshop shall aim to create agreement on about ten key boundary conditions, expressed in absolute terms and addressing not only the climate (350ppm) but also issues such as ocean acidity levels, biodiversity loss rates (i.e. geophysical and ecological indicators). Ideally, the workshop can communicate such "consensus messages" to the Forum audience. Ethical implications and various other aspects need to be addressed in this work, but the aim is also to provide a necessarily "simplistic" picture.[26]

Reference was also made to an existing "Zero paper" that was apparently on the table at the meeting on March 17, 2008. A later version of the paper, dated May 12, 2008, encompasses a provisional, yet fairly comprehensive, presentation not only of the ten boundaries – the last section of the Zero paper – but of the wider context of such an enterprise: to launch a set of science-based limits to human action on the planetary scale. At this point, the boundaries were ten (they would eventually become nine) and were in many ways different from the ones that later appeared in the *Nature* article. There was one on "fisheries," one on "cycles" (mentioning both nitrogen and phosphorous), one on the "Chemisphere," one on "Aerosols/Air quality." In the spring of 2008, a little over a year before publication, there remained quite a lot of refining and selection work to be done in building the eventual framework.

The Zero paper had a sizable section – more than half of the entire paper – called "The new understanding," discussing previous attempts to make future projections. Notably, this section discussed the nowadays little-remembered 1892 book, *Dans cent ans* (In a hundred years), a detailed and bold vision of the world in 1992 by the eccentric French physiologist/polymath Charles Richet, later to win the Nobel Prize in Medicine in 1913. Richet's projections, incidentally, failed on virtually all scores, which is precisely the main point made in the Zero paper: grand predictions and schemes tend to not turn out very well in the trial of history. They, very literally, rarely stand the test of time. The utopian tradition is also introduced, as are earlier attempts to suggest limits to the human enterprise, most prominently in the two 1972 reports on limits. One was *The Limits to Growth*, sometimes just "LTG," the famous

26 "The Tipping Points we cannot cross: Defining the Boundary Conditions for Planetary Sustainability." Notes from the planning meeting on March 17, 2008, with Johan (Rockström), Sverker (Sörlin), and Åsa (Persson). With Åsa Persson, SEI, and the authors (personal archives).

Club of Rome-commissioned computer-based projection of world development and environmental indicators for the twenty-first century.[27] The other was *A Blueprint for Survival*, Edward Goldsmith's small-scale decentralized vision of a future civilization based on self-subsisting communities, which was originally published in the January issue of *The Ecologist* and released later in the year as a Penguin book.[28] Much of the prose in this section of the Zero paper, and certainly some of the specifics – Richet being the prime example – cannot have been everyday parlance among many scholars regardless of background.[29]

A critical link between the previous literature and the message from Earth System science was the following paragraph in that section of the Zero paper:

> We now understand better why Richet and so many others went wrong: they dealt with phenomena which involved human behaviour and societal performance, which are in principle not foreseeable. One could contrast this tradition of projections with an example of a projection that was made on the earth system only, Svante Arrhenius' 1895 calculation of temperature rise as a result of the greenhouse effect. His calculation gave by and large the same result as the IPCC scenarios of 2007. From this we can learn a lesson: projections based on assumptions of how societies change, how humans may or may not act, and what kind of ingenuity they may be able to demonstrate to deal with challenges – such projections are extremely difficult, if not impossible, to project with any reasonable degree of certainty. However, if we focus on parameters which reflect the properties of the earth systems themselves, we move on safer grounds.[30]

This paragraph is telling because it justifies the novel approach of Planetary Boundaries thinking and explains why this version of a limits approach would be different from previous ones. Speculation on future projections and attempts to limit the human enterprise have, by and large, had a checkered history or been relatively futile. In order to take

[27] Donella H. Meadows, Dennis L. Meadows, Jørgen Randers & William W. Behrens III, *The Limits to Growth: A Report for the Club of Rome's Project on the Predicament of Mankind* (New York: Universe Books, 1972).

[28] Edward Goldsmith & Robert Allen, *A Blueprint for Survival* (London: Penguin 1972).

[29] Johan Rockström, "PREVIOUS PATTERNS OF THOUGHT. *Zero Outline High-Profile Boundary Paper 12th May 2008.*" Charles Richet, ed., *Dans Cent Ans* (Paris: Hachette, 1892, repr. 1920) was originally published as a series of four articles in *La Revue Scientifique* between December 1891 and March 1892. This part of the Zero paper was authored by Sörlin. Richet's utopian speculations had been introduced by him before, see Sverker Sörlin, *Naturkontraktet: Om naturumgängets idéhistoria* [The Nature Contract: Interaction with Nature in the History of Ideas] (Stockholm: Carlsson, 1991), 97.

[30] Zero paper (n.p.).

the risk to embark on yet another limit-setting project, something had to be done in order to make it credible and worth the while. Quantifying, Arrhenius style, was a possible way forward. In this regard, Earth System science sat on a new solution to an age-old problem. It could provide entirely novel, and stable, foundations for reawakening the limits project that the *Limits to Growth* debates had all but put to rest.

Another key idea of the Zero paper, in the same section, is that the politics of boundaries thinking can be less pessimistic than previous attempts had been.

> All these attempts, and many others here not mentioned, are different. What they share in common though is that they all represent a growing sense of unease with previous patterns of thought that in an unreflective way projected the future progress of humanity against a canvas of a bountiful and boundless natural world. They articulated the dark down side of the Anthropocene. In this new literature and in the computer-based projections that started to appear, Earth was a delimiting feature. However, the very way that nature was depicted in the models, and the choice of parameters, became, perhaps inadvertently, a constraining factor. By choosing parameters that were rooted in the supply side of the global metabolism – chiefly the availability of (cheap) natural and energy resources – there was an inbuilt pessimism in the prescriptions that followed. Recommendations tended to focus on modesty and restraint, which oftentimes turned global governance into a zero-sum game. Policy quickly turned into conflicts between the global haves and the have-nots. This was clear already in the aftermath of the Brundtland report and the Rio summit in 1992 and has not ceased since. As guidance for policy and governance, this was not very helpful.
>
> The approach to the issue of limits that we propose here is different. While we acknowledge the achievements from the past several decades, we use as a point of departure the stark and depressing fact that despite all intentions not much has happened to curb environmental decline and serious negative effects on social-ecological systems all over the world.[31]

The "we" assumed here is unclear, but the reasonable way to interpret it is to acknowledge the wider conversation among several scholars connected to SEI (Persson) and SRC (Folke, co-director, Steffen, affiliated, and Sörlin, a part-time employee during the SRC's first five years) or other parts of Stockholm University (Kevin Noone, who was also executive director of the IGBP from 2004 to 2008). It is also unclear to what extent the plus-sum game perspective was embraced by those engaged in the discussion of the paper. At any rate, it may be useful to observe that the idea articulated in this long quote stands in some tension to the quest for ethical concern that was also articulated in the paper, and

[31] Zero paper.

the previously cited words of caution based on earlier attempts to define limits. The tension in this early draft of the planetary boundaries idea contains elements of the debates that followed the publication of the finished paper. In the time that passed until the article appeared in print, the ethical and historical ideas largely fell by the wayside. The process honed in increasingly on refining and defining the boundaries themselves and securing a high-visibility publication. *Nature* was in fact already mentioned as a possible outlet at the meeting on March 17, 2008, a full year and a half before the paper appeared in print.

By May 2008, the plans had matured enough for a slightly larger team to assemble at Villa Brevik in Lidingö, the same island in the inner Stockholm archipelago where so many previous environmental meetings had taken place (see above, Chapters 3–5). In fact, the May 12 version of the Zero paper was intended for precisely this occasion. At this three-day gathering that started on May 14, those present were Will Steffen, Carl Folke from SRC and the Beijer Institute, Bo Ekman of the Tällberg Foundation, Uno Svedin – an SRC associate who worked for a major Swedish funding agency, and Rockström, Sörlin, and Persson, who had been the first to meet on March 17. During the May meeting at Villa Brevik, slides and other visuals were shown of the, at the time, ten boundary conditions, and deep, extensive discussions were conducted. One of the themes was how to avoid putting all the focus on climate and the 350 ppm boundary. In fact, much of the rationale of the whole enterprise was to build a strong argument for maintaining a wide range of Earth System parameters that were mutually dependent.

In late June, the tipping points meeting was held in Tällberg, with a range of international scholars present. Some showed up on fairly short notice, including UK science writer Mark Lynas and tipping points specialist Tim Lenton, a physicist at the University of East Anglia. Many of the attendees would later become co-authors of the article, a network of contributors that would continue to grow over the coming year. In preparation, yet another and much more ambitious and elaborated version of the paper had materialized, based on the Villa Brevik conversations the month before. It was now a massive piece of writing, 22,000 words over nearly sixty pages. Some of it was no doubt a result of Will Steffen joining the writing team. Much of the "Zero paper" remained. For example, long, verbatim reiterations of the quotes above and some of the key references to "previous patterns of thought," still a subheading in the June version, including the eccentric Richet and the neo-Malthusian precursors of setting planetary limits, this time also including the iconic

Swedish food and resources specialist Georg Borgström. The substantive expansion and elaboration were on the boundaries. They were now thoroughly penetrated, with references and visualizations. They were also called "Planetary Boundaries," a change that happened at some point between May 12 and June 18, when the Tällberg paper was finished for distribution to the impressive list of participants.

THE IDEA AND THE NARRATIVE

Limits, or boundaries, were a key idea, but perhaps even more seminal to the originality of the paper was some of the thinking around tipping points and regime shifts. The June 18 version articulated more than previous iterations the idea that humans, or "humanity" – still seen as a collective with little internal differentiation – had now, during the Great Acceleration, gained the capacity to move, or "tip" the "Earth" out of balance. The moral obligation not to do so, as put forward in the paper, was a fundamental principle of global environmental governance that became fully revealed and articulated by Earth System science. The Zero paper had also at this point been given a proper title: "Steering Away from Catastrophic Thresholds: Planetary Boundaries for Human Survival." After intensive debates at Brevik and again at Tällberg, "planetary" became the preferred framing of the boundaries, as "global" was deemed to be too vague and as carrying a fair amount of negative baggage.[32] The title summarized this moral idea. The limits were actually thresholds to spaces that were unknown to mankind. If humanity were to cross these thresholds and enter these spaces, even if inadvertently, the entire planet would be in peril, and humanity with it. Hence the word "survival,"

[32] Åsa Persson, personal communication. Persson adds that the group was surprised that so few attempts to quantify boundaries had been made. She was asked by Rockström to do a first screening of global environmental reports, like UNEP's GEO reports. She found they typically had only lists of key global environmental problems and unlike the PB concept did not, "1) attempt to be exhaustive (identifying ALL critical Earth system processes/services) 2) spring from a systematic framework/concept/criteria and 3) were not quantified into limits." Persson also mentions that she wrote much of the Supplementary Information (SI) to the more comprehensive *Ecology and Society* paper (Appendix 1 here www.ecologyandsociety.org/vol14/iss2/art32/appendix1.pdf). Concepts reviewed there were: guardrails/Tolerable Windows (by Schellnhuber), safe operating standards, limits to growth, and critical loads. Writes Persson: "My take-away from that exercise was that the PB concepts really filled a gap in terms of boundaries at the Earth system level, but it built on a standard approach...: scientists propose critical thresholds/loads/norms and policy-makers then set the (legally binding or not) boundary." Åsa Persson, e-mail to authors September 4, 2022.

harkening all the way back to William Vogt's 1948 diatribe, *Road to Survival*, or for that matter Edward Goldsmith's manifesto, *A Blueprint for Survival* from that formative year 1972, or why not the terminology of early evolutionary thought by Charles Darwin and Herbert Spencer in the nineteenth century?

Still, few thoughts during that June 2008 Midsummer planetary boundaries festival in Tällberg were drifting back to the twentieth-century classics in the self-flagellating neo-Malthusian canon of constraints. Probably because this new manifesto of sorts signaled so emphatically *the opportunities* inside the limits, certainly much more than it articulated the potential for disaster. Yes, there was "catastrophic" danger, but there was also ingenuity and innovation, and yes, there could, and should, be growth. Planetary boundaries "do not set limits to economic growth," the paper explicitly stated. So, here it was a quantified planetary apocalypse presented along with, in fact right next to, an evangelical narrative of Cornucopian transformation. This combination of realist pessimism in describing existential challenges for people and planet alongside a hopeful permissiveness for human agency, embedded within a comprehensive outlook for the future, fostered few enemies, at least in the Global North. And it likely goes a long way toward explaining the success, if not the triumph, of the PB idea.[33]

The thorough vetting at Tällberg 2008 gave the framework's ideas both additional stability and a kind of rubber stamp from a relevant community of expertise, albeit one slightly biased in favor of the general train of thought. Another key participant with long-standing Stockholm connections was Hans Joachim Schellnhuber, the director of PIK, the Potsdam Institute for Climate Impact Research in Germany. Schellnhuber was cited more than anyone else in the new long paper, chiefly for his thinking since the 1990s about the "switch and choke point paradigm."[34] He had broken down the fraught human–Earth

[33] Johan Rockström, Will Steffen, Kevin Noone, Carl Folke, Sverker Sörlin, Dag Broman, and Uno Svedin [Core writing team], "Steering Away from Catastrophic Thresholds: Planetary Boundaries for Human Survival." Background Paper for the Tällberg Pre-Forum Workshop, June 24–26, 2008. SEI and SRC. FIRST DRAFT, June 18, 2008. Mimeo. [Support in drafting and data compilations: Åsa Persson, Björn Nykvist, Louise Karlberg.] Pages 1–59, quote on 15.

[34] Rockström et al., "Steering Away," on 8. Hans J. Schellnhuber, "Coping with Earth System Complexity and Irregularity," In: Will Steffen, Jill Jäger, David J. Carson & Clare Bradshaw (eds.), *Challenges of a Changing Earth: Proceedings of the Global Change Open Science Conference* [Amsterdam, The Netherlands, July 10–13, 2001] (Berlin, Heidelberg, New York: Springer-Verlag, 2002), 151–156.

relationship to a set of tension zones, for example, the West African tropical rainforest, the Indian Monsoon transformation, the reversed Albedo in Himalayan glaciers, and the fate of the Greenland ice sheet. In total, there were about a dozen literal battle zones, indicated on a world map, where humanity's war with the Earth System was raging especially hard. It was pedagogical, and again underscored the importance of heuristic visualizations. But perhaps even more crucial, Schellnhuber had successfully entered the Earth System into a comprehensive narrative. There was a story out there already, and to launch a Big Idea on the scale of Planetary Boundaries, a trust-building narrative arc surely came in handy. The process could move on with reinforced self-confidence.

By September 2008, when the next iteration of the paper arrived, both the project and the process would carry the name of Planetary Boundaries. The "Tällberg consensus" branding of the previous year was gone; it had served as a starter engine for an activity that had now found a distinct direction and identity in its own right, and which had become increasingly refined in the process. Not least because the original broad brush sustainability ideas, Tällberg style, had by then gravitated much closer toward the home turf of Earth System science, with a sizable number of co-authors from that community who had not been part of the original consensus now becoming directly engaged in the paper.

Distancing the new project from past limits concepts was considered important for its success. At one of the preparatory meetings in his SRC office, Johan Rockström leaned toward one of the participants and asked: "Are you sure that this idea of boundaries is not the same as that of the Club of Rome in 1972?" The response was that, yes, you can be confident that it is different, and the reason was given. Still, the question lingered on. When Cambridge geographer Susan Owens, later an honorary doctor at the KTH Royal Institute of Technology (2012), came to Stockholm to take up a visiting Mistra Foundation Royal Carl XVI Gustav professorship in 2009, this was a central topic in conversations with her.[35] Yes, the PB idea was novel, but at the same time, there was reason to voice similar concerns about the problem of homogenization of Earth, Humanity, and Stability (the Holocene) that were already in circulation in the spring of 2009 as the *Nature* paper

[35] Susan Owens, personal communication on several occasions during spring 2009 and again on September 15, 2009 (Sörlin). Documentation in e-mail (with the authors).

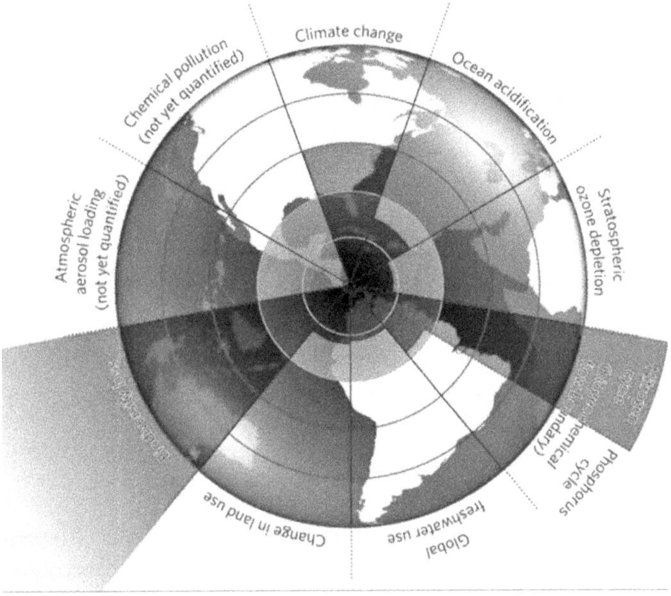

FIGURE 7.5 Planetary Boundaries diagram 2009. The diagram has evolved over the years. This is the original version. Courtesy of Nature Publishing.

started taking shape. Identical points that the humanities and social science critique would address in years to come, both with PB and the Anthropocene concept.[36]

The graphic representation of Planetary Boundaries was not obvious, and it came together late in the process (Figure 7.5). The original design was done by one of the paper's co-authors, Björn Nykvist at the SEI – "I was the data and visualization-person" – who was drawn in "half way" when the boundaries were given numbers and representations. He worked with a core writing team under the leadership of Johan Rockström, along with Åsa Persson, Will Steffen, and Kevin Noone. When they were finished, Nykvist drafted sketches that *Nature*

[36] The source for this paragraph is emails and recollections by one of us authors (Sörlin) and conversations over the years with several others involved in the winding PB process. See especially Sörlin to Rockström February 12, 2009, and Rockström to Sörlin February 14, discussing the status of the Holocene as a stability domain and homogenizing effects of concepts such as Earth and humanity, as well as tactical aspects of how to make the argument inclusive to the whole breadth of the scientific spectrum. E-mails with the authors. The Rockström quote is not documented but from memory by one of us (Sörlin), who also gave the affirmative answer.

then developed into their final form. He remembers that Rockström was clearly behind the visual idea and that several challenges arose, such as the wildly different scales of the boundaries.[37] The PB diagram quickly took on an iconic status and has been reproduced in countless versions, in textbooks and journal articles, on the internet, and in magazines such as *The Economist*. It underscores the role of visuals for communicating complex science-based relationships to wider audiences, both scientific and non-scientific, beyond the limited domain of "contributory" scientific expertise.[38] Surely, the striking, yet somewhat elusive, diagram that broke the rules of most science illustrations with its lack of comprehensive scaling helped the PB idea travel widely. It was a compelling demonstration of an idea rather than, strictly speaking, a diagram.

META-EXPERTISE

In combination, the planetary boundaries diagram and the article it appeared in did other things, too. The intervention reflected a particular role that Johan Rockström, as lead author, was now increasingly assuming: that of a "meta-expert," a torch bearer of "aggregative expertise" on the Anthropocene, resilience, Planetary Boundaries, and other conceptions and responses to global change. It was indeed a new way of looking at the world. The term, "aggregative expertise," has been used before to denote experts that do not limit themselves to speaking publicly on the particular, limited subjects that they have researched, that is, their own scientific specializations. They instead draw out the implications and produce, or circulate, narratives that also encompass results and concepts from others, for example, entire swaths of the Earth System science community and selected ideas from elsewhere to make a point that captures the attention and imagination of media, the public, and policymakers.[39]

While sometimes seen with skepticism by primary "micro" experts, "meta" (or "mega-") experts have been somewhat common in the history of "the environment." This type of actor has been extremely useful

[37] Björn Nykvist to Sverker Sörlin, e-mail April 6, 2018.

[38] P. Morseletto, "Analysing the Influence of Visualisations in Global Environmental Governance," *Environmental Science and Policy* 78(2017): 40–48. Birgit Schneider & Thomas Nocke, eds., *Image Politics of Climate Change: Visualizations, Imaginations, Documentations* (Bielefeld: Transcript, 2014). Harry Collins & Robert Evans, *Rethinking Expertise* (Chicago, IL: The University of Chicago Press, 2009).

[39] Warde, Robin & Sörlin, *The Environment*, 16, 25–46, 164–167, 173.

in advancing their cause beyond the scientific community. We can think of Rachel Carson (*The Sea Around Us* in 1951 and *Silent Spring* in 1962), Barry Commoner (zoologist, author, and eventually a United States presidential candidate), Bill McKibben (activist and author of numerous declensionist books since *The End of Nature* in 1989), Paul Ehrlich (Stanford biologist and author of the 1968 bestseller *The Population Bomb*), Garrett Hardin (the tragedy of the commons article from 1968), or Canadian geneticist David Suzuki (author, activist, and beloved television communicator). Some, like Ehrlich, had connections to Stockholm. Georg Borgström, originally Swedish and with a long career in the United States, appeared as an outspoken neo-Malthusian in the 1950s, a Scandinavian version of the arch-pessimist ecologist and contraception activist William Vogt, whose bestselling *Road to Survival* (1948) opened the discussion for ideas on "the environment." Remarkably, many of the post-war generation of meta-experts had a demographic outlook, wanting to restrict or reduce the world population, sometimes even the number of people in their home countries. A related focus was on resources. Borgström, an alarmist in word and deed, engaged millions of people in the specter of food scarcity, applying rhetorical skills he inherited from his preacher man father.[40]

A very different personality, but also matching the meta-expert profile, was the atmospheric scientist Bert Bolin, who pursued computer-based climate change research from the 1950s and became the founding father of the IPCC in 1988 as the climax of a long career as a world-leading science organizer (see Chapters 3 and 6). He segued discretely but importantly from active research to also assume the role of science diplomat, stitching together teams of scientists and huge collaborative programs, orchestrating their efforts to reach common goals. He was tirelessly presenting the meta-science of climate physics and chemistry to politicians and other leaders, signing off on reports and papers rather than putting in all the labor of collecting data and crafting equations. Bolin had learned a great deal from his mentor, Carl-Gustaf Rossby, a visionary genius and avid organizer of science (Chapter 3). Here was a rhetorical tradition going back to the famously outspoken, enticing, and versatile Svante Arrhenius (incidentally a distant relative of climate activist Greta Thunberg). Rockström was during the PB process taking on a similar kind of role, twenty-first century style, complete with a Netflix

[40] Björn-Ola Linnér, *The Return of Malthus: Environmentalism and Post-War Population–Resource Crises* (Cambridge: White Horse Press, 2003).

documentary (2021). He transformed, or the media turned him, into a meta-scientist, knowingly making himself dependent on many colleagues in order to achieve the "higher good" of establishing a synoptic, policy-friendly view of Earth System science.

This Swedish tradition of environmental meta-expertise, and the PB process, speaks to the same basic properties of Stockholm as a leading hub of environmental governance, namely the four "cons" of conceptualizing, connecting, convening, and contributing (Chapter 1). When the PB article was published in September 2009, it was the fruit of an over two-year-long process that had started with a pair of cons: contacts and conviviality at the Tällberg Forum in June 2007. It continued within a very small group of three or four Stockholm-based scholars that, at Tällberg in 2008, grew to more than twenty, mostly from other countries. In the end, twenty-nine co-authors had contributed to the framework's final formulation. It was a stocktaking process built on confidence, to name another con word.

Another thing that the article did was to communicate certain values that can be linked to Sweden. Then con-way was a Swedish way, of seeking consensus and common ground, and also to work on a level that was not overly esoteric and philosophical but robustly earthy and empirical. "A safe operating space for humanity" may be seen as a formula for the Swedish *Folkhem*, or welfare state, writ large in a planetary version. We may recall that in the 1990s, Swedish Prime Minister Göran Persson had, as part of the comprehensive and austere reform package to curb the country's fiscal crisis, launched the *Green Folkhem* as an optimistic, even cheerful, path forward in the midst of a political situation that was in other respects quite gloomy. In essence, Swedish policy, Folkhem-style, was a balancing act, navigating between guard rails. It was a policy of avoiding extremes, de-risking, and bringing everyone on board. A bit like Volvo cars, perhaps boring and not the sexiest vehicles on the road, but capable of safely taking you to your destination. It worked well for a good deal of the twentieth century, during which Swedish environmental policy was also, after all, pragmatic rather than radical, after protracted debates almost always settling for some modest but efficient version of mainstream modernism, grey rather than green (or even red).[41]

[41] Erland Mårald & Christer Nordlund, "Modern Nature for a Modern Nation: An Intellectual History of Environmental Dissonances in the Swedish Welfare State," *Environment and History* 26 (2020):4, 495–520.

The early Swedish Green Deal was perhaps the Folkhem's last hurrah before a more polarized politics of the 2000s set in (more on this in Chapter 9). The 1990s were Rockström's formative years as a scientist. The PB article is also at pains in not making nods in any particular political direction, probably intentionally. It is open to business and capitalism, and certainly comes across as one of the most outspoken attempts to launch the entire Earth as a "level playing field," where any forces – certainly including economic growth and development – can operate, as long as they help heal the planet.[42] But the planetary boundaries framework and its creators do not say no to socialism, either. The approach, especially if seen over the full fifteen years since the PB process started, is quite eclectic, to say the least. In later work, Rockström has spoken highly of "pragmatism," of state-led innovation (neo-Keynesian style, with reference to economist Mariana Mazzucato), and the necessity of equity in moving the PB concept forward. His latest book includes a foreword written by Greta Thunberg, a fellow Swede, but also a staunch radical, whose ideas Rockström has occasionally distanced himself from, but whose persona and presence on the world – or Earth – stage he has wholeheartedly welcomed. Rockström has occasionally talked about the need not just for transformation but "revolution."[43]

In this respect, the PB project can itself be seen as a convening enterprise, an initiative to build consensus around a new understanding of the Earth that may encompass some kind of politics down the road, although it is not entirely clear what type. It was a reconciliation of many opposites, not by design but a kind of tacit acceptance: between capitalism-socialism, environment-development, Global North-Global South, Holocene-Anthropocene. Coming out of the fields of development and agricultural policy, Rockström was rooted in expert management and could well use a firm guiding hand for development. Demonstrating the durability, or perhaps long gestation time, of certain ideas, the institutional and intellectual platform upon which Rockström gained much of his recognition, even fame since the late 2000s is built

[42] Based on email correspondence between Sörlin and Rockström during 2009 (with the authors).

[43] All examples and quotations from Rockström & Gaffney, *Breaking Boundaries*. For one of several statements in recent years about the need for "revolution," see an interview (in Swedish), "Rockström on Climate: Don't Know if We Will Pull This Off," *Syre* 2020: 2, https://tidningensyre.se/2020/2-januari/rockstrom-om-klimatet-vet-inte-om-vi-kommer-att-klara-det-har/.

upon a scientific theory from the early 1970s that was decades later mobilized through an international network of global change experts.

RESILIENCE – THE CONCEPT AND THE ALLIANCE

The Resilience Alliance was established in 1999 as an international network of natural and social scientists whose research revolved around the resilience concept first put forward by C. S. Holling in 1973. In addition to the core idea of *resilience* – the ability of a system to withstand disturbances while maintaining its core structure and functions – a suite of ancillary concepts and normative beliefs support "resilience thinking" among Alliance members.[44] The most fundamental of these is that human and non-human systems are inextricably linked, constituting coupled *socio-ecological systems* that can be scaled up to the level of the Earth System.

In the resilience worldview, interactions between different scales in space and time are of critical importance, with systemic changes understood as part of perpetual *adaptive cycles* taking place within a multilevel framework that the concept's followers call *Panarchy*.[45] To generate what are deemed to be positive outcomes, the Resilience Alliance promotes *adaptive management* as a scientifically informed method of environmental governance and embraces *transformation* of societal systems to increase resilience and achieve sustainability. Like the links between social and ecological systems, the coupling of science and governance is thus a fundamental aspect of the resilience concept as understood by its adherents within the Alliance.

Stockholm-based actors would, both before and after the founding of the Resilience Alliance, make a significant contribution to the institutionalization and popularization of the concept. This process began in earnest in the early 1990s with a series of workshops that facilitated international networking and resulted in interdisciplinary collaborations that helped further develop Holling's original concept, which was already twenty years old at that point. The institutionalization of resilience would expand significantly in the following decade

[44] Carl Folke, Stephen R. Carpenter, Brian Walker, Marten Scheffer, Terry Chapin & Johan Rockström, "Resilience Thinking: Integrating Resilience, Adaptability and Transformability," *Ecology and Society* 15(2010):4, 20.

[45] Lance Gunderson & C. S. Holling, eds, *Panarchy: Understanding Transformations in Human and Natural Systems* (Washington, DC: Island Press, 2002).

FIGURE 7.6 Brian Walker, Simon Levin, Karl-Göran Mäler, and Partha Dasgupta (left to right) in 2002 at the Beijer Institute's annual gathering of ecologists, economists, and other environmentally concerned scientists on the island of Askö, where Stockholm University manages a Baltic Sea marine laboratory. That same year, Mäler and Dasgupta – close friends and long-time scientific collaborators – were together awarded the Volvo Environment Prize. The Askö meetings have played a significant role in the development of the field of ecological economics since the early 1990s. Photo: Anna Sundbaum/ Courtesy of the Beijer Institute.

with the establishment of a well-financed scientific research institute in Stockholm that was mandated to inform policymaking both in Sweden and internationally. Moreover, the first major international conference dedicated to the concept of resilience would be convened in Stockholm shortly after the founding of the new institute.

Resilience and social-ecological systems had been recurring themes at the Beijer Institute's annual Askö workshops, held in the seaward skerries outside Stockholm in the years leading up to the founding of the Resilience Alliance in 1999 (Figure 7.6). Ecological economics, the scientific focus of the second incarnation of the Beijer Institute, would serve as an interdisciplinary bridge for studying interlinked social and environmental systems that underpinned resilience thinking. In summarizing the proceedings of the first Askö meeting in 1993, Beijer's deputy director Carl Folke concluded that ecologists and economists could indeed find common ground despite the significantly different perspectives on the human environment that had to date characterized their respective professions.[46]

As an essentially integrative concept, resilience offered the kind of common scientific ground on which experts from diverse disciplines could collaborate in pursuit of solutions to complex problems associated with sustainability. An early example was an article resulting from the 1995 Askö workshop, "Resilience in Natural and Socioeconomic Systems." Its seventeen natural and social science co-authors, some of whom would become founding members of the Resilience Alliance, included economists Karl-Göran Mäler, Partha Dasgupta, William Baumol, and Ing-Marie Gren; ecologists C. S. Holling, Simon Levin, Ann-Mari Jansson, and Bengt-Owe Jansson; biologist Paul Ehrlich, climate scientist Bert Bolin, and Carl Folke, whose scientific background encompassed both ecology and economics.[47] Like the Planetary Boundaries articles fifteen years later, and even the Tällberg Foundation's 350 advertisement, the programmatic paper from the Askö meeting in 1995 was an example of the big tent science that reflected the convening power of Stockholm. Not simply presentations of the results of scientific research, such pieces also served as proclamations for emerging intellectual movements that aimed to inform policy and influence the governance of socio-ecological systems at all levels of abstraction.

Many of the members of the expanding Beijer-Askö network would continue to collaborate across a range of projects and publications on related topics, as resilience became an increasingly prominent paradigm for understanding and addressing the challenges of the human environment during the 1990s and 2000s. With the retirement of Karl-Göran Mäler in 2006, Carl Folke became director of the Beijer Institute of Ecological Economics, which was at that point embarking upon a new collaboration to shape the institutional home for the Stockholm Resilience Center, bringing together Beijer and thereby the Royal Swedish Academy of Sciences, Stockholm University, and the SEI under the banner of resilience.

[46] Carl Folke, "Ecologists and Economists Can Find Common Ground," *Bioscience* 45(1995):4, 283–284.

[47] Simon Levin et al., "Resilience in Natural and Socioeconomic Systems," *Environment and Development Economics* 3(1998):2, 222–235.

The Stockholm Resilience Centre was inaugurated in 2007 with the unusual arrangement of Beijer's Carl Folke and Johan Rockström of SEI co-leading the new institution. In their capacity as science organizers and mediators of environmental knowledge, they would carry the mantle of Bert Bolin and Gordon Goodman, both of whom would pass away, each at the age of eighty two, in the year following the establishment of the SRC. While Folke had been engaged with the Beijer Institute's work on socio-ecological systems and the seminal Askö ecological economics workshops since the early 1990s, Rockström's specialization in hydrology and water management in semi-arid and tropical areas had seen him conduct extensive field research and development work in sub-Saharan Africa. His 2004 return to Stockholm from Zimbabwe came following a phone call from Bolin, who, on Folke's recommendation, recruited Rockström to become the executive director of the SEI – a position he would continue to hold until 2012, even after he took on the same title at the SRC in 2007.

Substantial support (the grant was around 25 million US dollars over ten years) for the SRC was provided by the Mistra Foundation (independent like all foundations, founded by the Swedish state in 1993) that finances strategic environmental research, usually of longer duration. Soon after its founding, the SRC would profile itself as a major international hub for the study of social-ecological systems (SES). Based on Mistra's call for an interdisciplinary research institute that could support sustainable governance in the wake of the Millennium Ecosystem Assessment, the strategic mission of the SRC would also include the production of science that was relevant for policy formation in Sweden and beyond. Hence, some thirty-five years after Holling's groundbreaking article, a cadre of Swedish and international resilience devotees would help establish the ideas for which they received handsome backing. Both directors had paid their dues for quite some time and performed impressively on most scores. This was a new step, with much broader support. Resilience and SES science had come of age, with Stockholm as a major hub within an extensive international network of experts.

However, despite the decisive gesture of establishing what was from the outset intended to be a world-class institution backed by long-term funding, staffed and supported by a network of renowned experts in their respective fields, the scientific niche and strategic purpose of the SRC were in its early years somewhat opaque to outside observers. The ambitious mission that combined producing cutting-edge research with ongoing engagement in societal decision-making processes, together with the Centre's complex organizational structure and the wide array of stakeholders its leaders were answerable to, generated a fair amount of uncertainty and confusion, internally as well as externally.[48] Moreover, resilience was initially a relatively unknown and ambiguous concept outside a community of sustainability specialists, with the term often being associated with infrastructure, health, and civic organizations, rather than socio-ecological systems or sustainable development. A successful rollout and embrace of resilience thinking, both within and beyond scientific circles, was therefore far from assured at the time of the Centre's founding.

THE FRIIBERGH MANOR WORKSHOP – AND SUSTAINABILITY SCIENCE

These early challenges were elaborated in a Start-up Review commissioned by Mistra two years into the Center's first funding period. The review was conducted precisely by Harvard ecologist William C. Clark, who had been a key participant in the Villach-Bellagio process during the 1980s and thus had long-standing ties to Stockholm. Clark was also a leading figure in another network centered on a subject known as sustainability science. It was a similarly comprehensive scientific paradigm concerned with nature-society interactions, which had developed in parallel to the rise of resilience during the decade leading up to the founding of the SRC. In problematizing what its practitioners consider modern society's unsustainable development pathway, the new synthetic discipline was based upon the application of scientific analysis to increase understanding and devise progressive solutions to the myriad complex challenges endemic to the twenty-first century human environment.[49] In fact, and ironically, the very concept of sustainability science, often attributed to Clark, was one that the Centre actively distanced itself from in its first phase, largely as it fought hard to define itself as different from the very start. Little would Centre leadership know that Clark was going to be their first evaluator, and even larger was the relief when the outcome was so reassuring.

[48] William C. Clark, *Start-up Review of the Stockholm Resilience Centre* (Stockholm: Mistra Foundation, 2009).

[49] William C. Clark & Nancy M. Dickson, "Sustainability Science: The Emerging Research Program," *Proceedings of the National Academy of Sciences (PNAS)* 100(2003):14, 8059–8061. William C. Clark & Alicia G. Harley, "Sustainability Science: Toward a Synthesis," *Annual Review of Environment and Resources* 45(2020):1, 331–386.

The formative moment for the international sustainability science movement, which has since spawned dedicated research programs, a scholarly journal, and an entire academic discipline – and has also become a core aspect of the SRC's scientific profile – took place at the turn of the millennium at a seventeenth-century mansion an hour outside Stockholm. The five-day Sustainability Science Workshop at Friibergh Manor in October 2000 was hosted by the Swedish Research Council and sponsored by various Swedish and American governmental organizations and private foundations.

The workshop's organizing committee included Clark, Bert Bolin, marine biologist Jane Lubchenco, and Uno Svedin, a trained physicist who spent much of his career engaged with research policy in the realms of environment and sustainability. As the director of research at the Swedish Council for Planning and Coordination of Research (*Forskningsrådsnämnden*, FRN) from 1981 to 2001 and international director at the funding agency Formas until 2010, Svedin has played a crucial role in connecting research communities and organizing international scientific events, including an earlier conference at Friibergh Manor in 1986. Such meetings were instrumental in the development of emerging concepts and disciplines like resilience, sustainability science, and ecological economics. Svedin had also during his years at FRN been actively engaged in the funding of interdisciplinary leaning environmental scholars such as Carl Folke in their very early career phases in the 1980s and early 1990s.[50]

Sustainability science explicitly entailed transcending particular disciplinary boundaries to bring diverse areas of expertise to bear in addressing practical, yet seemingly intractable, problems at the intersection of environment and society. In addition to the range of experts representing the Earth and life sciences, the convergence of research communities at the 2000 Friibergh Manor workshop also included specialists in the human and societal aspects of socio-ecological and planetary systems. Among the founders of sustainability science in attendance at Friibergh were geographers Akin Mabogunje and Tim O'Riordan and economist Sylvie Faucheux. Also there was Jill Jäger, the Villach-Bellagio veteran who was at that point the executive director of the International Human Dimensions Programme on Global Environmental Change – the humanities and social science component of the Earth System Science Partnership that had been founded in 1990.

[50] Personal communication, Uno Svedin.

The new discipline would be widely promoted within international scientific circles in the months following Friibergh. In the tradition of producing multi-author manifestos to publicly pronounce new scientific and policy-relevant standpoints, the twenty-three workshop participants signed the Statement of the Friibergh Workshop on Sustainability Science and reiterated their adherence to the sustainability science approach to real-world problems in an April 2001 article in the journal *Science*.[51] Momentum continued to build in the lead-up to the July 2001 Challenges of a Changing Earth Conference that was organized under the auspices of the Earth System Science Partnership. At the Amsterdam Conference (earlier, note 15), William Clark presented sustainability science to the global change community in what amounted to a coming-out party for the new paradigm that combined the rigors of science with the concerns and convictions of social and environmental activism. Sustainability science thus resembled resilience in its aspiration to serve as a norms-based transdisciplinary tool for the governance of socio-ecological systems, with substantial overlap and cross-fertilization between the two communities of activist scientists.

CONVENING POWER: RESILIENCE 2008

The convening power of Stockholm demonstrated by the Friibergh workshop – one of many milestones in a long tradition that dates back to the science organizing efforts of Carl-Gustaf Rossby and Bert Bolin in the 1950s – was further amplified with the SRC's founding in 2007. In his Start-up Review of the Centre's first two years, Clark asserted that the concept of convening power could extend beyond physical meetings to also include orchestrating and providing intellectual leadership for international scientific initiatives such as the Planetary Boundaries project:

> Measured by convening power, the Centre's *Resilience 2008 Conference* [in April that year] brought together a substantial fraction of the world leaders in the field, and was referred to by several senior scholars I contacted as one of the most exciting research conferences they could remember. The SRC's convening power is also reflected in the forthcoming (and soon to be legendary) *Nature* paper on "Planetary boundaries: Exploring safe operating space for humanity." The co-authors of this paper include a Nobel Laureate, directors of most of SRC's top competitor institutions, and sundry luminaries from the natural and

[51] Robert W. Kates et al., "Sustainability Science," *Science* 292(2001):5517, 641–642.

social science ... in short, people that any scientist would be proud to have as collaborators. Several of them that I contacted readily acknowledged the leading role played by first author Johan Rockstrom and other SRC scientists in conceiving and drafting this paper. I heard similar stories about other collaborative efforts driven by the Centre.

More than a strictly scientific event, "Resilience 2008: Resilience, Adaptation and Transformation in Turbulent Times" was a celebration of a concept that at times seems to transcend science to take on the trappings of an ideological or philosophical framework (Figure 7.7). "A central message of the Resilience 2008 Conference," according to its host institution, "was that resilience is not just an ecological issue – it involves ecological, economic, cultural, ethical and other social dimensions and values."[52] In its ambition to order earthly activity around the ideal of robust SES, resilience thinking had developed into something more than a scientific theory. It bears traits of a secular belief system, replete with a founding father in ecologist visionary C. S. Holling, a literary canon and cosmology in Holling's writings and his Panarchy framework, the Holy Ghost *qua* Hysteresis curve, the specter of Armageddon in ecological and planetary tipping points, and the promise of salvation in the case of societal transformation toward sustainability.

The April 2008 Conference was convened under the auspices of the Resilience Alliance, the Royal Swedish Academy of Sciences, and International Council for Science (ICSU). The latter was at that time led by Thomas Rosswall, a Swedish biologist and founding director of the International Geosphere Biosphere Programme, it too located in the Academy of Sciences building on the Stockholm University campus north of the city, just like so many other institutions and actors we have come across in the Stockholm story. Some 600 natural and human scientists would congregate in Stockholm for four days of what the SRC described as "a dense trans-impulsive program including science, art-exhibitions, music, culture, social events, and a high-level science-policy event."[53] In terms of scientific content, the Resilience 2008 Conference's call for papers announced nine overall themes encompassing a wide variety of human–environment interactions to be explored through the lens of resilience and its range of subsidiary concepts. The themes – inscrutable to

[52] SRC *Annual report* 2008.

[53] SRC *Annual report* 2008. On one of the art pieces presented, see Sverker Sörlin, "The Vulnerable Volvo," In: Jennifer Newell, Libby Robin & Kirsten Wehner, eds., *Curating the Future: Museums, Communities and Climate Change* (London & New York: Routledge, 2016), 215–218.

FIGURE 7.7 Concluding panel discussion of the Resilience 2008 Conference hosted by the Stockholm Resilience Centre. A milestone in the institutionalization of a scientific concept that had by then been in circulation for several decades, Resilience 2008 and the six panelists portrayed here embody the four fundamental con-words of this book: convening, conceptualizing, contributing, and connecting. From left to right: Sverker Sörlin, co-founder (with Nina Wormbs) of the KTH Environmental Humanities Laboratory; Elinor Ostrom, Nobel laureate in economics for her work on common pool resources; C. S. "Buzz" Holling, former director of IIASA who coined the resilience concept in ecology in 1973; Uno Svedin, an accomplished science organizer and research financer, as well as, along with Sörlin, a co-author of the 2009 Planetary Boundaries paper; Line Gordon, director of the Stockholm Resilience Centre since 2018; and Carole Crumley, a founder of Historical Ecology and former director and founder of the Integrated History and Future of People on Earth (IHOPE) community of researchers. All six held close associations with the Stockholm Resilience Centre from its founding in 2007.
Photo: Jerker Lokrantz/azotelibrary.com.

those uninitiated into resilience thinking – included "Traps, regime shifts, and transformations," "Knowledge management and social-ecological learning," "Urban social-ecological system challenges," and "Adaptive governance and multilevel challenges."

The convening power of Stockholm, reflected in the success of Resilience 2008 and embodied by the Planetary Boundaries project, would culminate in the convergence of two major scientific movements for which the Swedish capital had become a leading international node. The PB project was a collaboration between the SRC and the Earth System Science Partnership, with the Tällberg Foundation providing a policy perspective.[54] In line with this joint effort, the twenty nine authors of the 2009 Planetary Boundaries piece in *Nature* were drawn from international networks of resilience and Earth System scientists. The article's lead authors – Johan Rockström and Will Steffen – had held leadership positions in Stockholm-based institutions representing each of the respective scientific fields. What is more, three members of the 2000 Friibergh workshop's organizing committee – Uno Svedin, climate scientists Robert Corell, and the above-mentioned Hans Joachim Schellnhuber – were among the Planetary Boundaries co-authors, indicating the inclusion of a sustainability science angle inside the PB project as well as within the wider nexus of resilience and Earth System science.

In addition to the fundamental goal of producing knowledge on global change and aspects of the Earth System, the three intertwined networks were united in their ambition of creating concepts, tools, and frameworks for advancing sustainability and facilitating environmental governance at all levels. Corresponding to the holistic and interconnected outlook of Earth System science, the preferred perspective was planetary, with a common denominator in the idea that the global environment represents an indispensable life support system for human survival and civilization. Since the establishment of the IGBP in the mid-1980s and accelerating around the turn of the century, a variety of planet-level paradigms for environmental governance have been proposed by natural and social scientists associated with the resilience, sustainability, and Earth System science movements. A nonexhaustive inventory of maximalist concepts that have sought to render the Earth a governable object through synthesizing environmental science and politics includes "planetary management," "planetary stewardship," "Earth System stewardship," "Earth System governance," and, perhaps most prominently, the "Planetary Boundaries framework." Most of these emanated from Stockholm-based institutions, or from contexts that involved such institutions.

More than half of the Planetary Boundaries authors had SRC affiliations or other close connections to Stockholm. Published less than three

[54] SRC *Annual report* 2008.

months before COP15, the *Nature* article made specific reference to the highly anticipated UNFCCC Conference in Copenhagen that would take place in December 2009. Although the authors acknowledged that momentum was building in support of the "2°C guardrail" approach for containing climate change, the article advocated a limit expressed in atmospheric CO_2 concentration – a measurement pioneered by the Bert Bolin-collaborator Charles David Keeling – rather than the less precise benchmark of global mean surface temperature. In one of seven expert commentaries that accompanied, and in some cases criticized, the *Nature* article, the 350-ppm planetary boundary was questioned by climate scientist Myles Allen. In the critical lead-up to COP15, Allen saw the ppm-based boundary as a distraction to the more immediate 2°C target that had already been adopted by the European Union as its basis for climate policy.[55] Planetary Boundaries was thus, by design, embedded in environmental governance debates from the outset, and soon after the article's publication would serve as the foundation for a series of policy-oriented interventions by political scientists and other scholars at the SRC and elsewhere.

The 350 ppm climate threshold was one of three planetary boundaries that had already been transgressed at the time of the framework's publication. Despite the best efforts of many of the Planetary Boundaries authors to convince politicians on the imperative of decisive action on climate, and the fact that atmospheric CO_2 concentrations were in late 2009 well on their way to 400 ppm, COP15 concluded without reaching a binding global agreement on greenhouse gas emissions. The widely

[55] Richard S. J. Tol, "Europe's Long-Term Climate Target: A Critical Evaluation," *Energy Policy* 35(2007):1, 424–432; Samuel Randalls, "History of the 2°C Climate Target," *WIREs Climate Change* 1(2010):4, 598–605; Carlo C. Jaeger & Julia Jaeger, "Three Views of Two Degrees," *Regional Environmental Change* 11(2011):Suppl 1, 15–26; Sabine Höhler, "Two Degrees: A Global Climate Accord and Its Disparities," in *Creative Commensuration: Histories of Scaling in Science and Society*, Workshop Center History of Knowledge, Zurich, July 7–8, 2016. Eric Paglia, "The Socio-scientific Construction of Global Climate Crisis," *Geopolitics*, 23(2018):1, 96–123. Westermann & Höhler, "Writing History in the Anthropocene," esp. 590–591, 594, observing how scaling and global numbers tend to obscure historical actors, geographical particulars, and economic practices. Eric Paglia & Erik Isberg, "On Record: Political Temperature and the Temporalities of Climate Change," In: Anders Ekström & Staffan Bergwik, eds., *Times of History, Times of Nature: Temporalization and the Limits of Modern Knowledge* (New York: Berghahn, 2022), 277–301. See also Frank R. Rijsberman & Rob J. Swart, eds., *Targets and Indicators of Climatic Change* (Stockholm: SEI, 1990), produced by the SEI under the auspices of the Advisory Group for Greenhouse Gases, AGGG, discussed in Chapter 6, as a seminal early proposal for temperature targets as a basis for policy.

perceived failure in Copenhagen, which would demoralize much of the climate movement for months if not years after COP15, did however succeed in introducing the 2°C target into UNFCCC proceedings through the signing of the Copenhagen Accord. Although not immediately apparent amid the prevailing disappointment of COP15, the institutionalization of 2°C – which would become the foundation of the celebrated 2015 Paris Agreement – opened a new pathway for international climate politics. It also represented a relative success for one of the Planetary Boundaries authors, Hans Joachim Schellnhuber, who had advocated for a 2°C temperature target since the time of the first COP meeting in Berlin in 1995.

FAULT LINES IN THE STOCKHOLM CONSENSUS

While Schellnhuber is sometimes referred to as the father of the 2°C target, another Planetary Boundaries co-author – the Stockholm University-trained Paul Crutzen – could be regarded as not only the progenitor of the Anthropocene concept but also the leading proponent of the controversial idea of geoengineering. In a seminal 2006 editorial essay in the journal *Climatic Change*, Crutzen, who passed away in 2021, described the injection of sunlight-reflecting sulfate particles into the stratosphere as a potential "escape route" in the case of climate policy failure, which he considered likely given its track record, and the eventual rise of global temperatures above 2°C. His support for taking initial steps, such as modeling and eventual experiments, into exploring geoengineering options was reiterated in the heavily cited 2007 article in *Ambio* that he co-authored with Will Steffen, who passed in 2023, and John McNeill.[56] Crutzen's interventions on geoengineering as a last resort tool of climate governance punctured a perceived taboo on the topic, leading to a wave of new research and discussions – as well as recriminations – on the scientific, engineering, and ethical aspects of intentional climate modification.[57] International interest in geoengineering as a fallback option was

[56] Will Steffen, Paul Crutzen, & John R. McNeill, "The Anthropocene: Are Humans Now Overwhelming the Great Forces of Nature?," *Ambio* 36(2007):8, 614–621.

[57] Clive Hamilton, "Geoengineering and the Politics of Science," *Bulletin of the Atomic Scientists* 70(2014):3, 17–26. Mark G. Lawrence & Paul J. Crutzen, "Was Breaking the Taboo on Research on Climate Engineering via Albedo Modification a Moral Hazard, or a Moral Imperative?," *Earth's Future* 5(2017):2, 136–143. We have found no evidence, however, that Crutzen's co-authors, McNeill and Steffen, in earnest approved of Crutzen's predisposition toward climate geoengineering.

further enhanced following the COP15 failure to forge a new emission reduction regime, and the science of geoengineering is ongoing.[58]

Schellnhuber, for his part, was not as sanguine on the specter of solar radiation management as an escape route or silver bullet solution to the climate crisis. In his famous 1999 *Nature* article "'Earth System' analysis and the second Copernican revolution," he had listed geoengineering as one of five possible paradigms for proactive global environmental management or Earth System control. By 2011, however, Schellnhuber had signaled his opposition to solar radiation management, using geopolitical arguments based on the Cold War metaphor of mutually assured destruction. Yet both Schellnhuber and Crutzen had that same year, 2011, participated in a scientific working group on climate commissioned by the Pontifical Academy of Sciences that concluded it "may be prudent to consider geo-engineering if irreversible and catastrophic climate impacts cannot be managed with mitigation and adaptation." A degree of ambiguity, even ambivalence, can thus be discerned surrounding the personal convictions and eventual policy prescriptions that some Earth System scientists would, in the case of failed mitigation efforts, be willing to embrace on one of the most consequential global environmental governance questions of the twenty-first century.

PROMETHEUS VS. SOTERIA

No such ambiguity exists in the positions taken over the past decade by Mark Lynas and Clive Hamilton. The British and Australian, respectively, authors and public intellectuals have been two of the more prominent popularizers and outspoken supporters of Earth System science, the Anthropocene concept, and the Planetary Boundaries framework. Lynas and Hamilton have both employed the same scientific paradigms as points of departure for scholarly articles, media polemics and popular science books, and reports on the powers and prerogatives of humanity in confronting climate change and other environmental challenges. Nonetheless, they have come to see the relationship of people and planet in starkly different lights. Like Crutzen and Schellnhuber, both Lynas and Hamilton strongly support strict anthropogenic CO_2 emission reduction regimes under the UNFCCC as the primary political means for mitigating climate change. Yet in terms of the societal shifts required

[58] Jeroen Oomen, *Imagining Climate Engineering: Dreaming of the Designer Climate* (Abingdon & New York: Routledge, 2021).

to create an ecologically sound safe operating space for humanity, and the options that could be considered – or ruled out – in the case of an incipient climate catastrophe, the sustainability pathways they have espoused point toward dramatically divergent futures for the planetary human environment. Their contradictory perspectives and prescriptions reflect an underlying tension in what might otherwise be considered a Stockholm consensus on scientifically informed global environmental governance.

Lynas made his mark in the mid-2000s with a pair of popular science contributions to the climate debate. *High Tide: The Truth About Our Climate Crisis* and *Six Degrees: Our Future on a Hotter Planet* drew heavily on IPCC reports and helped amplify the growing public and political perception of climate change as an emerging global crisis. A frequent visitor to Stockholm, Lynas' connections inside the Swedish scientific community led to his invitation to attend the planetary boundaries planning session ahead of the 2008 Tällberg Forum. The idea of quantifiable planetary boundaries as elaborated by the scientists on hand at the closed-door meeting was a revelation for Lynas, and it was there in Tällberg, where he also consulted with Johan Rockström, that he decided to structure his next major work around the framework.

The God Species: How the Planet Can Survive the Age of Humans would serve to popularize the planetary boundaries concept beyond the rather restricted realm of scientific literature, where the articles in *Ecology & Society* and especially *Nature* had already made a major impact.[59] Published in 2011, *The God Species* marked a personal turning point for Lynas, who would from that point forward publicly embrace the ethos of ecomodernism and promote controversial technological solutions for combating the climate and environmental crises. The god-like planetary agency that humans had acquired with the advent of the Anthropocene should, according to Lynas, not be wasted but instead wielded for the worthy purpose of increasing welfare while

[59] An October 2, 2010, email sent to the Planetary Boundaries authors by one of the twenty nine co-authors, who had personal contact with Lynas, asked if anyone in the group would be interested in reading and commenting on a draft of the *God Species* manuscript. Parts of the email read as follows: "Inspired by our Planetary Boundaries paper (and, as I understand it, with encouragement from Johan), Mark has written a new book on global environmental change currently titled: *The God Species: How our planet can survive the Age of Humans* ... he has now finished the draft of the book and asked me if I would approach the 'team' to see if any of you are willing to read and comment on the draft. I have taken a quick look and it is a lively read – he certainly has an angle that will attract attention to the issues (and indirectly to our papers!)." E-mail with the authors.

keeping the Earth and its inhabitants within the scientifically determined safe operating space, where economic growth and most aspects of modern life could continue unabated. In his estimation, the vast majority of environmentalists – himself included – had been misguided in their opposition to powerful technologies like nuclear energy and the genetic modification of plants that would facilitate climate mitigation and adaptation efforts. If all else failed, Lynas' anointed God Species could as a last resort consider exercising its planetary agency by pursuing climate geoengineering options.[60]

By contrast, the Australian philosopher and professor of public ethics Clive Hamilton sees the increasing capacity of the human species to intervene in the climate system as a singularly frightening situation. Unlike the exaltation of human agency that underpins *The God Species*, the title of Hamilton's 2013 book *Earthmasters* is purely pejorative in its critique of the network of influential individuals and institutions – including certain scientists and research laboratories – that possess a "Promethean" urge to dominate the planet through geoengineering. Rather than a rational response to the prospect of runaway climate change, research on solar radiation management and the eventual implementation of large-scale climate modification schemes, such as spraying sulfate particles into the stratosphere, would, in Hamilton's view, constitute "the most dangerous experiment ever conceived."[61] Invoking the Greek goddess Soteria, who bestowed safety and preservation from harm to humanity, Hamilton – whose previous writings had criticized consumerism, climate denialism, and the fetish of economic growth – instead advocated a "Soterian" approach toward human interventions in the Earth System as a counter to the hubris and high-risk technological solutions of the Prometheans.[62]

Hamilton's outlook on geoengineering is consistent with his outright rejection of ecomodernism and the idea of a "good Anthropocene." Yet in some cases, on matters that are related but separate in terms of solutions, he also sides with some of the leading geoengineering advocates. Defending the prerogative of Earth System scientists such as Paul Crutzen to define and date the Anthropocene, Hamilton describes the onset of the human-dominated epoch as a very recent rupture in Earth history that

60 Mark Lynas, *The God Species: How the Planet Can Survive the Age of Humans* (London: Fourth Estate, 2011).
61 Hamilton, "Geoengineering," p. 25.
62 Clive Hamilton, *Earthmasters: Playing God with the Climate* (London: Allen & Unwin, 2013).

cannot be ascribed to activities and technologies prior to the modern period, as some ecomodernists have claimed. The de facto destruction of the relatively stable Holocene at the hands of humans – a radical historical break that Hamilton attributes to "*homo faber*, the technological man of modern Western civilization" – represented a phase shift in the functioning of the Earth System, which has made the latter a more volatile and potentially hostile operating space for humanity.[63] Hence, in a no-analogue state of the planet, the past has become a less reliable proxy for the future, and the ability of previously resilient ecosystems and human societies to adapt to environmental change, which might become sudden and severe rather than gradual and foreseeable, can no longer be taken for granted. Ecomodernist expectations of a good or even great Anthropocene therefore represent, according to Hamilton, a "perverse reading of Crutzen's conceptual innovation."[64]

Geoengineering can thus be seen as a subset of a larger and deeper social and political, as well as technological and even theological, struggle between Promethean and Soterian perspectives on environmental governance and sustainable development. As demonstrated by the antithetical prescriptions of Lynas and Hamilton, a shared scientific point of departure for diagnosing the environmental and climate crises does not necessarily create a consensus view on how societies, economies, and the international system should adapt, invest, and transform to address the governance challenges associated with global change. This fundamental divide can be illustrated by two diametrically opposed documents released in the spring of 2015, a landmark year in global environmental governance that would witness the signing of the Paris Accord and the adoption of the UN Sustainable Development Goals.

Mark Lynas was one of the eighteen authors of *An Ecomodernist Manifesto*, a Promethean treatise released in April 2015 that celebrated the social, economic, and technological power of humanity to shape its own planetary operating space. While acknowledging that significant damage had been inflicted upon the biosphere in reaching this stage of development, the *Manifesto* rejected imposing restraints on human agency and argued that technology has not only made humanity far less dependent on the natural world (a positive for ecomodernists), but that

[63] Clive Hamilton & Jacques Grinevald, "Was the Anthropocene Anticipated?," *The Anthropocene Review* 2(2015):1, 59–72, on 67.

[64] Clive Hamilton, "The Anthropocene as Rupture," *The Anthropocene Review* 3(2016):2, 93–106, on 99.

it can also enable environmental protection and extend prosperity to greater numbers of people on the planet. In placing their full faith in technology to empower humanity and avert environmental crises, the *Manifesto's* authors consider the prospect of a great Anthropocene to be well within reach.[65]

In stark contrast to the Promethean perspective of the ecomodernists, Pope Francis' encyclical letter on the environment, *Laudato Sí': On Care for Our Common Home*, attributes environmental deterioration and endemic poverty to precisely the same forces that *An Ecomodernist Manifesto* identified as the basis of success and salvation for the human species: free market economics and the widespread application of science and technology in modern society. Published in June 2015, *Laudato Sí* singles out climate change, water pollution, and biodiversity loss as symptoms of a consumerist culture that degrades quality of life and social cohesion while also fostering global inequities. Underpinning these problems of modernity is what Pope Francis sees as misguided and excessive anthropocentrism, and a lack of care and appreciation for God's earthly creation. The Pope's encyclical letter, for which Hans Joachim Schellnhuber served as chief scientific advisor, represented a de facto repudiation – in a vein similar to the Soterian convictions of Clive Hamilton – of the ecomodernist idea that a human-dominated Anthropocene world could ever be considered "good."[66]

TOWARD HUMANIZING THE FRAMEWORK – PLANETARY BOUNDARIES 2.0

The divergent pathways put forward by Prometheans and Soterians, and a spectrum of positions in between, are evidence that science is not independently equipped or capable of determining the structure and direction of global environmental governance. Even politics are in a sense downstream from some of the more deeply held beliefs surrounding how humans ought to interact with the biosphere and the Earth System as a whole. Despite the epistemological power of the Stockholm scientific nexus – resilience thinking and Earth System science, Planetary Boundaries and the Anthropocene, ecological economics, and sustainability science – reaching a broad Stockholm consensus on a range of

[65] John Asafu-Adjay et al., *An Ecomodernist Manifesto* (2015), www.ecomodernist.org.

[66] Pope Francis, *Praise Be to You – Laudato Sí: On Care for Our Common Home* (San Francisco, CA: Ignatius Press, 2015).

basic principles, priorities, and practices for governing the global environment is therefore likely to remain elusive. Given the highly contested social and political aspects of contemporary sustainability politics, guidance, rather than technocratic forms of governance, is perhaps a more reasonable expectation that scientists and the results of their research can be called upon to provide.

As it were, that was precisely how Planetary Boundaries 2.0 would be framed, with an updated version of the framework setting the stage for the turning point year of 2015. In the interim period since the original Planetary Boundaries publications in 2009, scholars associated with the SRC had produced a body of work intended to demonstrate the utility of the framework as an instrument for illuminating, conceptualizing and supporting environmental governance imperatives.[67] Planetary Boundaries would also underpin British economist Kate Raworth's acclaimed doughnut economics model, introduced in 2012, which combined the framework's Earth System limits with a set of social boundaries to specify "a safe and just operating space for humanity."[68] Other initiatives, such as the Future Earth-funded "Seeds of a Good Anthropocene" project based at the SRC, sought to leverage science in real-world situations at the interface of environment and development. The network of Stockholm-affiliated experts also brought Earth System science to bear on international political processes by organizing the 2012 Planet Under Pressure Conference in London. The event, held under the auspices of the IGBP and the Earth System Science Partnership, was timed to contribute to the Rio+20 sustainability summit, where the new global agreement – the eventual Sustainable Development Goals of Agenda 2030 – to succeed the Millennium Development Goals would be at the top of the agenda.

67 See, for example, an array of articles in special issues of the journals *Ecological Economics*, 81(2012), and *Current Opinion in Environmental Sustainability*, 4(2012), issue 1.

68 Kate Raworth, "A Safe and Just Operating Space for Humanity: Can We Live Within the Doughnut?," *Oxfam Discussion Papers*, 2012. Raworth's model contains clear echoes of Barbara Ward's early 1970s conceptualization of "inner limits" of fundamental human needs and "outer limits" of the Earth's physical integrity. See, for example, the Ward-written Coyococ Declaration of October 1974, reprinted in the *Journal of the Atomic Scientists* 31(1975):3. The inner/outer limits conceptualization was also advanced by key Cocoyoc (and Stockholm Conference) participants Maurice Strong and Marc Nerfin in conjunction with an international conference organized by the Dag Hammarskjöld Foundation in Täljövik, Sweden, in June 1974. Dag Hammarskjöld Foundation, "50 Years Dag Hammarskjöld Foundation," *Development Dialogue* 60(2012).

By the beginning of 2015, as anticipation and anxiety mounted over the upcoming COP20 meeting in Paris and with the UN General Assembly scheduled to decide upon the Post-2015 Development Agenda that had emerged in Rio in 2012, a window of opportunity had opened. The time was ripe for providing additional precision on the contours of humanity's safe operating space, and for contributing updated guidance on how to remain inside it. Within a week of the January 2015 publication of "Planetary Boundaries: Guiding Human Development on a Changing Planet" in the journal *Science*, the new version of the framework would make its debut beyond scientific circles before global elites at the World Economic Forum in Davos.

8

"Listen to the Scientists"

Yes, But What Is the Message?

A week after the January 2015 publication of the new Planetary Boundaries article in *Science*, the updated version of the framework made its public debut before the global financial and political elites that had descended upon Davos for that year's installment of the World Economic Forum (WEF). The aristocratic resort of Davos has long been internationally known as a site of mystique and the worship of health and wealth, and the setting at Schatzalp of the European sanatorium overlooking the village below in Thomas Mann's novel *The Magic Mountain* (1924) that won him a Nobel Prize. A century later, the small town nestled in the Swiss Alps is primarily known as a playground of elites and mega-influencers, and the home of the WEF, arguably one of the central players of the world in its globalized era – and of the small subspecies of Homo sapiens that has been baptized "Davos Man."[1]

Aiming high in terms of potential political impact, and surfing on previous waves of success in both its science and its communication, the Stockholm Resilience Centre (SRC) hosted no less than seven sessions at the 2015 WEF meeting in Davos. Representatives of the SRC were there to promote Planetary Boundaries 2.0 as well as a popular science book by Johan Rockström (Figure 8.1) and the renowned Swedish nature photographer Mattias Klum. Rockström, Klum and the SRC framed *Big World, Small Planet: Abundance within Planetary Boundaries* as a contribution to a "new narrative of world growth within Earth's limits."[2] The wider

[1] Peter S. Goodman, *Davos Man: How the Billionaires Devoured the World* (New York: Custom House, 2022).

[2] Johan Rockström & Mattias Klum, *Big World, Small Planet: Abundance within Planetary Boundaries* (Stockholm: Max Ström, 2015). www.stockholmresilience

FIGURE 8.1 Johan Rockström at the 2008 Tällberg Forum, just over a year before the publication in *Nature* of the first Planetary Boundaries paper, for which he was the lead author. At the time director of the Stockholm Environment Institute and co-director, with Carl Folke, of the SRC, Rockström went on to lead the Potsdam Institute for Climate Impact Research and, as a "meta-specialist" on a wide range of environmental issues, has become a prominent public voice for sustainable development. Photo: Courtesy of the Tällberg Foundation.

context was the upcoming Paris climate conference and the adoption of the UN Sustainable Development Goals (SDGs), scheduled for later in 2015, "the most important year of environmental decisions in decades." Like the conceptual framework it drew upon, *Big World, Small Planet* combined scientific data with the deployment of vivid, often tropical, nature imagery – this time in the form of Klum's photography – in evoking a planetary human environment. Despite having been pushed to the brink of critical boundaries for safe habitation, this environment still held the hope of rehabilitation and a return to a Holocene-like state of stability.

Rockström and Klum also employed a variety of by then well-established SRC themes and concepts straddling science and governance, including "planetary stewardship", "reconnecting to the biosphere", and the idea of a "good Anthropocene." The latter was an ostensibly attainable outcome,

.org/publications/publications/2016-05-06-big-world-small-planet-abundance-within-planetary-boundaries.html

as implied by the book's quasi-Cornucopian title. The word "abundance" also echoed contemporary works of techno-optimism, signaling a more Lynasian than Hamiltonian outlook on the prospect of expansion for the human enterprise on a planet that was certainly not without limits, but that remaining within them could offer more to many, provided it was done in the right way. According to Rockström and Klum, "Abundance within planetary boundaries requires a deep mind-shift. Not growth without limits. Not limits to growth, but growth within limits."[3]

PLANETARY METAPHYSICS?

The outreach efforts at Davos and beyond had a background. The massive impacts and lively debates that followed the 2009 Planetary Boundaries paper turned the PB research initiative into a more or less permanent field of activity at the SRC. Soon after the article's publication, Sarah Cornell, a British marine and atmospheric chemist, was hired by the SRC to coordinate Planetary Boundaries-related research, which became a de facto line of research in its own right. An early area of focus was to translate the PB framework into potential policy practices, again taking advantage of Stockholm's convening power to organize workshops and launch multiple-author papers, occasionally bringing the *Nature* article co-authors into the scheme. A great deal of additional research was pursued in Stockholm to refine the work, make the quantifications more robust, and introduce data for possible new boundaries that were not part of the original framework.[4] After five years, the time was considered right to take a major step forward and publish a new, comprehensive version of the framework. This initiative, including the publication and the thinking behind the updated version of planetary

[3] Also emblematically cited on the SRC website, www.stockholmresilience.org/publications/publications/2016-05-06-big-world-small-planet-abundance-within-planetary-boundaries.html (accessed July 21, 2024).

[4] Several papers have followed up on different aspects on Planetary Boundaries, often connected to global environmental governance, for example, Victor Galaz, Frank Biermann, Carl Folke, Måns Nilsson & Per Olsson, "Global Environmental Governance and Planetary Boundaries: An Introduction," *Ecological Economics* 81(2012):1, 1–3; Tiina Häyhä, Paul L. Lucas, Detlef P. van Vuuren, Sarah E. Cornell & Holger Hoff, "From Planetary Boundaries to National Fair Shares of the Global Safe Operating Space – How Can the Scales be Bridged?," *Global Environmental Change* 40(2016), 60–72; D. Collste, S. Cornell, J. Randers, J. Rockström & P. Stoknes, "Human Well-being in the Anthropocene: Limits to Growth," *Global Sustainability* 4(2021), E30. https://doi.org/10.1017/sus.2021.26. Sarah Cornell on Planetary Boundaries +10: www.stockholmresilience.org/research/research-news/2019-11-06-a-scientific-story-that-matters.html

boundaries, reflected many important aspects of the ongoing shifts in environmental governance, both before and after the 2009 *Nature* piece.

Coupling evocative concepts with big data and powerful imagery has proven to be an effective science communication strategy as demonstrated by the extraordinary diffusion of the Planetary Boundaries framework, not only in scientific journals but also in the lexicon of environmental politics and activism. The release of the 2015 edition of the framework coincided with the publication of another scientific article that, like the iconic Planetary Boundaries diagram, encompassed dramatic, data-derived visualizations of humanity's mounting impact on the Earth system. Named after a concept that emerged at a 2005 conference on the history of the human–environment relationship in Dahlem, Berlin, a revised version of the so-called Great Acceleration graphs[5] was published in *The Anthropocene Review* the day after the Planetary Boundaries paper appeared in *Science*.

In "The Trajectory of the Anthropocene: The Great Acceleration," two sets of twelve hockey stick-style graphs depicted an array of socio-economic drivers. They included population, real GDP, energy consumption – and selected Earth system disruptions – atmospheric CO_2 concentration, ozone depletion, and ocean acidification – that had been rapidly rising in virtual lockstep with one another during the decades following World War II. That period, which some consider co-temporaneous with the onset of the Anthropocene, was dubbed The Great Acceleration by Dahlem participants John McNeill, Paul Crutzen, and Will Steffen in their 2007 article in *Ambio* (Chapter 7) where the concept first appeared in print. It has since become a key conceptual tool for explaining and periodizing the recent past – as well as assigning blame for environmental crisis – among members of the global change community.

Will Steffen (Figure 8.2), an SRC fellow and former director of the International Geosphere Biosphere Programme (IGBP), was the lead author of both January 2015 articles that updated the Planetary Boundaries framework and the Great Acceleration graphs – the latter described as a "planetary dashboard" by the heuristically attuned science communicators at the SRC. Two of the other co-authors of the Great Acceleration article were Stockholm-based expatriates who, like Steffen, held long-term affiliations with the IGBP. Marine chemist Wendy Broadgate was deputy director of the Programme for fifteen years before becoming founding director of the global hub in Stockholm for Future

[5] These had originally been published in a slightly different version in Steffen et al., *Global Change and the Earth System*, 132–133.

FIGURE 8.2 Climate scientist Will Steffen (right) on stage in the Beijer Hall at the Royal Swedish Academy of Sciences during the tenth anniversary celebration of the Planetary Boundaries framework in 2019. Executive director of the Stockholm-based IGBP from 1998 to 2004 and a world-class science communicator, Steffen, who passed away in 2023, was instrumental in the development and popularization of Earth system science and the Anthropocene concept: He was also the lead author of the second Planetary Boundaries article, published in *Science* in 2015. Also depicted, left to right: Line Gordon, Belinda Reyers, and Johan Rockström. Photo: Courtesy of the Stockholm Resilience Centre.

Earth, which in 2015 supplanted the IGBP and the other member organizations of the Earth System Science Partnership. This institution, too, found its home in the Academy of Sciences building adjacent to the Stockholm University campus.

Owen Gaffney was the communications director at IGBP and went on to lead the Resilience Centre's international media strategies, while also holding an affiliation with Future Earth. His dual role – mirroring in a sense the public engagement of scientists such as Johan Rockström – as a media expert and global sustainability analyst reflects the high degree of integration between science and communications at leading Stockholm institutions like the SRC and IGBP. Some of Gaffney's other activities and affiliations can, moreover, provide additional insight into the ways in which scientific products have been conceived, perhaps even packaged, for communication purposes. He is for instance an associate of the non-profit organization Globaïa that specializes in data visualizations combining art, science, and narrative – including concepts like the Anthropocene, the Ecosphere, and Big History – to "inspire a 'big picture' understanding of our home planet." Through its connections to Stockholm, Globaïa was entrusted to produce the opening "Welcome to the Anthropocene" film for the 2012 Rio+20 Conference and provided data visualizations for the Netflix documentary "Breaking Boundaries" (2021) based on the Planetary Boundaries framework. Gaffney also co-authored its accompanying volume with the same title (see Chapter 7).

Indicative of a decidedly positive outlook on high-technology solutions to sustainability challenges among some key members of the Stockholm network, Gaffney is also on the faculty of Singularity University. This institution, not precisely a "university" as usually understood but rather a global think tank, was founded by the prominent futurist Ray Kurzweil and the engineer-entrepreneur Peter Diamandis. The latter is also well known for being the co-author of *Abundance: The Future Is Better Than You Think*, a 2012 bestseller on the potential of exponential technologies such as artificial intelligence and quantum computing for solving many of humanity's most intractable problems.

It could also be expected to find this eco-modernist line of exegesis percolating in the wake of Planetary Boundaries thought. As we have seen, Mark Lynas had already preached the eco-modernist gospel in *The God Species* (2011). What was needed to deal with the Anthropocene challenges was not less technology or less modern civilization, less urbanization, less nuclear power, or, for that matter, less energy – just smarter versions of all of the above and more in order to create a cybernetic super-sphere operated by a much more capable and well-equipped humankind. This kind of futurist thinking was far from novel – it has roots in Francis Bacon's *Solomon House* utopia (1625), or in Charles Fourier's super-rational *phalanstères* (1816) – the word itself a mix between the French words *phalang*, a long military flank, and *monastère*, a monastery, perhaps a congenial place for finally escaping the ordinary world of contradictions and conflict. Or, for that matter, in utopias by Edward Bellamy and H. G. Wells in the decades around the turn of the twentieth century.[6] More recently, the Cold War version of technological ultra-optimism was suggested by the wizard mathematician John von Neumann at the Institute for Advanced Study in Princeton. He spectacularly argued that global warming could be increased by reducing

[6] Frank E. Manuel & Fritzie P. Manuel, *Utopian Thought in the Western World* (Cambridge, MA: Harvard University Press, 1980).

Arctic ice sheets with a combination of nuclear blasts and soot. There may be downsides – "to make a new ice age in order to annoy others, or a new tropical, 'interglacial' age in order to please everybody, is not necessarily a rational program" – but all stones must be turned.[7] Certainly Promethean, not at all Soterian.

ENTER GAIA

There was a metaphysical ring to much of these future anticipations, perhaps to be expected in Early Modern thought, but still lingering on in the era of "the environment." It did not go away just because science had become more solid and numbers introduced. On the contrary, after initial work in the 1960s, James Lovelock, who died at the age of 103 in 2022, proposed his hypothesis of a self-regulating "Gaia," named after the Greek goddess of the Earth, in the early 1970s. Based on ideas developed over a long career as a missile scientist, thinking deeply about feedback loops in ballistic trajectories, Lovelock, in collaboration with biologist Lynn Margulis, author of *The Symbiotic Earth* (1998), conceived of the entire planet as a huge living organism, evolving under Darwinian laws.[8] Their first common piece was published in 1974 in *Tellus*, Carl-Gustaf Rossby's Stockholm-based meteorology journal that had expanded its remit to also encompass broader ideas of biogeochemistry.[9] On this Earth, humans were just another species, no more or less special than termites, and Gaia would simply shrug them off if they were threatening her long-term survival.

Gaia thought is vast and complex and has strong scientific strands, but it also includes streaks of teleology in the eyes of many of its critics over the years, of whom Richard Dawkins, the Oxford biologist, and Paul Crutzen have been the most vocal. There were also many well-respected defenders of the idea, among them UK physicist Tim Lenton, a PB

[7] Jacob David Hamblin, *Arming Mother Nature: The Birth of Catastrophic Environmentalism* (New York: Oxford University Press, 2013). John von Neumann, "Can We Survive Technology?," *Fortune* (June 1955), reprinted in von Neumann, *Collected Works*, ed., A. Taub, vol. 6, *Theory of Games, Astrophysics, Hydrodynamics and Meteorology* (New York: MacMillan, 1963), 504–519, on 513.

[8] Lynn Margulis, *Symbiotic Planet: A New Look at Evolution* (London: Weidenfeld & Nicolson, 1998).

[9] James E. Lovelock, "Gaia as Seen through the Atmosphere," *Atmospheric Environment* 6(1972):8, 579–580. James E. Lovelock & Lynn Margulis, "Atmospheric Homeostasis by and for the Biosphere: The Gaia Hypothesis," *Tellus*. Series A. Stockholm: International Meteorological Institute 26(1974):1–2, 2–10.

co-author who had specialized in research on tipping points and planetary equilibrium models.[10] In response to the critique, Lovelock and Margulis sharpened their thinking and revised the Gaia concept in several publications, with Lovelock emphasizing the hypothesis framing of the Gaia concept in 1979 and subsequently shifting towards a considerably more mathematized "theory" formulation in 1990.[11]

Central to Gaia thinking are ideas of homeostasis, a wonderful equilibrium that can be repeated on ever higher levels of complexity and self-awareness, or spirituality. Earlier expressions of the idea can be found in Russian scientific thought, interestingly by proto-Anthropocene biogeochemist Vladimir Vernadsky, and similar works, as outlined at that time in a book by Helan Jaworski, *Le Géon ou la Terre vivante* (1928); the Earth was a living organism, a *Terre vivante*.[12] In an ultimate instance, this line of thought can be traced back to versions of idealism and not least ideas presented by spiritual Russian visionary Anna Helena Blavatsky, founder of the International Theosophical Society in 1875. A far cry from science-based global environmental governance, it may seem, but big, often controversial, ideas have a tendency of "always creeping in," as the French historian Lucien Febvre put it in his very early and insightful analysis of the concept of "environment," *La Terre et l'évolution humaine: Introduction géographique à l'histoire* (1922), translated as *A Geographical Introduction to History* (1925).[13] For Febvre, environment meant freedom, in contrast to the predominant geographical determinism of the day, where environment typically meant hard natural barriers, a steering force for both individual humans and entire societies.[14]

Perhaps it is also precisely in philosophical and political concepts such as "freedom" and "determinism" that we should seek the roots of the

[10] T. M. Lenton & H. T. P. Williams, "On the Origin of Planetary-scale Tipping Points," *Trends in Ecology and Evolution* 28(2013):7, 380–382.

[11] Tim Lenton, "Gaia," In: Steffen et al., eds., *Global Change and the Earth System* (2004), 70–71. A more critical version was also published, in the same IGBP volume, likely to indicate that the Gaia controversy was far from resolved. Paul Crutzen, "Anti-Gaia," *ibid.*, on 72.

[12] Helan Jaworski, *Le Géon ou la Terre vivante* (Paris: Librairie Gallimard, 1928).

[13] The full sentence is worth quoting: "Between man and his natural environment, ideas are always creeping in and intervening." Lucien Febvre, *A Geographical Introduction to History*, The History of Civilization Series (1925), new ed. (London: Kegan Paul, Trench, Trubner & Co., 1932), on 363.

[14] On Febvre and his anti-determinism, Sverker Sörlin, "Environmental Times: Historical and Scientific Temporalities from *Annales* to Anthropocene, ca 1920–2020," In: Anders Ekström & Staffan Bergwik, eds., *Times of History, Times of Nature: Temporalization and the Limits of Modern Knowledge* (New York: Berghahn, 2022), 64–101.

protracted fighting over Planetary Boundaries. What the eco-modernists wanted was actually not the boundaries. They by and large sided with the critical opinion that rose up against the *Limits to Growth* report from the Club of Rome in 1972 – one of several earlier attempts to introduce limits into the politics of resources and environment and arguably the most famous. *Limits* was distinctly neo-Malthusian in its perspective and was heavily criticized by economists for disregarding the inherent capacity of economies to innovate, increase productivity, and thereby improve resource efficiency. The following decades seemed to justify their view. Wealth kept growing and resource scarcity did not materialize. But the economists were only partly right.[15] On the planetary scale, impacts did in fact grow. Hence the Anthropocene, hence the unbroken Keeling curve of CO_2 emissions, and hence the volatility and geopolitical criticality of oil, gas, and strategic minerals, involved in literally every major armed conflict on earth. Hence, the massive spread of natural resource extractivism during the Great Acceleration, which propelled strong linkages to the rapidly expanding financial sector, constantly on the lookout for investment objects upon which to build capital.[16] The eco-modernist "good Anthropocene" was essentially a new version of the 1970s anti-Limits to Growth argument. Let technological capitalism reign, and it will once again make sure that we do not need to restrict economic growth, reduce consumption, or temper consumerism.

This is an argument that requires a great many models and mathematics as well as some reassuring guesswork. It nonetheless risks being unconvincing. It is more effective rhetorically to provide a consistent narrative that shows how the trade-off between resource depletion and wealth production is constantly taking humanity to a better position. Ideas are needed. Eco-modernism itself is an ideology that feeds upon the seeming absoluteness that the Planetary Boundaries provide. The boundaries are not political products; they are, in a sense, God-given, properties of the Earth, and to observe them makes a lot of sense. Even more so for what follows: the freedom of capitalism to operate within these boundaries – the "level playing field," a recurring turn of phrase in the communications of some, but not all, of the PB spokespersons.

[15] For an insightful review of the scarcity concept, see Fredrik Albritton Jonsson & Carl Wennerlind, *Scarcity: A History from the Origins of Capitalism to the Climate Crisis* (Cambridge, MA: Harvard University Press, 2023).

[16] Martín Arboleda, *Planetary Mine: Territories of Extraction under Late Capitalism* (London & New York: Verso, 2020).

THE NEOLIBERAL SPECTER

Another strand of the PB debate used the very same idea to critique the framework. The boundaries may be reasonable and scientifically sound, yet for some – including the green movement, climate activists, many environmental scientists as well as some scholars in the social sciences and humanities, ethicists and theologians – it was the lack of a level playing field that was the problem. This entailed the same old Darwinist view of competition where the richer countries, and in each country, the wealthy, could continue to take advantage of their privileges and skew politics in their favour. The level playing field is a metaphor, and also an idea. In reality, it can be defined in many ways. The level playing field was a way to suggest that inside the boundaries, societies were free to choose. In fact, the global community was also free to allocate operating space to actors, including states and firms. But the critics in this camp, politically more to the left or "liberal" in US parlance (eco-modernists would typically lean to the right, neoliberal or conservative), were skeptical. Would this be governance with enough clout, they asked? Would it support directionality, or would it just provide a green light for the same actors – states, firms, technologies, fuels, and resources – that had created all the problems? What guarantees were there that these actors would actually change course or be ejected from the competition?

When PB2 was being prepared, this embrace from the right and critique from the left had sunk in among the core team in Stockholm and their associates around the world. They may have perceived the debate differently, with some of the authors engaging in technical and scientific debates surrounding individual boundaries and their scientific foundation. That was one line of defense. But the other, and potentially more problematic issue, was over the politics of the boundary concept. The fact that the Global South and the developing countries were skeptical of the libertarian/Darwinist slant of the approach was a drawback. It didn't help that both Rockström and Gaffney, and other PB advocates, signed appellations to include limits thinking within the UN SDGs.[17] Will Steffen, the lead author of the follow-up article in *Science*, took the critique seriously enough to engage in an in-depth revision. The *Science* article included a section on equity issues, or rather the lack of them in the framework. It was also now carefully stated that the

[17] D. Griggs, M. Stafford-Smith, O. Gaffney, et al., "Sustainable Development Goals for People and Planet," *Nature* 495(2013), 305–307.

framework did not offer any advice on how to reach the goals. The following excerpt could be seen as a wider framing and a caveat, but also as a signal of a desire to escape from the ideological stalemate that Planetary Boundaries found itself in.

> The PB approach is embedded in this emerging social context, but it does not suggest how to maneuver within the safe operating space in the quest for global sustainability. For example, the PB framework does not as yet account for the regional distribution of the impact or its historical patterns. Nor does the PB framework take into account the deeper issues of equity and causation. The current levels of the boundary processes, and the transgressions of boundaries that have already occurred, are unevenly caused by different human societies and different social groups. The wealth benefits that these transgressions have brought are also unevenly distributed socially and geographically. It is easy to foresee that uneven distribution of causation and benefits will continue, and these differentials must surely be addressed for a Holocene-like Earth-system state to be successfully legitimated and maintained. However, the PB framework as currently construed provides no guidance as to how this may be achieved [although some potential synergies have been noted (54)], and it cannot readily be used to make choices between pathways for piecemeal maneuvering within the safe operating space or more radical shifts of global governance (93).[18]

This move was probably necessary to build and maintain trust among the large portions of key audiences for which equity issues and distributional politics were of major concern. That Will Steffen was keen to explore these issues he had already demonstrated by co-authoring a paper with Mark Stafford-Smith, an Australian behavioral ecologist, on the relationship between PB and equity, arguing that the rich countries in fact had much to win by increasing global equity. Stafford-Smith had been vice chair of the IGBP Scientific Committee and also co-chair of the 2012 *Planet Under Pressure: New Knowledge Towards Solutions* conference on global environmental change during the lead up to Rio+20 in 2012. These were contexts that Steffen too worked within. It was also

[18] Steffen et al., "Planetary Boundaries." Reference 93 in the quote is a piece that had been published the year before by humanities and social science scholars with a constructive critical approach to PB ideas, emphasizing especially the need to bring society and equity into any scheme of the scale and scope of Earth system boundaries, Noel Castree, W. M. Adams, J. Barry, D. Brockington, B. Büscher, E. Corbera, D. Demeritt, R. Duffy, U. Felt, K. Neves, P. Newell, L. Pellizzoni, K. Rigby, P. Robbins, L. Robin, D. B. Rose, A. Ross, D. Schlosberg, S. Sörlin, P. West, M. Whitehead & B. Wynne, "Changing the Intellectual Climate," *Nature Climate Change* 4(2014):9, 763–768. In this piece, too, there were Stockholm based members of the team of contributors (Robin, also at ANU in Canberra, and Sörlin).

in the same period that Oxford economist Kate Raworth's "Doughnut Economics" ideas began penetrating the PB framework, suggesting that social and economic sustainability should be given due attention in the struggle taking place within the boundaries.[19]

This debate may seem slightly off the mark; wasn't it supposed to be about saving the planet? If so why all this tinkering on the sidelines? Other aspects became so central because they were directly related to the position of the "governance" in environmental governance. The Stockholm tradition had always been about keeping science at the core, just as the PB framework insisted. So far so good. But it was also a Stockholm tradition to deliver advice for policy and for the common good. This entailed turning to politics and society, and raising a call for action, or at least an invitation to a discussion about what could be done about the problem at hand. In this regard, PB was a different kind of animal. It shifted the focus to an entirely new arena: a commercial one, and possibly a civic one. It was no longer a consensus that was sought. Rather the opposite. As the 2015 *Science* article honestly stated: "... the PB framework does not as yet account for the regional distribution of the impact or its historical patterns. Nor does the PB framework take into account the deeper issues of equity and causation."[20] This may also explain why the developing world remained unconvinced, and why they in the end vetoed the proposed inclusion of PB and Anthropocene-related language in the UN SDGs.

In that respect, the 2015 *Science* article served a function of reconciliation and modification. Partly, the issue was one of communication. PB was a vague enough framework as to be presented in very different ways. The "level playing field" language was one that signaled a business approach, but perhaps also more generally a neoliberal perspective. Neoliberalism isn't often talked about in the context of sustainability and environmental protection. That is a mistake. Just like in almost any other policy area in the last two decades of the twentieth century, environment and climate were also affected by these ideas.

Since the 1980s, a range of literature has arrived proposing new measures in environmental protection. Many of the mechanisms were derived from economics: the polluter pays principle, road tolls and congestion fees

[19] W. Steffen & M. Stafford Smith, "Planetary Boundaries, Equity and Global Sustainability: Why Wealthy Countries Could Benefit from More Equity," *Current Opinion in Environmental Sustainability* 5(2013):3–4, 403–408.

[20] Steffen et al., "Planetary Boundaries," on 743.

to reduce traffic, public-private partnerships to share costs and increase efficiency in infrastructure, carbon offsetting schemes, monetarized ecosystem services, market-based certification schemes, economic compensation to species-rich countries for not cutting down rainforests, and many others. In structure and logic, these ideas and steering instruments can be described as a set of specialized applications *of* the basic tenets of market liberalism *on* the environmental realm, where in the past civic engagement and political intervention had been the dominating modes of action, in both cases typically with strong links to science. In Sweden, such tying together of actors and interests had suited the Swedish model well.

There is no comprehensive history of the neo-liberalization of environment, just as there isn't any proper, full-scale history of environmental governance.[21] But we must surely consider the former as a part of the latter, alongside more state-centric and less market-oriented versions of environmental governance. Lacking a standard version of this history, we can still look back upon the Dutch political scientist Maarten Hajer's very influential book from 1995, *The Politics of Environmental Discourse: Ecological Modernization and the Policy Process*. It includes a brief historical account of earlier versions of environmental governance and is worth revisiting, since it was for several years the most cited work that at the same time both launched and attempted to explain why and how environmental politics are also a politics of modernization. It explains convincingly what a major shift it was that began in earnest in the 1980s. A core element of the book is the view of "ecological modernization" as a new way of talking about environmental challenges. Language is important, not simply a choice of words, but as clearly reflecting a different way of thinking about environmental challenges. Above all, Hajer did not pit the environment against economic growth, development, or capitalism, but rather put forward the de facto argument that environmental management can be seen as a "positive sum game."

[21] Which does not exclude that several modern global and regional environmental histories invoke neoliberalism as a key factor in environmental policy and politics. Some good recent examples are Mark Stoll, *Profit: An Environmental History of Capitalism* (Cambridge: Polity, 2023) and Carl A. Zimring, *Clean and White: A History of Environmental Racism* (New York: New York University Press, 2015). Geographers debated the neoliberalization of nature since the beginning of the 2000s, for example, Noel Castree, "Neoliberalising Nature: The Logics of Deregulation and Reregulation," *Environment and Planning A* 40(2008):1, 131–152. Karen Bakker, "The Limits of 'Neoliberal Natures': Debating Green Neoliberalism," *Progress in Human Geography* 34(2010):6, 715–735.

Hajer devoted a full chapter to the historiography of this line of thought. He depicted a breakthrough moment in the early 1970s when the environment gained traction with the 1972 Stockholm Conference and the simultaneous experience of two emblematic publications that year, the Club of Rome's *Limits to Growth* report, and *Blueprint for Survival*, juxtaposed in Hajer's narrative as respectively elitist and modernist (LTG), and radical with streaks of civilizational critique (BFS). He suggested that in the 1970s and 1980s, these positions converged under the influence of major institutions like the Organisation for Economic Co-operation and Development (OECD) and the United Nations Environment Programme (UNEP) taking up the modernist paradigm.[22] *The World Conservation Strategy* (1980) and, above all, the OECD International Conference on Environment and Economics of June 1984 are presented as milestones in this evolution. The WCS report was the joint product of "moderate NGOs," such as the International Union for the Conservation of Nature and the World Wildlife Fund, together with UNEP, and was written in collaboration with the FAO and UNESCO. The major conclusion of the 1984 OECD Conference was that "the environment and the economy, if properly managed, are mutually reinforcing; and are supportive of and supported by technological innovation."[23] A characteristic summary statement in Hajer's presentation reads thus: "The positive-sum game format that became characteristic of ecological modernization was simply born out of necessity: for environmental policies to survive they needed to work with the grain of the time. What is more, environmental politics was not only positioned as non-contradictory to economic policy, it was also suggested as a potential instrument for economic recovery."[24]

The convergence, Hajer argued, was further facilitated by the fact that both 1972 reports departed from the same gloomy computerized predictions, and also the fact that both sides in this modernization debate had concrete ideas and solutions that gained ground in policy.

[22] On OECD, see Matthias Schmelzer, "'Born in the Corridors of the OECD': The Forgotten Origins of the Club of Rome, Transnational Networks, and the 1970s in Global History," *Journal of Global History* 12(2017):1, 26–48 and *The Hegemony of Growth. The OECD and the Making of the Economic Growth Paradigm* (Cambridge: Cambridge University Press, 2016). On UNEP, Maria Ivanova, *The Untold Story of the World's Leading Environmental Institution: UNEP at Fifty* (Cambridge, MA: MIT Press, 2021).

[23] Organisation for Economic Co-operation and Development, *Environment and Economics: Results of the International Conference on Environment and Economics, 18th–21st June, 1984* (Paris: OECD, 1985), 1.

[24] Maarten Hajer, *The Politics of Environmental Discourse: Ecological Modernization and the Policy Process* (Oxford: Oxford University Press, 1995), quote on 97.

The Blueprinters argued for public transport, smarter technologies, and an ecotax, all ideas with a future. The Club of Rome argued, back then, that in the ultimate instance, continued growth and consumerism were necessary; it was a matter of avoiding the negative consequences. The World Commission on Environment and Development (the Brundtland Commission), whose report *Our Common Future* appeared in 1987, crowned this evolution of compromises with a politically convenient conceptualization of how to save the environment with continued development, including for the developing countries.

This third way, *so zu sagen*, of avoiding the dilemma of having to choose between a safe and healthy planet and a prosperous future for mankind comes across as the central feature of Hajer's version of environmental modernization. He does not talk about neoliberalism at all. Ideologically, he instead locates this line of modernization in a reformist, social-democratic tradition of crisis solving and responsible progressivist thought. He mentions mid-twentieth-century economists such as John Maynard Keynes and the welfare- and equity-oriented Dutch Nobel Laureate Jan Tinbergen as influential background thinkers, and draws deeper intellectual roots to nineteenth-century economic and environmental philosophers John Stuart Mill and George Perkins Marsh. He also cites the multiple North-South development commissions and reports from the late 1970s and early 1980s, chaired by social democrats like Willy Brandt and Olof Palme to which Brundtland and her 1987 report can be seen as a successor. It may be mentioned that Hajer worked within the world of Dutch politics, had government commissions, and co-authored a political program for the Social Democratic Party.[25]

It is hard not to notice the incompleteness of Hajer's historiography, and writing within the framework of a PhD dissertation at Oxford University in the early 1990s obviously set some limits for what was possible. Several lacunae stand out. The first is the rise of neoliberal policy instruments, not least in economics, that receive no attention from Hajer. Likewise, ideas and results from Earth Systems science are entirely absent, unsurprisingly, since it was early in their ascent at the time Hajer was writing. Still, his insightful analysis shows that a middle-way trade-off between growth/development and environment was also considered possible, although as neoliberalism expanded, its toolbox of market-oriented instruments followed suit.

For Sweden, this analysis should be no surprise. There had always been an element of pragmatism to Swedish environmental policy. It can

[25] https://en.wikipedia.org/wiki/Maarten_Hajer.

be traced to debates on forestry in the nineteenth century and to heated discussions about water pollution around 1900. In both cases, the interests of industry were satisfied first, and for forestry, production-oriented conservation legislation was introduced in 1903. Nature reserves and other such restraints were kept at a minimum, except in high mountain areas above the tree line. The fact that national parks in Sweden established early and extensive in size was rather because the largest parks could be located beyond areas of productive forestry and agriculture.[26]

This pattern has been consistent throughout the twentieth century. Landmark decisions such as the repeated expansion of hydroelectric power, harnessing cherished river rapids, were continued up until the 1980s. Nuclear energy was expanded by parliament after a split referendum in 1980. The decision was to phase nuclear energy out by 2010, but over 60 percent of the capacity remains in operation well into the 2020s. The current plan is to keep nuclear power as a significant part of Sweden's energy output until mid-century, and several political parties argue that it should be re-expanded to facilitate the green transition. Right-wing parties in fact won the 2022 election on a distinctly pro-nuclear platform (albeit with the narrowest possible margin; more on this in Chapter 9). In an attempt to present "Environmental Dissonances in the Swedish Welfare State," two environmental historians have recently identified this pattern of erring on the side of least friction with industry and growth as being predominant in Sweden, and in this they are in line with earlier research (see also Chapter 2).[27]

EXTENDING PLANETARY BOUNDARIES – DOUGHNUT THINKING

This backdrop is essential to understand the evolution of the new influential turn of Stockholm-based environmental science and policy after 2000. The need to link scientific work with the new economic and political trends that prevailed under globalization was real, as was the

[26] Sverker Sörlin, *Framtidslandet: Debatten om Norrland och naturresurserna under det industriella genombrottet* [Land of the Future: The Debate on Norrland and Its Natural Resources at the Time of the Industrial Breakthrough] (Stockholm: Carlsson, 1988), new revised edition, *Framtidslandet* (Gothenburg: TEG Publishing, 2023). Lars J. Lundgren, *Vattenförorening: Debatten i Sverige 1890–1921* [Water Pollution: The Swedish Debate, 1890–1921] (Lund: Gleerup, 1974). Lars J. Lundgren, *Staten och naturen: 1869–1935* [State and Nature: 1869–1945], 2 vols. (Brottby: Kassandra, 2009–2010).

[27] Erland Mårald & Christer Nordlund, "Modern Nature for a Modern Nation: An Intellectual History of Environmental Dissonances in the Swedish Welfare State," *Environment and History* 26 (2020):4, 495–520.

increasing institutional density of think tanks and university-based centers that struggled to make their voices heard. In this era of stiffening competition for influence and resources, some groups and institutions opted to stay *intra muros* and get on with their scientific work. Those who wanted to remain interactive and influential, or were newcomers with such ambitions, had to comply with the rules of the game. The latter certainly applied for the SRC, which in that regard became not just a new arrival on the Swedish scene but also one with unusually high expectations to take on an outreach and policy-oriented position. This was surely facilitated by the fact that the Centre started out in alliance with two well-established players on the Swedish scene: the Academy of Sciences, which had made its own environmental turn in the early 1970s, and the Stockholm Environment Institute. The SRC was also from the beginning well connected with an array of international networks and institutions, including within the UN system, and was keen on rapidly expanding its reach into the realms of business and policy.

This is rewarding and demanding at the same time, and it requires several skills that conventional academic institutions do not always possess. The SRC did, or quickly developed them. In fact, it was one of the few points of critique in William Clark's 2009 start-up review (Chapter 7): that the SRC needed to find a strategy for its work with stakeholders and policy. Outreach would within a few years become an important aspect of the Centre's activities. With scientists working in close collaboration with media specialists in producing knowledge adapted for both scientific and public consumption, skillful science communication at the SRC and other Stockholm institutions has been a significant factor in the popularization of concepts such as Planetary Boundaries, the Great Acceleration, and the Anthropocene. What is more, the emphasis on visualizations has not only made the scientific and policy-oriented messages contained in these and other concepts more media-friendly and accessible to non-specialist audiences, but it has even also facilitated new synthetic paradigms that have expanded their applicability to other disciplines and societal sectors.

Developments surrounding PB played out internationally, which is not to say that other Swedish scientists or policy actors were silent. All along, there was significant research taking place at universities and institutes across the country, some of it with considerable influence. The physical resource theory group at Chalmers University of Technology, linked in turn to several other groups at Gothenburg University (economics, climate science), was among those. Other important natural science research environments engaged with PB included the ecology departments at the universities of Lund, Uppsala, and Umeå. Political scientists contributed, with Stockholm University as a key player. Work in history and other strands of the humanities was taking shape at the KTH Environmental Humanities Laboratory in Stockholm. A stronghold for climate-oriented social science was established at Linköping University. If anything, environment and climate gained academic ground in many institutions, but did so in increasingly diverse ways. Since "environment" started to appear on university curricula in the 1960s, the academic setting has become more uniform, and almost exclusively disciplinary-based. In the new century, academic specializations and departmental boundaries have meant increasingly less. Issue orientation and capacity for dialogue to raise institutional and economic support from *extra muros* have become ever more important. Unavoidably, what it means to be a Swedish or a Stockholm voice in the world of global environmental governance has not remained static. In fact, it could be argued that many institutions were not able to make that transition, while others did so successfully. Various actors have navigated these changes in society and academia differently, and with different degrees of impact.

The most prominent and impactful of the conceptual modifications to the PB idea has undoubtedly been "Doughnut Economics," a framework derived from the Planetary Boundaries diagram by the abovementioned Kate Raworth (Figure 8.3). The self-described "renegade economist" added a set of socio-economic boundaries to the environmental limits of Rockström and colleagues to create an aspirational operating space for humanity that was both safe and just. Planetary Boundaries has also in recent years converged with the Club of Rome's pioneering scenario analysis paradigm, the Limits to Growth, generally considered the first attempt to systematically model human–environment interactions at the global level by using the World3 computer model developed by researchers at MIT in the early 1970s. For the Club of Rome's fiftieth anniversary in 2018, the SRC – together with the Norwegian Business School's Jörgen Randers, one of the original co-authors of the 1972 LTG report – was commissioned to perform a scenario analysis employing a global system model called Earth3. Similar to the juxtaposition of Planetary Boundaries with the idea of abundance in *Big World, Small Planet*, the 2018 SRC-Club of Rome report regarded the nine boundaries as a set of non-negotiable environmental limits within which the seventeen UN SDGs of Agenda 2030 could be realized. Like Raworth's Doughnut Economics framework,

FIGURE 8.3 A strong critic of conventional economics, "renegade economist" Kate Raworth created the Doughnut Economics model that combines a set of human development indicators – the inner ring of the doughnut – with outer environmental limits determined by the Planetary Boundaries framework. The space in between is what she calls a safe and just space for humanity. Photo: Arbeid & Milieu/Flickr.

the Earth3 model links socio-economic and biophysical processes in mapping pathways for achieving the SDGs by 2030 and 2050 while remaining safely inside planetary boundaries.

The SRC-Club of Rome collaboration was an example of a growing engagement with structural economic issues and an emerging entente with institutions concerned with the future of global capitalism among some sustainability scholars associated with Stockholm. This came about during the tenure of Anders Wijkman as the Club of Rome's co-president between 2012 and 2018. A former politician and Secretary General of the Swedish Red Cross and the Swedish Society for Nature Conservation, Wijkman has since the 1970s been an influential voice in Sweden on issues related to the environment and sustainable development. In 2012, he co-authored with Johan Rockström a Report to the Club of Rome called *Bankrupting Nature: Denying Our Planetary Boundaries*. In the report, Wijkman and Rockström applied a Planetary Boundaries perspective to traditional Club concerns such as resource depletion and population growth, while calling for transforming the prevailing model of economic growth based on what they considered excessive consumption as measured by the environmentally suspect metric of Gross Domestic Product.

It was also during this period that Rockström became increasingly involved with the World Economic Forum, thus gaining further visibility that helped catapult him into the TIME Magazine 2023 list of the world's 100 most influential people, following Greta Thunberg's inclusion in 2019.[28] The WEF is another organization of economic and political elites that, like the Club of Rome, was founded in the years before the 1972 Stockholm Conference – a time when the global economic system that emerged during the postwar era began to be called into question, due in part to the mounting environmental impacts of the Great Acceleration. Recent research on the WEF has pointed to its combined role of think tank and ideological platform for a particular form of opinion-making. Not fully a view from nowhere, what can be considered a "Davos consensus" of financial and political elites typically favors free market-oriented agendas. But it also reflects an increasing engagement with contemporary concerns. By the 2010s, the Stockholm story extended even to this frosty and lofty place, high above the muddy terrain of ordinary politics.[29]

Rockström became a regular participant in the work of the World Economic Forum, a "voice of science" at the annual Davos meetings, and a recurring contributor to the WEF website. In a column posted in conjunction with the 2016 WEF, Rockström conditionally endorsed the concept of the Fourth Industrial Revolution – the theme of that year's Davos meeting. The idea of 4IR, conceived as a moment of historic transformation based on rapidly increasing interconnectivity and the integration of exponential technologies in all aspects of society, had been popularized in 2015 by WEF founder and Executive Chairman Klaus Schwab in an article in *Foreign Affairs*. Rockström's endorsement of 4IR was predicated on the possibility that the dramatic enhancement of human agency it entailed could be enacted equitably and sustainably. Planetary Boundaries was to serve as a benchmark for ensuring an ecologically resilient safe operating space for humanity in an Anthropocene world that was increasingly dominated by algorithms and an array of advanced technologies such as automation and the Internet of things.

[28] https://futureearth.org/2023/04/13/johan-rockstrom-named-in-time-100-most-influential-people-2023/. https://time.com/collection/100-most-influential-people-2019/5567758/greta-thunberg/ (accessed July 22, 2024).

[29] Christina Garsten & Adrienne Sörbom, *Discreet Power: How the World Economic Forum Shapes Market Agendas* (Stanford, CA: Stanford University Press, 2018). Goodman, *Davos Man*.

In *Breaking Boundaries*, the 2021 book by Rockström and Owen Gaffney that the Netflix documentary was based upon, a cutting-edge technology at the core of the 4IR concept – artificial intelligence – was classified as one of a number of "novel entities." According to the authors, these had been introduced into humanity's operating space and posed a potential risk to people and planet if not properly managed. Yet Gaffney and Rockström are in general sanguine on the critical role of technology, citing the fourth industrial revolution and calling the technological revolution of recent years a potentially positive tipping point that could countervail the trajectory toward catastrophic Earth system tipping points embarked upon during the Great Acceleration. In the chapter "Taming the Technosphere," they even explore a range of geoengineering options, describing measures such as cloud seeding, deploying giant sunshades, and atmospheric sulfur injection as extremely high-risk technological interventions – some of which, they argue, will likely have to be considered, perhaps a decade into the future, because of the unmanageable risks the Earth will soon face.

GRETA – AND "DAVOS MAN"

A less extreme way forward is put forth in the foreword of *Breaking Boundaries* by the Swedish climate activist Greta Thunberg (Figure 8.4). Her method for addressing the climate crisis centers on science communication: educating oneself on the issue and spreading knowledge to others in order to create a critical mass of awareness that will in turn lead to broad-based action. In her first in-person public address – at a climate action concert in Stockholm on September 7, 2018, three weeks into her famous school strike – Thunberg began her short speech by referring to Johan Rockström's warning that the world had to drastically lower the trajectory of CO_2 emissions by 2020 if the 2°C target of the Paris Agreement was to be met. A week later, as her international notoriety continued to expand exponentially, Thunberg posted an English-language video message on Twitter in which she strongly criticized Sweden for its outsized ecological footprint and invoked planetary boundaries in demanding that developed countries reduce their excessive consumption levels to keep the Earth within habitable limits.

Sweden's political class was in fact the initial target of Thunberg's intervention in the climate debate. Her public writing debut in May 2018 – several months before she sparked the Fridays for Future global youth movement – came in a debate article in the daily newspaper

FIGURE 8.4 Climate activist Greta Thunberg started her School Strike for Climate by the steps of the Swedish parliament on August 20, 2018. Following a summer of record-breaking heat and wildfires and three weeks before national elections in Sweden, her solo school strike soon became a global youth movement under the banner of Fridays for Future. Thunberg is a distant relative of Svante Arrhenius, the Swedish chemist who in 1896 calculated the effect of carbon dioxide emissions on global surface temperature. In demanding political action to combat climate change, Thunberg has consistently called for decisionmakers to "listen to the science," particularly that of the IPCC. Photo: Gustav Delhilage/Flickr.

Svenska Dagbladet. In the article, she condemned Swedish climate policy and challenged Sweden's self-conception, perhaps even national conceit, as a progressive and pioneering country (*föregångsland*) due to its high per capita carbon emissions. Several months later, in the lead up to the September 2018 national elections in Sweden, following a summer of record-breaking heatwaves, Thunberg on August 20 initiated without fanfare her school strike for the climate on the steps of the Swedish parliament.

Thunberg continued to champion climate science, particularly the work of the Intergovernmental Panel on Climate Change (IPCC), as her public profile as the world's leading climate activist soared during the autumn of 2018. Her perhaps most impactful speech was delivered at the September 2019 UN Climate Action Summit in New York, where she wielded IPCC data in admonishing world leaders for ignoring climate science and continuing to pursue business as usual pathways predicated on the idea of eternal economic growth. Several days earlier, in testimony to the US House Committee on the Global Climate Crisis, she had in lieu of a written statement submitted the 2018 IPCC *Special Report on Global Warming of 1.5°C*, telling the Committee, "I don't want you to listen to me, I want you to listen to the scientists." Her inclination to communicate science to power has been consistent across a range of high-profile gatherings of global elites. But, after all, what is it that the scientists say? Having come so far in this book, we have noted that their solutions can vary, and that they sometimes touch upon fundamental divides on values and political philosophy. What is it that "scientists" recommend politicians (ultimately: voters) to do?

In addition to polemical speeches at consecutive COP meetings and other climate events, Thunberg also followed Rockström's lead in speaking at the World Economic Forum in 2018, 2019, and 2020 for an audience that was unlikely to share Thunberg's commitment to climate justice and transformative change of global capitalism. Whether or not the inclusion of voices such as those of Thunberg and Rockström – both listed as WEF contributors on the Forum's website – signals a genuine shift in thinking among the Davos corporate and political elite is at this point unclear. Are the "Davos men," perhaps also some of its women, sincere in their oft-espoused allegiance to trendy "Environment, Social and Governance" ideals, or are they simply attempting to coopt critical perspectives on climate, environment, and sustainability issues and thereby sidestep transformative changes to prevailing economic structures? Whither Davos, whither the world?

OUR PLANET, OUR FUTURE – NOBEL LININGS

Publicly communicating science to a broad, if not elite, range of government and private sector stakeholders has, since approximately the turn of the millennium, become a core activity for a cadre of environmental experts connected to Stockholm. The epistemological commitments and conceptual repertoires of these activist Earth System and sustainability scientists have, in parallel, led to a predominantly planetary framing of the environmental governance challenges facing humanity as a whole in the near and longer-term future. Leveraging Sweden's arguably most prominent international brand in connecting science to policy imperatives, some leading members of the Stockholm network organized the first Nobel Prize Summit in 2021 to promote the idea of planetary stewardship and address what they identified as the interlinked "supranational crises" of climate, ecology, information, and inequality.

The "Our Planet, Our Future" summit was convened in April 2021 by the Nobel Foundation in association with the US National Academy of Science, the Stockholm Resilience Centre, the Beijer Institute, and the Potsdam Institute of Climate Impact Research, where Johan Rockström had several years earlier taken over as director after Hans Joachim Schellnhuber. Originally meant to take place in Washington, DC, the year before, the summit was rescheduled and moved online due to the coronavirus pandemic. The purpose of the meeting was to advance a transformative agenda at the outset of the 2020s, a decade the organizers considered critical for combating the climate crisis and implementing the UN SDGs. The three-day virtual event brought together Nobel laureates and other international experts, as well as representatives of business, government, and advocacy groups. It featured interventions by Al Gore, Mexican climate activist Xiye Bastida, and the Dalai Lama of Tibet.

The intellectual foundation for the summit took the form of a seventy-two-page white paper published by the Beijer Institute. Synthesizing the broad spectrum of Stockholm's twenty-first century conceptual canon, *Our Future in the Anthropocene Biosphere: Global Sustainability and Resilient Societies* emphasized the embeddedness of humankind in an Earth system that was increasingly prone to extreme events, a situation exacerbated by human activities that have reduced resilience in social-ecological systems. Ranging well beyond the realms of resilience and Earth System science, the Beijer report – authored by Carl Folke with support from some twenty colleagues affiliated with Stockholm – not only outlined the problems of climate change and biodiversity loss but also touched upon non-scientific topics. These comprised inequality and inequity, social media and social innovation, and even the notion of narratives as instruments of societal change. An overarching theme and ambition of the white paper, a condensed version of which was published in *Ambio*, was the promotion of transformation – a somewhat elusive and ambiguous idea that over the past decade has spread from sustainability discourse into general political language, particularly in the European Union.[30]

This tendency to invoke transformation – and the sweeping societal and technological changes it implies – without taking the realities of local and international politics explicitly into account points to an inherent weakness within the genre of ostensibly policy-oriented scientific literature that Stockholm has specialized in. While diagnosing drivers of environmental change and leveling general critiques of, for example, CO_2 emissions and excessive consumption in industrialized countries, the lack of precision in specifying concrete policy alternatives, in particular political contexts, has obscured to some extent the inevitable sacrifices and economic tradeoffs entailed by the transformation of societies and the global system. Furthermore, the cacophony of concepts marshaled in papers like the Beijer report risks rendering the manifold implications and imperatives of transformation less, rather than more, intelligible for the range of stakeholders that experts associated with Stockholm have otherwise sought to reach with their scientific insights. Perhaps William Clark in his start-up review saw a pattern that goes even deeper than he himself might have thought. There may still be reason to think carefully about directionality and the choice of stakeholders when a wider societal impact is on the agenda.

IS THE GREEN STATE HOLLOWING OUT?

Already well into the Agenda 2030 decade, the end of which should ideally see the SDGs achieved, it is clear that one of the environmental policy innovations that still draws the most attention to Stockholm and Sweden is the Planetary Boundaries concept. Like the Anthropocene idea that preceded it, Planetary Boundaries is now in its third decade, and both concepts continue to extend their reach well beyond scientific circles. The other high-profile innovation originating in Stockholm this century

[30] Carl Folke, with contributions by Stephen Polasky, Johan Rockström, Victor Galaz, Frances Westley, Michele Lamont, Marten Scheffer, Henrik Österblom, Stephen Carpenter, F. Stuart Chapin III, Beatrice Crona, Gretchen Daily, Partha Dasgupta, Owen Gaffney, Line Gordon, Holger Hoff, Simon Levin, Jane Lubchenco, Will Steffen, Brian Walker, et al., *Our Future in the Anthropocene Biosphere: Global Sustainability and Resilient Societies*, Beijer Discussion Paper 272 (Stockholm: Beijer Institute, 2020). Carl Folke, Stephen Polasky, & Johan Rockström, et al., "Our Future in the Anthropocene Biosphere," *Ambio* 50(2021):2, 834–869.

is the aforementioned Fridays for Future movement initiated by Greta Thunberg as recently as 2018. Interestingly, although these scientific and social interventions share many features and ideas, they each represent the far end of a spectrum. One is a singular planetary framework with little attention paid to details on the level of what German philosopher Jürgen Habermas, precisely in opposition to the large "system," called "the life world."[31] The other end is a loosely structured popular movement rooted in the frustration, pain, and longing for justice of millions of young people feeling a personal responsibility for not just their own personal life world but for those of all others as well. Both positions at either end of the spectrum are rooted in traditions that run deep in Swedish history as we have seen throughout this book.

Perhaps most interesting, however, and somewhat worrisome, is the nagging observation that there seems to be little common ground between these two peripheral phenomena, systemic meta-expertise, and on the ground activism. Much of the middle ground where most day-to-day politics takes place, and which used to be the strength of the Swedish green model when it first emerged in the post-war decades, has in recent years been on a path of incrementally hollowing out. PB is at its core a scientific approach on the planetary level, drawing its insights and policy ideas from an understanding of the operations and constraints of the Earth System, a massive abstraction that we can only speak to through the intermediary of scientific instruments, data, and equations. On the other end, we find a youth movement, vocal and committed but with an unclear relationship to the general opinion of the societies where it developed a following, largely in Europe. Between these very active frontlines, we used to find capable social and political institutions that could absorb the energies from various edges of society. Over the last few decades, these institutions have certainly become more numerous, at times suggesting the appearance of what some scholars have called a "green state." This is of course encouraging. But are they potent, do they

[31] Habermas' lifeworld concept, in his *Theorie des kommunikativen Handelns* (1981), rests on our human abilities to establish relations between each other and with our lifeworlds in their totality. The title of its volume II is *Lifeworld and System*, and it reflects an attempt to reconcile these two spheres of action and understanding. Jürgen Habermas, *Theorie des kommunikativen Handelns*, vol. 1 *Handlungsrationalität und gesellschaftliche Rationalisierung*, vol. 2 *Zur Kritik der funktionalistischen Vernunft* (Frankfurt am Main: Suhrkamp, 1981), English translation, Thomas A. McCarthy: *Theory of Communicative Action*, vol. 1 *Reason and the Rationalization of Society*, vol. 2 *Lifeworld and System: A Critique of Functionalist Reason* (Boston, MA: Beacon Press, 1984–1987).

deliver, and are they functional? Are we currently witnessing an emerging green twist on "the hollowing out of the state," an epithet and phenomenon typically linked to the rise of neoliberalism and the expansion of markets at the expense of the state and its institutions?

This "hollowing out of the green state" is not only true for Sweden but it is *also* true for Sweden, which makes the appearance of Sweden as an environmental Great Power very different now from what it was half a century ago. In Chapter 9, we will take stock of what has been learned in this book so far, and attempt to draw out some implications for the remainder of the Agenda 2030 decade and beyond. This entails taking into account the very bleak global situation, with a recent global pandemic, war and conflict, economic volatility, trends toward less democracy and more autocracy on all continents, and, sadly, most environmental and climate indicators still moving in the wrong direction. Perhaps above all, it means widening the scope and examining what seems to have been an enduring weak spot in most places, Sweden not excepted. Namely the difficulties in productively linking the deepening insights into how the human–Earth relationship is actually playing out on different scales, from the micro and local levels to those of the planetary and Earth system. An understanding of such interlinked systems is of little use if it is not applied by those with the capacity to implement far-reaching change.

9

Conclusion

An Environmentalism of the Rich?

On June 2, 2022, Swedish Prime Minister Magdalena Andersson addressed heads of state, ministers, diplomats, and other elites from around the world.[1] They had assembled in Stockholm to celebrate a half-century of global environmental governance (GEG), fifty years after the UN Conference on the Human Environment. The semi-centennial gathering is a large-scale international event, but it spans only two days, compared to almost two weeks in Stockholm in June 1972. Leading up to the 2022 meeting, there had been lingering concern among the world's activists, researchers, engaged companies, and citizens that so little was being heard about Stockholm+50, which, in comparison to the original Stockholm Conference, lacked energetic leadership and an extensive preparation period. By the eve of its opening, many questioned the level of ambition that lay behind an event that should have been a milestone in the half-century struggle to tackle the environment and climate crisis.

Stockholm 1972 had sparked a phenomenal development. The Conference had represented the birth of an entirely new arena of international politics, a set of issues of grave importance for people and planet. However, after fifty years, the Swedish prime minister had to concede that things were still not moving in the right direction. It was time – once again – to turn words into deeds.

[1] Swedish Prime Minister Magdalena Andersson, opening speech at the Stockholm+50 conference, June 2, 2022, Älvsjö fair, Stockholm, www.government.se/speeches/2022/06/stockholm-50-opening-address/ (retrieved August 2, 2022). All quotes in the following are from that speech.

Heat waves, water scarcity, droughts, storms, floods, wildfires, melting glaciers, rising sea levels, warming oceans – soon containing more plastic than fish.

We are hurting our planet.

We must start delivering on the declarations we have signed, and the pledges we have made.

We have already talked the talk,
Now it's time to walk the walk.

She speaks in a cavernous auditorium at Stockholmsmässan, a hangar-like facility on the outskirts of the city center typically used for massive marketing events, dog shows, and trade fairs. Andersson cites her predecessor, Olof Palme, who opened the Stockholm Conference fifty years earlier with a speech at the Royal Swedish Opera, a grand nineteenth-century structure located opposite the Swedish Parliament, the Foreign Office, and the Royal Palace right in the heart of Stockholm.

Let our work be guided by the words spoken at the 1972 Conference by the Prime Minister, Olof Palme,

"There is no individual future, neither for human beings, nor for nations.
Our future is common. We must share it together.
We must shape it together."

Before rounding off her remarks, the Swedish prime minister also attempts to ignite a degree of optimism. She does not speak about great achievements on the path toward global sustainability, but rather provides examples of ostensible environmental progress that has been made in Sweden. Specifically, about – cars and steel…

And yesterday, right outside these doors, the very first vehicle built using fossil-free steel was presented. This project is the result of a public-private partnership supported by the Swedish Government.

We see that it's possible to combine reduced emissions with economic development.

One can reflect on whether or not these remarks, and the convention center setting of the event, contributed to lifting the spirit and ambition of GEG and its practitioners. Perhaps what came across was the pragmatic, down to earth voice of a disillusioned politician who spoke in less than lofty terms, compared to the grand visions, and scathing critiques, articulated fifty years earlier.

Palme had spoken about the need to stop "ecocide," a word he is credited for putting into diplomatic and political circulation in

FIGURE 9.1 The two-day Stockholm+50 Conference was hosted by Sweden on June 2–3, 2022, in association with the government of Kenya, where, in Nairobi, the secretariat of the United Nations Environment Programme – a main outcome of the 1972 Stockholm Conference – is based. Front row, left to right: Sweden's Crown Princess Victoria and King Carl XVI Gustaf, UN Secretary General António Guterres, Kenyan president Uhuru Kenyatta, and Swedish prime minster Magdalena Andersson. The back row includes senior UN officials and members of the governments of Sweden and Kenya. Photo: Jessica Gow/TT.

June 1972.[2] The prime minister of 2022 talks about a Swedish automobile, in the wake of the demise of Saab and the Chinese acquisition of Volvo. But Andersson's statement should, perhaps, rather be interpreted as a gesture to the everyday character of the sustainability transition that is not only necessary but, she wants to suggest, has already started in Sweden. She adds that it plays out in the north of the country,

[2] A concept Sweden itself does not officially endorse, and passed by with silence at Stockholm+50 in June 2022. Also, the large majority of political parties in Sweden, except the Left Part and the Green Party, are not willing to move on the issue. Lisa Röstlund & Alexandra Urisman Otto, "Allt fler länder stödjer lag mot miljömord – men inte Sverige" [Ever More Countries Support Ecocide Law – But Not Sweden], *Dagens Nyheter*, May 29, 2022. At the same time, a large number of Swedish NGOs, including WWF, SSNC, the Church of Sweden, and the Olof Palme International Center, want this policy to change: "26 Swedish NGOs Demand Ecocide Law at Stockholm+50," *Stop Ecocide International*, May 17, 2022, www.stopecocide.earth/new-breaking-news-summary/26-swedish-ngos-demand-ecocide-law-at-stockholm50 (retrieved December 24, 2022).

a region with some of the biggest challenges, both environmentally and economically, where salvation may lie in the so-called new green industrial revolution. Most of the international excellences listening would not know the details of this alluring future vision, but those of us who do are acutely aware that there are formidable obstacles to overcome even along this pathway of progress.

Andersson's narrative is reminiscent of the Soterian-Promethean dichotomy (see Chapter 7). She largely adopts the Promethean perspective in making the case for a Big Technology approach to the future. But, as we noted earlier (Chapter 2), Swedish Social Democratic political rhetoric also has a Soterian side, a position we can see in Palme's 1972 opening speech at the Stockholm Conference. On the other hand, Palme was by and large a strong proponent of the standard technology-modernity narrative. He was for instance an advocate of nuclear power, even using climate change as a motivating factor long before such arguments were commonly employed.[3] In the Social Democratic Party, these categories were also always fluid. Hans Palmstierna (Chapters 4 and 5), for example, was first pro-nuclear, but by the early 1970s, had turned against the technology at a time when he also became increasingly critical of industry – an unwelcome development for some members of the Social Democratic government as well as business leaders, who started to see him as a threat to the economic growth that underpinned the Swedish welfare state.

Fifty-plus years after Stockholm, progress on problems of the human environment is not nearly rapid enough, and the GEG endeavor embarked upon in 1972 is still far from complete. Where will Sweden's green transition stand a few decades from now? The truth is that the Swedish prime minister does not know, because it is indeed impossible to know. Had Andersson invoked Sweden's aboriginal people, the prime minister would have had to concede that the Sami typically do not like the mines that she mentions as part of the new green future in their part of the country. There is a vision for industry, perhaps even a sustainable one, but the road to take us there is just as disputed as it has always been.

This is Stockholm in the 2020s. What are we supposed to think? Andersson ends her speech by asking everyone, upon departing Stockholm, to speed up their work to reach all the important goals they have agreed upon; commitments that were in some cases made many years ago. Will they finally follow through this time?

[3] Kristoffer Ekberg & Martin Hultman, "A Question of Utter Importance: The Early History of Climate Change and Energy Policy in Sweden, 1974–1983," *Environment and History*, 29(2023):3, 399–421.

Only months after Stockholm+50, Sweden held general elections. Environment and climate were largely absent from the election campaign, which revolved around the rising energy crisis in Europe following Russia's invasion of Ukraine in February 2022. The right-wing opposition, of which the ultra-right "national conservative" Sweden Democrats were part, won the election with a tight margin. Soon after, the Conservative government made a radical shift of environmental and climate policies. It dismantled the Ministry of Environment and brought environmental and climate issues into the Ministry of Climate and Enterprise. It withdrew a series of policy measures to reduce the use of fossil fuels in the transport sector, lowered taxes on, and reimbursed households and companies for, rising prices on fossil fuels and electricity. It also cut the environment budget by more than 50 percent over a few years.[4] Signals of weakening steadfastness on the green front had been seen for some time in Swedish politics. With the shift of government in the fall of 2022, the floodgates opened and raised eyebrows in Sweden as well as abroad. Where was the green leadership going?[5]

IS IT PROGRESS?

As historians, we will of course answer that it is too early to tell. Here at the end of this book, it is, however, time to see how both long-term trajectories and more recent changes have worked together to shape the arc of "the *human* environment" – with the italicized *human* – as articulated in the Stockholm approach to GEG since the mid-twentieth century and the Agenda 2020 decade. Might there be more dimensions to "human" than the benign, humane, and social ones that we discussed at the outset? Are there perhaps more anthropocentric and nationalist interests – perhaps also deeply human self-interest – at play in the Swedish position, after all?

We began this book by raising progress as a theme. Much history writing about the environment in general, and GEG in particular, is built upon narratives of gloom and decline. We wanted to reverse the optics and put

[4] *Dagens Nyheter*, April 17, 2023: www.dn.se/sverige/regeringen-storsatsar-pa-kontroversiell-klimatkompensation/.

[5] The official government statement on its policy for the coming four years was presented on October 18, 2022: www.government.se/speeches/2022/10/statement-of-government-policy/. Some reactions to the new politics are here: www.euronews.com/2022/10/18/devastating-consequences-as-new-swedish-government-scraps-environment-ministry, and here: www.bloomberg.com/news/articles/2022-10-19/climate-role-model-sweden-s-new-leader-axes-environment-ministry?leadSource=uverify%20wall.

the focus on the role of a city, Stockholm, and a country, Sweden, and ask if we could find a progressive story to tell about a small state assuming leadership on a big issue. We opted to hone in on Stockholm and Sweden partly because we have a long experience in that geographic and social context. One of us, Sverker, was born in Sweden and has lived and worked there over the course of a long career as a historian, public intellectual, and policy advisor, but has also spent periods outside of Sweden, long enough to see the country from an outsider's perspective. Eric was born and raised in the United States, came to Sweden well into adulthood, and went to university in both countries. He has spent the better part of the past thirty years as a resident of the Swedish capital, working with foreign policy, sustainability, media, and now as a historian specializing in the history of international governance and geopolitics.

What have we found? Were we correct in assuming that there is a story to tell, and that a small country and its capital city can, under certain circumstances, make a useful, perhaps in particular moments, even decisive contribution to world affairs on issues that literally span the entire planet? The reader will have already formed her or his own opinion by now, but arriving to this concluding chapter, we as investigators have found it reasonable to retain our initial assumption. We do think that the Stockholm story in fact provides added value both to the understanding of GEG and, just as importantly, to the wider community of the science and politics of sustainable development, as well as to historians and policy thinkers like ourselves. At the very least, we hope that we have contributed some insights on how environmental governance actually takes place and plays out in a concrete empirical setting and over a considerable amount of time, in our case the long half-century since the UN Stockholm Conference. In reality, an even far longer timespan, given the attention we have paid to the several years of Conference preparations within the UN system in New York and Geneva. To this may be added the several decades of scientific work at Stockholm University and other Swedish institutions after World War II, and the formation of Sweden's modern democratic political culture earlier in the twentieth century. What Marquis Childs called "the middle way," with a focus on dialogue, democracy, compromise, the embrace of science and technology, and firm belief in international institutions, especially the United Nations.

Our conclusion, however, does not come without a few caveats. The first is that neither a city nor a nation is a fixed entity that passes through time unaffected. We have become much more aware through this project of just how different Stockholm, and Sweden, are today compared to the time of the Swedish initiative at the United Nations in 1967–1968.

Sweden is no small, inert monolith working to improve the relentlessly changing world outside. The country itself changes just as quickly as any other part of the world, and in no unified and predictable direction. Overall, as a global environmental player, Sweden's significance, while still probably punching a bit above its weight, is today not anywhere near where it was at the beginning of this period in the late 1960s and in the final decades of the 1900s. Some of its early strengths have weakened, while in the most recent years, few new strengths have emerged to offset the loss. On the contrary, a new cynicism has entered the environmental register, a shrugging of shoulders as past green achievements are abandoned, manifested for example when the government decided to reduce the level of mandatory biofuel mixed into gasoline and diesel, ostensibly disregarding goals set in Sweden's climate law of 2017. Reviewing the government's 2023 Climate Plan, the Swedish Climate Policy Council delivered a scathing critique, declaring that under current policy, Sweden will not meet either its own or the European Union (EU)'s 2030 emission targets, nor long-term net zero for 2045 and 2050.[6]

The Stockholm+50 Conference in June 2022, officially a major UN event like the 1972 Stockholm Conference, illustrated the sea change in Sweden's position in the international system and the structural changes in global affairs in general. Stockholm 1972 was a long, intense, well-prepared, and innovative meeting that produced landmark institutions like UNEP and pointed the world in a new direction of hope and collective commitment to a better managed human environment. Stockholm 2022 was a brief event with a short preparation period and few surprises (perhaps except for the new car…) that generated little excitement before, during, or after the two-day gathering. The circumstances were certainly not optimal, given the preceding and still ongoing Covid-19 pandemic and a horrifying new war in Ukraine that deeply disrupted international relations. On the other hand, fifty years down the road, the world should have been infinitely better equipped to organize a major event on the fate of the planet and its soon eight billion inhabitants, and deliver at a critical moment for the international system. Sweden, as the initiator of what has come to be called GEG, would have been expected to provide strong leadership for such a mission. Neither proved to really be the case.

Another caveat has to do with the political commitment of the country itself. There have been major shifts in the circumstances enabling progressive politics in Sweden, just as in Europe at large. Populist politics with

[6] *Swedish Climate Policy Council: Report 2024* (March 2024). www.klimatpolitiskaradet.se/wp-content/uploads/2024/04/klimatpolitiskaradetsrapport2024.pdf.

a strong nationalist orientation has eaten its way into the very fabric of political life in many countries. Since the financial crisis of 2008–2009, and in particular the tumultuous and ominous past decade – with Brexit, the Trump presidency, the deepened autocracy in China, Moscow's war of aggression against Ukraine, and the worsening wars and crises in the Middle East as some of the most disheartening manifestations – this has inserted mental and material friction into all potential sustainability policies. Sweden is no exception. Even without that friction, the speed of progress has been slower than what most would have expected a half century ago. If participants at the Stockholm Conference had been told in June 1972 that most of the major problems would still be unresolved in 2022, and that some of the biggest, including climate and biodiversity, would only accelerate over the next fifty years, they would hardly have thought it plausible.

Stewart Brand's famous sojourn in Stockholm (Chapter 6) included an entourage of not only Hopi, Mohawk, and Navajo Native Americans that gave talks, organized rituals, and performed music but also a small team of journalists equipped with recording devices. Their tour of the Swedish Environmental Protection Agency is available on YouTube. At one point, the camera stops in front of the desk of a young Swedish civil servant wearing a quintessentially bureaucratic outfit, a summery light suit and a tie. He responds to a question by saying, without the slightest hint of irony (or any indication of being under the influence of hypnotic substances ...), that he expected industrial pollution to be eradicated in Sweden by 1973! His response reflected the mood of the day. In the same film, a representative of Rotary International tells the same journalist that he is, alas, too old to be reformed. He has committed environmental sins all his life, implying that he simply can't stop now. The new generation will be entirely different, he says. They have the spirit his cohort never learned and will take society in a new direction. Like the young Swedish bureaucrat, the much older Rotary man was also wrong.[7]

Needless to say, the general lack of environmental progress to date cannot be blamed on Stockholm or Sweden in particular. Had there been no Stockholm Conference and had the Swedish capital not been a proactive nexus for green world affairs since 1972 and before, the global environment would in all likelihood not have fared any better, and the same hockey stick curves would have climbed upward anyway.

7 Youtube URL, www.youtube.com/watch?v=TFBDXm8DaYc. A longer playlist from Soundings Mindful Media is here, "Long Live Life – Feature Documentary on Eco Politics – 1972 UN Conference on the Human Environment," www.youtube.com/playlist?list=PLECo3BNUfXE8288XStZadEOGqqLDTjVq- (retrieved August 5, 2022).

But the very fact that these curves are so hard to bend back down and the processes of progress are so painstakingly slow, or even negligible, should give us pause. If we want to be serious about the long-term improvement of the global environment, we should take into consideration the fact that the only progress we can reasonably talk about here is *relative*. Relative, that is, to the even more disastrous earth story that might have played out without the emergence and evolution of GEG. What we may call a counterfactual nihilism, or denialism, scenario, which is clearly not what we have experienced over the past half century (although it may at times have felt that way). The work that we have taken on in writing this book has been to identify the factors in the case of Stockholm and Sweden that have made a difference. Looking at progressive elements in the modern history of the global environment is not the same as arguing that the efforts of an array of individuals and institutions have also resulted in real progress on the ground at various levels of abstraction. An engaged analytical discourse on progress toward a more sustainable human environment at the global or planetary scale must be possible, even if much of that (potential) progress still lies in the future.

IS STOCKHOLM'S ROLE OVERSIZED?

The reader will have noted that we started the chapters with an emblematic vignette as a preamble to the theme and historical period covered. Many of these, like the extract from Magdalena Andersson's speech at Stockholm+50 at the start of this final chapter, were from or about *meetings*. We did this with a purpose: to demonstrate that a primary method of GEG was adopted from diplomacy, where dialogue through face-to-face meetings has always been considered the best working method. Parallels are apparent with the evolution of the in-person scientific conference – still defended by many academics, while heavily critiqued by others as reproducing hierarchies and for being environmentally unsustainable because of the inevitable air travel involved. Progress, the argument goes, starts with agreement, which requires negotiation.[8]

[8] Charlotte Bigg, Jessica Rheinisch, Geert Somsen & Sven Widmalm, "No amount of technology can replicate in-person conferences," *Times Higher Education Supplement*, August 4, 2022. Astrid Eichhorn, Magnus Breitholtz, Valerie Domcke, Jan Hladky, Debbie Hopkins, Agnes Kreil, Sverker Sörlin, & Diarmuid Torney, *Towards Climate Sustainability of the Academic System in Europe and Beyond* (Berlin: Alliance of European Academies, ALLEA, 2022).

We also wanted to demonstrate that certain evidence of progress can already be gleaned from the progression of topics encompassed within environmental governance. Issues like toxic chemicals and pollutants have actually been, if not solved, at least cast as a set of problems where agreements were made and the rates of dangerous change have slowed down.[9] In the case of population growth, once a fundamental aspect of neo-Malthusian environmental crisis discourse, research on demographic transition and the science communication of the late Swedish physician Hans Rosling – also, like Johan Rockström and Greta Thunberg on the *TIME Magazine* 100 list (in 2012)[10] – have, for instance, alleviated concerns to some extent. Such issues have moved down the list of urgency. Our later chapters focus on the significance of Stockholm in enabling scientific work and fostering international cooperation on the most complex issues, where point sources of pollution and other problems can't be identified. For such intractable problems, it is rather the sum total of human activities that have caused the negative effects, as in the case of climate change, biodiversity loss, and across a wide set of Earth system indicators. An arc of at least some progress is built into our narrative.

Have we been able to demonstrate that Sweden has played an oversized role in the realm of environmental governance? It is hard to measure such a thing, and we have not attempted to do so. We opted not to count the number of conferences and agreements nor to quantify participant statistics. We have instead tried to provide a narrative where a prominent role of Swedes and other Stockholm-associated actors in the sequence of meetings, the establishment of institutions, and the conceptualization of environmental imperatives has been a feature, following in a rich tradition from diplomacy, science, and international environmentalism.[11] We hope that we have been convincing. Sweden's

[9] Again with many exceptions, such as the worrying finding that PFAS pollutants have now been detected in literally every corner of the planet, like the Antarctic ice cap and the Himalayan glaciers, indicating that they will stay with humanity forever. Jack Garnett, Crispin Halsall, Holly Winton, Hanna Joerss, Robert Mulvaney, Ralf Ebinghaus, Markus Frey, Anna Jones, Amber Leeson, & Peter Wynn, "Increasing Accumulation of Perfluorocarboxylate Contaminants Revealed in an Antarctic Firn Core (1958–2017)," *Environmental Science & Technology*, 26 July (2022). https://doi.org/10.1021/acs.est.2c02592. Epub ahead of print.

[10] https://sv.wikipedia.org/wiki/Time_100.

[11] The importance of meetings and of networks of expertise, formed and maintained during meetings, often taking place in sequences, is visible, although not always explicitly theorized in the literature on the formation of international environmentalism.

contribution to this process should also be weighed against the small size of its population and gross domestic product (GDP) relative to world, or Organisation for Economic Co-operation and Development (OECD), totals.

Is there a pattern in Sweden's engagement? We do think there is. The more common and international an issue is, and the less the response centers on the actions of an individual government, the more active Sweden tends to be. In Swedish domestic politics, it has sometimes been said that Sweden has wanted to serve as a "world conscience" (*världssamvete*). The expression in fact became a stereotype for Sweden and is often used by critics of the country's engagement in global issues.[12] A perpetual target of this critique has been Sida, the Swedish International Development Cooperation Agency, which has since the early 1970s been funded at the level of 1 percent of GDP, certainly a growing budget over time and among the highest in the world along with Norway and Luxembourg.[13] For the first time, the 1 percent goal has now been abandoned. Along with it, the Sida-funded program of development research was slashed by the government, all designed to hollow out the position of Sweden as the "progressive small state" with an outsized role in world affairs, and apparently succeeding as far as international publicity can tell.[14]

For an example of an approach that emphasizes the combination of meetings, individuals, networks and institutions, see Matthias Schmelzer, "'Born in the Corridors of the OECD': The Forgotten Origins of the Club of Rome, Transnational Networks, and the 1970s in Global History," *Journal of Global History*, 12(2017):1, 26–48. On the usefulness of the concept "espistemic communities" in this context, see Peter M. Haas, "Introduction: Epistemic Communities and International Policy Coordination." *International Organization* 46(1992):1, 1–35. www.jstor.org/stable/2706951, on page 3, note 4. Haas cites especially literature from the history of science. In science, just as in diplomacy, meetings have been institutionalized as tools of cohesion building and organized progress.

[12] Here it is used as such in the *New York Times* in 1998: Warren Hoge, www.nytimes.com/1998/08/10/world/sweden-once-the-world-s-conscience-now-drifts.html.

[13] Annual Report 2021 to the European Council on the EU Development Aid Targets – Council conclusions (June 14, 2021). https://data.consilium.europa.eu/doc/document/ST-9549-2021-INIT/en/pdf (retrieved March 21, 2023). Sunniva Engh, "The Conscience of the World?: Swedish and Norwegian Provision of Development Aid," *Itinerario* 33(2009):2, 65–82. https://doi.org/10.1017/S0165115300003107.

[14] The 1 percent target was officially abandoned in the budget of the center-right government that was formed following the September 2022 general election. David Nilsson, "Enprocentmålet avskaffas: Sverige måste hitta en ny roll i världen" [One Percent Target Abandoned: Sweden Must Seek a New Role in the World], *Omvärlden*, November 21, 2022, www.omvarlden.se/opinion/debatt/enprocentmalet-avskaffas-sverige-maste-hitta-en-ny-roll-i-varlden (retrieved December 24, 2022). The drastic dismantling of Swedish development research in June 2023 – see: www.vr.se/english/just-now/news/

Some of that critique has for a long time been adopted by the rising populist right, whose more cynical "Sweden-first" nationalism is consistent with a low profile on the international scene. It has been the hallmark of Swedish foreign policy to engage in conflicts and concerns that are shared by many. This has in part been rooted in national priorities. Robust multilateralism and a strong UN have been deemed to be in Sweden's best interest. It also follows from the concept of *solidarity*, once a word of high prestige, originating in the trade union movement, which has always been strong in Sweden.[15] During the Cold War, the term took on a new meaning associated with supporting the cause of national independence and the struggle against colonialism – what Alfred Sauvy in 1952 called "tiers-mondism," in favor of the Third World and the Non-Aligned Movement that took shape with the 1955 Bandung Conference of newly independent Asian and African states, as well as China.[16]

The solidarity principle, in its Swedish version, was about protecting the weaker countries of the world. In a certain sense, Sweden itself was part of this cohort; it has never been a major power since the demise of its Baltic empire in the eighteenth century. But it was "weak" only because of its size. In most other respects, it was a nation replete with natural as well as societal resources (Chapter 2). In science, medicine, and technology, Sweden was punching far above its weight; a small superpower, which took the position that using its strengths in the interest of global solidarity and cooperation would be both good for the world in general and, especially, for the weak. For Sweden, too. This is the most reasonable way to read the words of Olof Palme from his 1972 speech: "There is no individual future, neither for human beings, nor for nations." There is a future only if there is cooperation. As Magdalena

news-archive/2023-06-27-no-new-grants-in-development-research.html – led to massive protest among scientists and negative international publicity: www.nature.com/articles/d41586-023-02239-8; www.swedev.dev/the-suspension-of-development-research-grants/#; www.thelocal.se/20230628/academics-in-uproar-after-sweden-cancels-research-funding.

15 A. Wildt, "Solidarity: Its History and Contemporary Definition," In: K. Bayertz, ed., *Solidarity: Philosophical Studies in Contemporary Culture*, vol. 5 (Dordrecht: Springer, 1999). https://doi.org/10.1007/978-94-015-9245-1_11.

16 Leslie Wolf-Phillips, "Why 'Third World?': Origin, Definition and Usage," *Third World Quarterly* 9(1987):4, 1311–1327. Christoph Kalter, "A Shared Space of Imagination, Communication and Action: Perspectives on the History of the 'Third World'," In: Samantha Christiansen & Zachary A. Scarlett, eds., *The Third World in the Global 1960s* (New York & Oxford: Berghahn, 2013), 23–38. Christopher J. Lee, ed., *Making a World after Empire: The Bandung Moment and Its Political Afterlives* (Athens, OH: Ohio University Press, 2010).

Andersson stated in her opening address at Stockholm+50, quoting Palme's 50 years earlier: "our future is common." Hence, the Swedish commitment precisely to cooperation and the types of *con*-activities that we have elaborated throughout this book.

This may be the most fundamental component in explaining the oversized Swedish engagement in environmental politics at the global level. The inbuilt rationale of Swedish foreign policy – in all of its areas – can be distinguished in the rising issue of "the environment." It is in fact a particularly relevant policy object, and perhaps even the quintessential case of Sweden's international relations in the late-1960s. When it became clear in autumn 1967 (see Chapter 4) that transboundary pollution was a real and present danger that threatened countries downwind – such as Sweden in the case of British sulfur emissions that generated acid rain affecting Swedish soil and lakes – it was the decisive piece of evidence, in combination with the emergence of "the environment" as a framing narrative, that ignited what was already an ideology of solidarity rooted in Swedish political culture. The mobilization of the Swedish state and members of its foreign office to go to the UN and propose global action is, in that light, not as sensational as it might otherwise seem – and as it *has* seemed to some commentators on the Swedish initiative in the scholarly literature and, indeed, on Swedish environmental leadership in general.[17] On the contrary, it could be seen as a playbook example of an intervention in international politics for the rational, solidarity-minded small state and its conscientious leadership. It was just that this was the first time that a high-stakes political gambit in the realm of the environment took place at the global level, and it was this that made it historical.

The rise of Sweden as an environmental vanguard nation has sometimes been presented as largely an accomplishment of naturalists, popular movements, media figures, and young radicals as part of a major mind shift in the second half of the 1960s.[18] Just as important, however,

[17] Annika Kronsell, "Sweden: Setting a Good Example," In: Mikael Skou Andersen & Duncan Liefferink, eds. *European Environmental Policy: The Pioneers* (Manchester: Manchester University Press, 1997); Martin Jänicke, "Trend-setters in Environmental Policy: The Character and Role of Pioneer Countries," *European Environment* 15(2005):2, 129–142. https://doi.org/10.1002/eet.375; Steven Sarasini, "Constituting Leadership via Policy: Sweden as a Pioneer of Climate Change Mitigation," *Mitigation and Adaptation Strategies for Global Change* 14(2009): 635–653. https://doi.org/10.1007/s11027-009-9188-3.

[18] Andrew Jamison, Ron Eyerman, Jacqueline Cramer & Jeppe Læssøe, *The Making of the New Environmental Consciousness: A Comparative Study of Environmental*

were the strong ties within Swedish society and domestic politics, in which government, industry, and organized labor enjoyed a robust institutional alliance based on a long-standing tradition of close collaboration to promote modernization and growth. Environment became woven into this well-established concept of soft corporatism at around the same time as the Swedish initiative in the UN.[19] This was in part due to the realization among employer organizations, labor union federations, and Sweden's Social Democrat-led government that environmental exploitation and despoilment could negatively affect Sweden's focus on industry as a driver of economic growth, which in turn was a foundation of the welfare state. To industry, taking the environment seriously also meant a safer license to operate. The new area of collaboration was made manifest with the founding in 1966 of the Institute for Air and Water Research (IVL), co-funded by the large, influential forest industry and a Swedish government that was becoming increasingly interested in the environment.

What is more, in a country where innovation, resource exploitation, and high levels of industrial activity have led to large concentrations of capital, the involvement of industrialists and entrepreneurs in institution-building at the intersection of science and environment – following in some respect the tradition established by Alfred Nobel – represents another important aspect of the Stockholm story. One example (Chapter 5) is Kjell Beijer's donation to the Academy of Sciences that led to the creation of the Beijer Institute in 1977. Building on the Beijer Institute, the establishment of the Stockholm Environment Institute (SEI) in 1989 (Chapter 6) was in turn reminiscent of earlier efforts by the Swedish state to promote progressive global issues, such as the 1966 founding of the Stockholm International Peace Research Institute

Movements in Sweden, Denmark and the Netherlands (Edinburgh: Edinburgh University Press, 1990). David Larsson Heidenblad, "Mapping a New History of the Ecological Turn: The Circulation of Environmental Knowledge in Sweden 1967," *Environment and History* 24(2018):2, 265–284. Anna Kaijser & David Larsson Heidenblad, "Young Activists in Muddy Boots: Fältbiologerna and the Ecological Turn, 1959–1974," *Scandinavian Journal of History* 43(2018):3, 301–323.

19 Lennart J. Lundqvist, *The Hare and the Tortoise: Clean Air Policies in the United States and Sweden* (Ann Arbor: University of Michigan Press, 1980). Ann-Kristin Bergquist & Kristina Söderholm, "Green Innovation Systems in Swedish Industry, 1960–1989," *Business History Review* 85(2011): 677–698. Kristina Söderholm & Ann-Kristin Bergquist, "Firm Collaboration and Environmental Adaptation: The Case of the Swedish Pulp- and Paper Industry 1900–1990," *Scandinavian Economic History Review* 60(2012):2, 183–211.

(SIPRI). This was largely on the initiative of Alva Myrdal, who would also be instrumental in the creation of the Secretariat (now Institute) of Futures Studies in 1972.[20]

THE CON-WORDS

Given this background, it is also easier to contextualize the significance of the structuring analytical keywords that we proposed at the outset (Chapter 1). We called them con-words, because they all started with the prefix "con": connecting, convening, contributing, and conceptualizing. Over the chapters that followed, we added more such words as context required. If you are a small player, like a Nordic country with less than 2 percent of Europe's population, and if your ambition is to make the world come together around rational principles of action in order to protect the weak and make sure that solutions are fair and acceptable to all, it seems quite reasonable to think in con-terms. After all, "con" means "with," or "together." It also seems reasonable to be pro-active, because bigger powers may have better chances to protect themselves or take measures that others will have to adjust to. As a small actor, you approach the big powers, and the UN, with a language of collaboration. What you can offer is only valid if it can be carried out by many actors in concert. To not be active and demonstrate your capacity to be a relevant partner means that you would rather seek refuge under the wings of a bigger partner or jump on some bandwagon. To connect, convene, contribute, and conceptualize is to signal that among the weak you are in fact strong, strong enough to maintain an active role, for example, in scientific and environmental diplomacy, or in security policy.

This explains the rationale for *connecting* with others. In the chapters above, we have seen Swedish and Stockholm-based actors establishing relations with other progressive forces in the environmental arena, domestically and internationally. This has been facilitated by the early inception of an official Environmental Protection Agency, set up in 1967 as a government agency, combined with the distribution of various environmental missions to existing agencies responsible for forests, minerals, agriculture, public health, work safety, and other sectors. Such authorities could do connecting work alongside the foreign office.

[20] Jenny Andersson, "Choosing Futures: Alva Myrdal and the Construction of Swedish Futures Studies, 1967–1972," *International Review of Social History* 51(2006):2, 277–295.

After the early major initiative that shaped the UN meeting in 1972, this has been less of a driver in Swedish environmental engagement, perhaps even less so after the installment of a Minister of Environment (currently environment and climate) in the cabinet in 1985 and a separate government department for environmental matters since 1987 (until its dismantling in 2022).

Throughout this volume, we have witnessed the performance of connecting work, sometimes by government actors, and increasingly by other societal institutions as well as individuals. A number of these were scientific institutions, some of which had close ties to policy and were based in Stockholm. The need of staying connected also explains the effort to build institutions with a mission of providing research of relevance for policy and governance. The background stories to the institutes may vary, as did their management, funding, and mission. Their growing critical mass, their persistence and international orientation, and certainly their close relation to government through a range of intermediating partners, for example, research councils, can be best understood if institutional density is seen as a corollary to the solidarity logic of staying connected and therefore reaching out.

Convening can be seen as a logical continuation of connecting, when you actually both get in touch with others and start building relations and networks, and then move on and take a wider responsibility as host, organizer, facilitator, and sometimes also, importantly, as funder. These are time-consuming activities that require endurance and diplomatic patience. They also require trust from the wider world. Sweden's reputation as a convening actor was catapulted by the 1972 Conference, including the ambitious and skillful preparations in the years leading up to the "twelve days in Stockholm" (Chapter 5). Adding to this were the organizational abilities of Swedish authorities on site in June that year, and the innovative ideas and bold statements by Olof Palme, including the ecocide concept he deployed in the context of the then-controversial Swedish critique of the United States and its war in Vietnam. Although it annoyed some, it impressed many more, not least in the Global South. The credibility of Sweden as a convening power rested also on its excellent research institutions and the Royal Swedish Academy of Sciences, whose convening power as the conferrer of the Nobel Prizes in chemistry and physics was supplemented by the strong shift toward environmental issues in what was a comprehensive makeover of the venerable institution, founded in 1739. Add to this the strong popular commitment to nature and outdoor activities and domestic industries based on efficient processing of natural resources (despite including environmentally negative effects), and Sweden could in a fairly short period of time secure its position as one of the world's foremost conveners of meetings, workshops and negotiations with relevance for GEG (Chapter 6).

Of great importance for a connecting and convening power of any kind is that the rest of the world finds it meaningful to be connected to and convened by it. Reliability and a capacity to deliver over the long term are key factors. Since Sweden has not held much hard decision-making authority of its own, but instead stressed global (preferably UN) governance over (big power) government, what the Nordic country has sought to deliver has rather centered on the underlying science and ideas that can move policy work forward, or at least create new openings. Hence the need to be a *contributing* convener and not just a facilitator. From this stems another reason for remaining strong in science and for maintaining, on your own soil and preferably close at hand (in Stockholm!), a set of comprehensive institutions that work with you on behalf of your broader political objectives. In theory, these do not always have to be national institutions. In reality, for a small country, it greatly benefits the policy process if there is easy and natural access between the two spheres.

The UN Stockholm Conference proved to be a somewhat inadvertent strike of genius in that respect. It served as a catalyzing event for the series of science and policy institutions that have been a major part of the Stockholm story – such as MISU, Beijer, the Bolin Centre, SEI, SRC, and most recently the KTH Environmental Humanities Laboratory, founded in 2012 in the spirit of extending Stockholm's environmentally relevant expertise more decisively to also include the humanities.[21] These are, typically, centers and institutes with distinct profiles, taking positions at the intersection of research, policy, and activism. They have thus become discernably more pro-active and visible than the universities that they are in some cases part of, and are always reliant upon for staff, PhD students, and collaboration. Politics thus helped science, and science helped policy in a positive feedback loop that, as we have shown (Chapter 3), predated 1972 by several decades, and has carried on for the many decades since then.

[21] David S. Emmett & Robert E. Nye, *The Environmental Humanities: A Critical Introduction* (Cambridge, MA: The MIT Press, 2017). Sverker Sörlin, "Reconfiguring Environmental Expertise," *Environmental Science and Policy* 28(2013):1, 14–24. http://dx.doi.org/10.1016/j.envsci.2012.11.006. Sverker Sörlin & Graeme Wynn, "Fire and Ice in the Academy: The Rise of the Integrative Humanities," *Literary Review of Canada* 24(2016):6, 14–15.

A parallel may be drawn here with the Swedish Cold War security policy. The doctrine was to stay alliance-free in peace in order to remain neutral in the case of war. This doctrine was not completely honest. In reality, Sweden maintained a secret collaboration with NATO throughout the Cold War (Chapter 3). Even a compromised neutrality policy required, it was thought, a strong deterrence capacity to stave off potentially hostile states that might otherwise consider attacking and occupying Sweden. In fact, this capacity needed to be much stronger under a neutrality regime than under the umbrella of an international defense coalition such as NATO. As a result, Sweden had a strongly militarized economy up until the 1980s, with as much as 5 percent of GDP spent on defense. A similar logic applies to a civic, and an environmental, internationalist agenda. Our analysis supports the notion that strong institutions with relevant and innovative contributory expertise are the price that a small nation must pay to maintain its connecting and convening role. Sweden's geopolitical position as a neutral state was in itself a factor that contributed to its connecting and convening power, since it was one of the few developed countries that could be considered nonthreatening and thus acceptable for both Cold War blocs. Also, from a North-South perspective, Sweden's lack of a significant colonial past made it a suitable mediator for developing country concerns. Two of the first UN Secretary Generals were from the Nordics, Dag Hammarskjöld (Sweden) and Trygve Lie (Norway).[22]

Finally, *conceptualizing* work is the result of functional and cutting-edge scientific institutions and think tanks, or in some cases, organizations and civic movements. Ordinary science and scholarship – what Thomas Kuhn called "normal science"[23] – does not typically produce path-breaking ideas. Under normal science, the scientist adheres to the ruling paradigm. To become revolutionary, to break free from

[22] This may also explain why prominent Swedish and Norwegian politicians such as Olof Palme, Ingvar Carlsson, and Gro Harlem Brundtland have been called upon to lead global commissions. Brundtland chaired the Commission on Environment and Development which produced *Our Common Future* (1987), and Olof Palme the Independent Commission on Disarmament and Security Issues, which resulted in *Common Security: A Blueprint for Survival* (New York: Simon and Schuster, 1982). On Palme's vast foreign policy engagement with the Global South, see Andreas Mørkved Hellenes & Carl Marklund, "Sweden Goes Global: Francophonie, Palme, and the North-South Dialogue during the Cold War," *Histoire@Politique*, n° 35, mai-août 2018 [online, www.histoire-politique.fr].

[23] Thomas Kuhn, *The Structure of Scientific Revolutions* (Chicago, IL: The University of Chicago Press, 1962).

the ordinary, many things are required, one of them being luck. What is needed above all is the capacity to conduct the scientific work: the resources, the equipment, and the people, often from abroad, which needless to say means funding and material support. Concepts linked to Stockholm have made an impact, from Arrhenius's Greenhouse Effect to Rossby's Jet Stream to Crutzen and Stoermer's Anthropocene to Planetary Boundaries and Hothouse Earth, and why not Greta Thunberg's Fridays for Future, or "bla-bla-bla"!

MATERIAL REALITIES

In a country like Sweden, where universities and research institutes rely to a very large extent on public funding (direct funding or competitive funding from research councils), such creativity is only possible if there is continuous and reliable support, in reality through political decisions by parliament and government to concentrate and direct funding toward research on environment and climate. In addition, thanks to its particular tax legislation (e.g., allowing deductions for donations to foundations), Sweden has the world's largest concentration of private foundations, as well as an array of internationally oriented nonprofit organizations.[24] While most are typically small and supplementary to public funding schemes, a sizable number of the largest foundations channel their resources toward research in "do-good" areas such as health, technology, or environment, where they make considerable contributions.[25] In this way, private capital also consolidates trust, first because of the prestige that science enjoys – consistently highly ranked in surveys of public trust in Sweden; and second, since some of their fortunes are derived from what might be today considered "bad" environmental behavior. Similar in a sense, in the Swedish context, to the key role of the Rockefeller Foundations in the promotion of international climate initiatives in the 1980s (Chapter 6).

The Nobel Foundation is the obvious forerunner and analogue for Swedish institutions that have emerged to promote science and

[24] Like The Natural Step Foundation, which was founded in 1989 by the Swedish oncologist Karl-Henrik Robèrt and has carried out national awareness raising campaigns.

[25] Filip Wijkström & Stefan Einarsson, "Comparing Swedish Foundations: A Carefully Negotiated Space of Existence," *American Behavioral Scientist* 62(2018):13, 1889–1918. Sverker Sörlin, *Vad kan stiftelser göra?: Den privata stiftelsesektorn som forskningsfinansiär* [What Can Foundations Do?: The Private Foundation Sector as Funder of Research] (Stockholm: Kempestiftelserna, 2005).

progressive causes, as well as draw attention to good deeds and groundbreaking work, through awarding prestigious prizes. Established in 1980, the Right Livelihood Award is presented at the Swedish Parliament in December each year and is often referred to as the "alternative Nobel prize" (although it has no relationship with the Nobel Foundation). The Volvo Environment Prize is awarded by a foundation that is financed by the Swedish automotive company, with an independent scientific committee that is hosted by the Beijer Institute. Recipients of the Volvo Prize since its founding in 1990 have included many of the scientists and sustainability leaders portrayed in this book. One individual, the American physicist, environmentalist, and energy expert Amory Lovins – who has had a close relationship with Sweden since the 1970s – has received both the Volvo Environment Prize and the Right Livelihood Award. An additional benefit of bestowing generous awards, besides furnishing Sweden's reputation as a place that has the credentials (and the cash) to pass judgment on what counts as excellent and progressive, is the steady flow of experts that pass through Stockholm. Not just those receiving prizes, but, likely, those that want to draw attention and cultivate relationships with those that decide.

We have not in any of the chapters described in full how research funding has developed in Sweden, but hopefully we have offered enough information to demonstrate that this has been a priority. In the 1970s and 1980s, science faculties across the country were funded to conduct environmental research. However, it should be noted that several of the prominent Stockholm institutions that have figured in this volume were initiated through funding initiatives over and above regular faculty funding. As mentioned above, it was the government, expanding on the previously existing Beijer Institute, that started and financed SEI from 1989 as a legacy of the 1972 UN Declaration on the Human Environment.[26] When the state decided to endow a new generation of foundations in the early 1990s, two of these were devoted to environmental purposes. One for strategic environmental research, Mistra, which provided the cornerstone grant to found the Stockholm Resilience Center in 2007, and the other for the training of master students from all over the world. The KTH Environmental Humanities Laboratory was founded based on a 2011 donation from industrialist Carl Bennet. A 2016 government bill encompassed a forward-leaning

[26] See SEI strategy 2020–2024, www.sei.org/strategy/2020-24/chapter/who-we-are.php (retrieved August 6, 2022).

initiative to establish broad, thematic, interdisciplinary, and well-funded research programs for sustainability distributed by the research councils, not as direct allocations to universities. The funding portfolio for environment and climate has over time broadened to include medicine, technology, agriculture, the social sciences, and, increasingly, the humanities and theology.[27]

Hence, not only have we seen a great deal of empirical evidence to support our employment of the con-words to help understand and explain Sweden's policy ambitions for a global approach toward the environmental challenges and the rising climate crisis in the new century, we can also see how it is rooted in a broader foreign and security policy idea, adopted and articulated by a progressive small state in northern Europe, and how it was ideologically, pragmatically, and, not least, rhetorically held together by a *principle of solidarity* and a *doctrine of neutrality*. These fundamental political assumptions, in combination with the prevailing geopolitics of the Cold War, were instrumental in making Sweden's remarkable tour de force on the world stage possible and helped secure its position as a leading international environment and sustainability player over decades.

SWEDISH ENVIRONMENTAL EXCEPTIONALISM?

To the extent that the Stockholm road to, and from, 1972 was a useful one – was it "exceptionalist"? The term itself is most often used about the United States, "the first new nation," whose exceptionalism is legendary. It is typically seen to rest, at least partly, on its status as a global superpower, with the prerogatives that afford for being selective and self-serving.[28] Another famous exceptionalism is the German *Sonderweg*. Yet another is Nordic exceptionalism, a variation on the concept of the "Nordic model," which underscores the tendency that Nordic countries consistently place high on welfare indices and happiness ratings. Sweden

[27] Mats Benner & Sverker Sörlin, "Shaping Strategic Research: Power, Resources, and Interests in Swedish Research Policy," *Minerva* 45(2007):1, 31–48. Sverker Sörlin, ed. *"I den absoluta frontlinjen": En bok om forskningsstiftelserna, konkurrenskraften och politikens möjligheter* (Stockholm: Nya Doxa, 2005). Mats Benner, Göran Marklund & Sylvia Schwaag Serger, eds., *Smart Policies for Societies in Transition: The Innovation Challenge of Inclusion, Resilience and Sustainability* (Cheltenham & Northampton, MA: Edward Elgar, 2022).

[28] It should be noted, though, that American exceptionalism is a tremendously complex concept with a long and winding history.

as the ultra-modern European version of the United States, the "second new nation," is a standing trope in this discourse.[29]

Swedish environmental exceptionalism, we argue, belongs in the same class of phenomena, part reality, part trope: a national attribute that may only be possible, or at least is far less unlikely, among countries that are already exceptional in the way the Nordics are. We have already mentioned Peder Anker's *The Power of the Periphery* on Norway as a green nation (Chapter 1). It may be no coincidence that the two books, Anker's and ours, that single-out individual countries as environmental vanguards, both center on Scandinavian states.[30] Ironically, it can also be noted that despite their global contributions, Norway's and Sweden's activities on their respective home fronts over the long term are less convincing. Norway continues to practice whaling and is one of the largest oil producers in the world per capita, and in total, far ahead of countries such as Qatar, Libya, and Nigeria. Sweden, for its part, has harvested almost all of its old growth forest, dammed its great northern rivers for hydropower, and over the quarter century since 1999, managed to reach only one of its sixteen national environmental goals, which are set by the country itself (Chapter 1).

Hence, the two Nordic neighbors that produced the likes of Nobel laureate author-conservationist Selma Lagerlöf and passionate nature metaphysician Rolf Edberg (Sweden) and the world-leading eco-philosopher Arne Naess (Norway) are still today major exploiters of natural resources, and especially in the case of Sweden, highly industrialized. Such activities have to a not insignificant extent formed the basis of the prosperity that has enabled the Nordic model to flourish. Other Nordic countries differ markedly in their environmental history. Denmark, with few natural resources, has instead pursued the path of wind power and high-tech agriculture since the 1980s and stands at the top of global sustainability indexes. There is thus no unified "Nordic green model." As techno-modernist states of similar but distinct types, all Scandinavian countries have performed well on the environmental policy front. But they have done so following very different trajectories, conditioned by geographical factors, geopolitical position (Baltic,

[29] Arne Ruth, "The Second New Nation: The Mythology of Modern Sweden," *Daedalus* 113(1984): 53–96. Jenny Andersson & Mary Hilson, "Images of Sweden and the Nordic Countries," *Scandinavian Journal of History* 34(2009):3, 219–228.

[30] Denmark, a global leader in wind power, being the third; Iceland is also sometimes counted among the Scandinavian countries. The Nordic countries also include Finland (again ranked the world's happiest nation in 2023 for the sixth consecutive year!).

Arctic, North Atlantic), and political cultures. Norway and Sweden are wealthy nations, but they are nonetheless marked by their respective "resource peripheries" and varieties of the "resource curse" that affect countries that modernize on the basis of raw materials, be they forest, fish, minerals, or oil and gas.[31]

Again, it makes most sense to see Sweden's environmental exceptionalism not as the existence of a particularly abundant green genius that happened to hover close to the Arctic circle. Nor do we believe that it was an obvious byproduct of the country's vast forests, the popularity of mushroom and berry picking, and a widespread and deeply ingrained tradition of outdoor activities in Swedish nature, although these may have helped.[32] It is far more convincing to interpret a persistent and large social and political effort like the one we saw in Sweden during the second half of the twentieth century as a product of a political culture and a national strategy with internationalism as one of its central principles. In that respect, it may well be called an environmental exceptionalism that for a particular historical moment produced remarkable results and, indeed, brought progress, specifically in the emergent realm of GEG.

Arguing that the ultimate causes lie deeper, and are structural and strategic, is not to diminish the effort and brilliance of individual politicians, diplomats, activists, civic leaders, scientists, business leaders, and bureaucrats, who not only talked the talk but also often walked the walk. These individuals, operating through different domestic and international networks, undeniably helped foster Sweden's reputation as an environmentally progressive place. In this book, we have met a phenomenal array of such actors. But as a group of however skilled and forward-looking individuals, they would not have been able to pursue their ambitions and realize their visions had those not been in line with the broader interests of the country and rested upon what

[31] D. B. Carson, D. A. Carson, R. Porter, C. Y. Ahlin & P. Sköld, "Decline, Adaptation or Transformation: New Perspectives on Demographic Change in Resource Peripheries in Australia and Sweden," *Comparative Population Studies* 41(2017): 3–4. On Scandinavia's Arctic and North Atlantic resource connections, see Sverker Sörlin, ed., *Resource Extraction and Arctic Communities: The New Extractivist Paradigm* (Cambridge: Cambridge University Press, 2022). See also the contributions in Felipe Irarrázaval & Martín Arias-Loyola, eds., *Resource Peripheries in the Global Economy: Networks, Scales, and Places of Extraction* (Cham: Springer, 2021).

[32] Daniel Wolf-Watz, Klas Sandell & Peter Fredman, "Environmentalism and Tourism Preferences: A Study of Outdoor Recreationists in Sweden," *Scandinavian Journal of Hospitality and Tourism* 11(2011):2, 190–204.

governments and parliament could actually do. It was not (just) a stroke of genius. It was the general direction taken by an entire democratic nation. Seen in this coherent, retrospective light, it makes sense. It was a product of history.

DOES IT STILL WORK?

Here at the very end of our examination of the Stockholm story, we may ask: what happened to this coherent policy? Does Swedish environmental exceptionalism still exist today? What may happen to it in the future? These are questions, now broadly asked in the Swedish policy debate, which if answered in full would require another book.[33] What we can offer here in the final few pages is indeed only a sketch, drafted in the present about the recent past.

We have already indicated that Sweden's general position on the world stage is not the same now as it was in the last few decades of the twentieth century. *Stockholm* as an international hub of GEG thinking may not, however, have declined in status and visibility to the same extent. This may sound like a strange and somewhat paradoxical statement, but we think it is underpinned and supported by a certain logic. So far in this chapter, we have articulated the *policy rationale* for Sweden to take a proactive exceptionalist role. Let us now turn to the *external factors that can explain* the Swedish position and how these circumstances have changed since the Stockholm Conference. Although these changes started right after 1972, their implications have become more pronounced only in the last couple of decades.

National security and geopolitical context have been important drivers. There was also a set of skills that became embedded in the state apparatus for working within international fora, with Sweden continuing to learn through experience during the postwar period how useful international organizations could be.[34] Dag Hammarskjöld, a broad-minded liberal with a humanities and economics background – and an

[33] The future of Swedish exceptionalism is an issue that runs through several themes of politics at the same time, for example, foreign and neutrality policy (with increasing Swedish NATO collaboration, an application for membership in 2022 and full membership in March 2024) and the Swedish party structure (for a long time lacking a radical right wing populist party). One example, among many, discussing such themes in recent years: Jens Rydgren & Sara van der Meiden, "The Radical Right and the End of Swedish Exceptionalism," *European Political Science* 18(2019):3, 439–455.

[34] Ann-Marie Ekengren, *Olof Palme och utrikespolitiken: Europa och Tredje världen* (Umeå: Boréa, 2005).

avid nature lover[35] – rose to become Secretary General of the United Nations, killed in a still-today unresolved plane crash in Northern Rhodesia (today's Zambia). Sida became a pioneering agency of its kind that enmeshed multiple interests in Swedish society and was generally considered a success, both in the Global South and in other Western nations, although not always among Swedish opinion makers.[36] Despite – or perhaps because of – Sweden's advanced plans in the 1950s to develop its own nuclear weapon, the Swedish anti-nuclear and peace movement became a powerful political force, with many, especially female, social democratic leaders taking strong positions on the issue. Among them were the diplomats Inga Thorsson and Alva Myrdal, the latter also a long-standing member of the government. Both were active in articulating the rationale of Swedish internationalism, with an emerging feminist tinge, and played important roles in the Swedish initiative (Chapter 4).[37]

The institutional conditions for Sweden in developing this pronounced ideological version of internationalism were of central importance. In the postwar decades, there were no international alliances or functional arrangements that constrained the move to make Sweden engage in international cooperation on its own terms. This would change after 1972. The first set of changes concerned the deregulation of financial markets and banks in the early 1980s. In the same decade, the concept of "governance," with the diffusion of authority it implies, entered international discourse.[38] For a state that had relied heavily on its strong central government, the governance agenda reduced autonomy and ushered in a new world that Sweden needed to adapt to. Governance was a concept associated with the new neoliberal toolbox that came to be applied in most policy areas, with the environment as no exception (Chapter 6).[39] Economic crisis and currency pressures

[35] Hammarskjöld was an avid out of doors practitioner, vice president of the Swedish Tourist Association, president of the Swedish Mountaineering Club (Svenska Fjällklubben), and wrote poetry and essays on Swedish nature.

[36] Annika Berg, Urban Lundberg & Mattias Tydén, *En svindlande uppgift: Sverige och biståndet 1945–1975* (Stockholm: Ordfront förlag 2021).

[37] See for example Inga Thorsson, *Att internationalisera Sverige* (Stockholm: Tiden, 1971).

[38] Anne-Mette Kjaer, *Governance* (Cambridge: Polity, 2004), 1–7.

[39] A review article from 2008 demonstrates the proliferation of concepts and approaches, including "governance" that can be associated with the neoliberal influence on environmentalist discourse since the 1980s: Joseph Huber, "Pioneer Countries and the Global Diffusion of Environmental Innovations: Theses from the Viewpoint of Ecological Modernisation Theory," *Global Environmental Change* 18(2008):3, 360–367.

made Sweden question its distanced relationship to the European Community. When the government eventually applied for membership in 1991, it was not presented as a major policy change, rather a pragmatic adjustment, characteristically announced in a small passage in a crisis policy statement. In 1995, Sweden joined what had by then become the European Union, and although environment was national sovereignty, the European integration gradually reshaped the image, and self-image, of Sweden as a nation, and it became increasingly reluctant to take on the oversized role that it had played in the past. This development has sometimes been linked to political personalities, arguing that the charisma and leadership of Olof Palme and Alva Myrdal were lacking in the following, less illustrious generation of political leaders, across the political spectrum. There may be some truth to this. Palme was a spectacular presence on the international scene, easily outshining his predecessor Tage Erlander, who had served as Sweden's prime minister for the entirety of the postwar period up until Palme's ascent to the position in 1969.[40]

The relative weakening of Sweden's convening and connecting powers at the political level had less to do with individual Swedish politicians, however, or even Swedish environmental policies. Wider changes in European and global geopolitics were the main driving forces behind this relative decline in Sweden's authority in the realm of the environment. Agency instead shifted in the direction of the *Stockholm institutions* that could act, at least formally, independent of the state, although funding often came from the government's purse. This de facto deferring, even devolution, of governance should have already been visible in the way agency was presented in the chapters above. These demonstrate a historical arc, starting with almost exclusively state actors: diplomats, government representatives, and a limited set of scientists who were state employees or had close ties to political leaders and public officials. The state thus coordinated the early phase of GEG, which was in this sense distinctly more govern*ment* than govern*ance*. Its self-imposed remit covered the entire spectrum: from the moral high ground, speaking on behalf of humanity and the need for global cooperation, through to policy initiatives ranging from the local to the global. It comprised the funding and institutional infrastructure of science and knowledge, from public agencies to universities, research institutes, and high-profile centers like SIPRI and SEI.

40 Ekengren, *Olof Palme och utrikespolitiken.*

It also included the social mobilization of citizens, activists, and experts, many of which were funded by Sida and other Swedish government agencies. One prominent example is the Stockholm International Water Institute (SIWI), which bridges science, policy, and development work to improve water governance around the world. Since its founding in conjunction with the first Stockholm Water Symposium in 1991, SIWI has awarded the high-profile Stockholm Water Prize as part of the annual World Water Week in Stockholm, one of the largest international events on issues related to water and sustainability.[41] Other initiatives were funded through state expenditures for culture or education. The actual work in such cases was often carried out by strong and vital popular environmental and anti-nuclear movements, whose membership numbers have risen dramatically over the past fifty years. The Swedish Society for Nature Conservation (SSNC), which up through the 1950s was largely grounded in a combination of (male) naturalist expertise and nationalism, trebled its membership to 70,000 in the 1960s and 1970s as the environment went global and progressive, and trebled it again to reach 200,000 in the 1980s.[42] During this period, the environment became more civic; not less expert, but with a broadening understanding of what expertise could mean, and with less state mediation in the circulation of knowledge through the interaction of experts and civil society.[43]

DEFERRING FROM STATE TO SOCIETY

Under pressure from mounting constraints and new institutional and political commitments, the Swedish state has gradually deferred agency to a set of other actors that had been nurtured under the exceptionalist order. Research institutions, think tanks, academies, and their support structures in funding agencies and foundations continued to maintain a high profile. They increasingly stand out today as the inheritors of earlier state efforts and have served in many cases as a quasi-political task force for sustainability. One could think of this as a version of what has been called the politics of big issues playing out in (many) small places.[44] The

[41] https://siwi.org/ (retrieved August 6, 2022).
[42] Membership in the organization today stands at around 200,000.
[43] Sörlin, "Reconfiguring."
[44] Thomas Hylland Eriksen, *Small Places, Large Issues: An Introduction to Social and Cultural Anthropology* (1995), 4th ed. (London: Pluto Press, 2015).

Stockholm Environment Institute in particular has taken on this role. Officially declaring its origins to be part of the Stockholm Conference legacy, SEI is now a world-leading consultancy and think tank with offices on all major continents. This book has also paid attention to the Stockholm Resilience Centre, which has been especially active in conceptualizing work (e.g., Planetary Boundaries and Hothouse Earth) and has had strong ties from the start with the Academy of Sciences. The IGBP and Global Change programs – forerunners to a great deal of the work conducted at the SRC – had been based at the Academy since the 1980s and were succeeded there by one of the Future Earth global hubs following the transformation in 2015 of the international architecture for global change research. The Academy continues to host the Beijer Institute, as it has done since the first incarnation of the Institute was established there in 1977.

These are not just scientific institutes and consultancies that provide "policy advice" – itself a too-limited and slightly misconstrued way of understanding how the science-policy process plays out. Rather, such institutions have considerable policy agency of their own. Through initiatives, ideas and innovation, and action on the "con"-fronts, they replace some of the political clout that the state has either lost or deferred. Their alliance-building efforts have also been more dynamic outside Sweden, which is of course where most of the work on the Earth system scale takes place, conducting connective activities in order to catalyze larger actions rather than implementing change processes on their own.

Similarly, the profile of civic organizations has increased. The largest of those, SSNC, has evolved from a small, field-oriented expert organization in the middle of the twentieth century, to become a broad civic movement reaching hundreds of thousands of citizens. The organization's growth clearly reflects the far higher profile of nature and environment as areas of widespread popular concern. It has also engaged more actively in conservation and, more recently, climate as societal and political issues, providing consumer advice and ranking the "greenness" of political parties before general elections. SSNC and the Swedish/Nordic branches of Greenpeace and the WWF, all based in Stockholm, and several other small- and medium-sized civic organizatins and networks, have increasingly served as important nodes of global knowledge. This has in effect pluralized policy work, again shifting the nexus of agency from government to wider society. The work of civic movements has in recent years been crowned by the astounding success and worldwide fame of Greta Thunberg's Fridays for Future activism. Significant in its own right, this should also be seen as the tip of an iceberg that has grown slowly and silently, but surely. Something similar could be said about the municipal and regional networks across Sweden that have formed alliances of the willing to share ideas and experiences, establish benchmarks, and move forward – faster than the state – in addressing community concerns over issues related to people and environment.[45]

One could see this as a series of documented strengths of the institutional and civic capacities of Sweden and its capital, Stockholm, as a broker and major convening hub. Representatives of government should be credited for their formidable ingenuity in grasping the moment of opportunity that emerged in the late 1960s and early 1970s to raise the issue in the UN. They subsequently built the domestic institutions and international networks that have since then served as part of the infrastructure and architecture for GEG. Government, however, cannot be expected to perform all the necessary tasks, especially as the agenda expands – which it has, dramatically, in the case of the environment and sustainable development. The engagement of civil society tends to make commitment more resilient.[46] Civic movements also managed to effectively halt many of the attempts to modernize and transform – in the eyes of the critics: essentially destroy – Swedish city centers during the same period as Sweden's UN initiative, with the May 1971 "Battle of the Elms" in central Stockholm being the most famous example (Chapter 1). Partly impacted by such events, Swedish national and urban governments have been less prone to launch massive modernist state projects, and although some are still pursued, they are certainly not nearly as common, especially in urban centers.

This deferring of initiative and execution may over the long term have drawbacks, and the moment may in fact have passed when politics not only defers but also devolves its institutional creativity. In today's political landscape, with a marked shift toward nationalist/authoritarian conservative politics setting the agenda – with or without an actual majority in the electorate – there is a tendency that the government, in its ambition to win elections, withdraws the kinds of messages that

[45] One of these is Climate Communes (*Klimatkommunerna*), an association of some fifty cities and municipalities (designated communes out of a total of nearly 300). Including most of the largest Swedish cities, fewer rural communities, they wish to be a vanguard of climate neutrality. https://klimatkommunerna.se/

[46] Sonny S. Patel, M. Brooke Rogers, Richard Amlôt & G. James Rubin, "What Do We Mean by 'Community Resilience'?: A Systematic Literature Review of How It Is Defined in the Literature," *PLOS Currents: Disasters*. *1*. PMC 5693357 (2017).

risk losing swing voters, especially in the countryside and among the socio-economically less privileged.[47] These are among the groups that tend not to readily identify with the gospel of transformation and a dematerializing economy. Nor are their favorite media on the frontlines of climate activism; even in Sweden, the opposite is the case.[48] Recent Swedish governments, experiencing that kind of electoral calculus, have become more reluctant to accept the progressive environment and climate politics of the European Union, including its new "taxonomy" (of what is in line with EU climate and sustainability goals) and solidarity programs to assist the Union's poor. Sweden for one has taken a strong stand against minimum wages and opposed a climate-focused forest policy, instead defending voluntary agreements between firms and trade unions, and protecting the property rights of companies and foresting farmers and households. Sweden has also tried to protect its car manufacturing industry by slowing down the implementation of some of the tougher emission targets.

DEVOLUTION – THE ENVIRONMENTALISM OF THE RICH?

What used to be an official Swedish environmental foreign policy aligned with development and solidarity has drifted toward a more cautious navigation of domestic politics and international engagement, eager to not neglect opportunities for Sweden, which remains one of the richest countries in the world. To paraphrase the Catalan ecological economist José Martinez Alier's famous expression "the environmentalism of the poor" (2003), we may in the case of Sweden instead speak of a gradual swing toward an "environmentalism of the rich."[49] Such an "inverted solidarity environmentalism" does not threaten to provoke the more cynical and blatantly self-serving outlook of the anti-environmentalist populist right. On the contrary, it encompasses a more careful and selective determination of an environmental and climate agenda that does not violate national economic interests and, perhaps above all, the sentiments, or resentments, of the authoritarian right and its electoral base.

[47] Natascha Strobl, *Radikalisierter Konservatismus: Eine Analyse* (Berlin: Suhrkamp, 2021).

[48] Kjell Vowles & Martin Hultman, "Scare-Quoting Climate: The Rapid Rise of Climate Denial in the Swedish Far-Right Media Ecosystem," *Nordic Journal of Media Studies* 3(2021): 1.

[49] José Martinez Alier, *The Environmentalism of the Poor: A Study of Ecological Conflicts and Valuation* (Chaltenham: Edward Elgar, 2003).

Although this is not just a Swedish dilemma, it is a policy pathway one would not exactly have expected from "the first new green nation" with the added ambition, since the Paris Agreement, to become the "first fossil free welfare nation." What has likely happened is a forceful combination of two major processes. One is the shift in the political landscape since the financial crisis, which has provoked the discombobulating and perplexing reversals of center-left politics in Europe, including, after the 2022 election, in an unexpectedly pronounced way, in Sweden. The other, partly related, is a more long-term process captured by the 1980s catchphrase "the hollowing out of the state."[50] It referred at first to the lack of capacity of the state in Great Britain but became an established view of how states increasingly defined themselves under globalization and neoliberalism in the decades around the turn of the Millennium. States took steps back everywhere, lowering taxes, seeking market solutions, devolving agency to local and sectorial interests, and replacing political deliberation with various forms of competition. Another expression in line with this hollowing out of the state was "multi-level governance," which, also reflecting in some way the EU subsidiarity principle, declared that decisions and authority should stay on the level where the issues and the expertise are.

We think this is what has happened in Sweden, which can to some extent help explain the declining interest in green issues on the level of parliament and government, despite the activism and deep engagement among Greta-inspired young people and, for that matter, wide swaths of the citizenry. The current environmental and climate politics of Sweden – simultaneously puzzling, and disappointing for many – are not, we would argue, an isolated policy mistake made in 2022 and 2023 in the face of rising European crises such as Covid-19, war-generated energy shortages, increased migration, and climate-related phenomena such as worsening droughts, floods, and fires. Rather it is a phenomenon that has been developing gradually over a period of several decades, including the neoliberal tendencies that have been pronounced in Sweden since around 1990.[51] It is particularly important to note that it is not something that

[50] R. A. W. Rhodes, "The Hollowing Out of the State: The Changing Nature of the Public Service in Britain," *The Political Quarterly* 65(2005):2, 138–151. Bob Jessop, "Hollowing Out the 'Nation-state' and Multi-level Governance," In: *A Handbook of Comparative Social Policy* (London: Edward Elgar, 2013).

[51] Kristoffer Ekberg & Viktor Pressfeldt, "A Road to Denial: Climate Change and Neoliberal Thought in Sweden, 1988–2000," *Contemporary European History* 31(2022):4, 627–644. https://doi.org/10.1017/S096077732200025X.

is chiefly orchestrated by environment-oriented politicians or by civil servants, experts, or activists, who have often been forced to stand by, almost helplessly, as the insufficient and counter-environmental policies play out. Environmental politics has been the inadvertent object of changes that originated elsewhere, typically with reference to an unstoppable globalization, emblematic of the profound transformations in the role of the state in recent years.[52] What once seemed to be an almost inevitable, rational ascendency of a growing number of enlightened democracies to a reformed and ethically balanced position, *so zu sagen* within "planetary boundaries," has turned out to become a much more fraught and embattled terrain. As a result, some of the more utopian and normative versions of the emerging "Green State," heralded in the literature since the early 2000s, already seem hard to attain.[53]

In Sweden, long singled out as a green state vanguard, this devolution has been concealed by the fact that so many of the institutions, arrangements, and processes that were built in the postwar decades up until the end of the previous century, when the state provided more directionality and resources, have stayed vital and shouldered the agency that the state gradually deferred. To return to one of the central themes of this book, it is striking how well this sits with the observation that *the city* can play a significant role in policy. To put it very succinctly, Stockholm institutions have stepped in and filled some of the vacuum caused by the hollowing out and squeezing of the state by constraining external forces. This should perhaps come as no surprise. Cities also compete for attention, influence, and visibility in a globalized world. To successfully compete, they must also

[52] Significant examples in a vast literature are Arthur P. J. Mol, *Globalization and Environmental Reform: The Ecological Modernization of the Global Economy* (Cambridge, MA: MIT Press, 2000), and D. C. Esty & Michael Porter, "National Environmental Performance: An Empirical Analysis of Policy Results and Determinants," *Environment and Development Economics* 10(2005):4, 391–434. For an overview of the literature until 2008, see Frank Biermann & Philipp Pattberg, "Global Environmental Governance: Taking Stock, Moving Forward," *Annual Review of Environment and Resources* 33(2008): 277–294. https://doi.org/10.1146/annurev.environ.33.050707.085733.

[53] E.g., Lennart J. Lundqvist, "A Green Fist in a Velvet Glove: The Ecological State and Sustainable Development," *Environmental Values* 10(2001):4, 455–472; Robyn Eckersley, *The Green State: Rethinking Democracy and Sovereignty* (Cambridge, MA: The MIT Press, 2004); Andreas Duit, Peter Feindt & James Meadowcroft, "Greening Leviathan: The Rise of the Environmental State?" *Environmental Politics* 25(2016):1, 1–23. See also the careful overview of the literature by Annika Kronsell, & Karin Bäckstrand, "The Green State Revisited," In: *Rethinking the Green State: Environmental governance towards climate and sustainability transitions* (London: Routledge, 2015), 1–23.

learn to collaborate. We hear most about this global competition in areas such as finance, technology, art, and fashion (Chapter 1), but, as we have consistently argued, it is essential to see how similar processes take place in environmental governance. To be green is also, or has become since 1972, increasingly a game of urban competition and responsible branding.

This is a far-from-new observation. Alliances of proactive green cities have existed since the 1990s. One of the largest organizations of this kind, Local Governments for Sustainability, started in 1990 and now coordinates activities among urban and regional governments in more than 125 countries around the world. Similar initiatives exist in North America and Europe, including Sweden, where the international Transition Town movement and network have also taken root across the country. Indeed, some of the earliest environmental action in Stockholm and around Sweden was focused on the urban environment (Chapter 4). Stockholm has since the 1960s taken a prominent role in this area and received a great deal of international recognition, including being named the first European Green Capital in 2010 (Chapter 1). What is new, however, in Sweden and elsewhere, is the increasing mismatch between the different levels of multilevel leadership, with cities in many cases being the most pro-active. In Stockholm and other more or less progressive places around the world, the analysis is simple: if you are not green, you are not on the right side of change, and will soon be out of the game. While at the state level, government officials – bound by the constraints and compromises of national politics – often do not want, or are simply unable, to take on a leadership role, even now during the Agenda 2030 decade when the sustainability transformation was supposed to accelerate.

*

In the current crisis-ridden moment of history, what can we expect of GEG, and what can we expect of Stockholm? We should probably look ahead at, and embrace, continued critical involvement and active participation of civil society, young people, and the private sector in fostering GEG. Such involvement certainly cannot solve all problems, especially not those at the level of international relations and geopolitics, but there is evidence to suggest that engagement from below and from consciously acting citizens influences politics at the state level, and there is rising interest "to understand how individuals in affluent societies reason around their own actions" in relation to environment and climate change.[54]

[54] Recent work on Sweden in this vein is Nina Wormbs & Maria Wolrath Söderberg, "Knowledge, Fear, and Conscience: Reasons to Stop Flying Because of Climate Change,"

This speaks in favor of opinion-making and education, to continue building understanding and narratives that mobilize science in shaping a new kind of general education that encompasses the basic tenets of geo-anthropology and the Anthropocene. It also entails profound opportunities, and challenges, for educational systems at all levels. Again, this cannot take place without civic effort to overcome inertia and push environmental politics out of its current deferral mode.

We would likely see a continued renegotiation of the knowledge bases for "the environment." These have continuously widened since the emergence of the concept in its modern understanding. We are now in a phase where the implications and challenges for societies of environmental and climate changes are moving to the forefront on a large scale, also in affluent societies. To reach a more sustainable state of the world, economies will have to change and legal systems, inherited since the birth of capitalism and global trade, need to be reformed. The very criteria and parameters of what it means to lead meaningful lives in sustainable societies under a different kind of human–Earth relationship must be re-articulated. A particular challenge, especially since this requires democratic discussion and decision-making, is the speed with which the changes will take place. Multiple timescales need to be operating in tandem.

An increased role will likely be assumed by the human sciences, that is, scholarly disciplines that provide insights on, and in favor of, change in human and social organization. These are vast fields of knowledge, some of which have not been sufficiently mobilized for the environment in the past. In recent years, there has been a distinct rise in interest in the environmental, climate, energy, and other humanities that speak to these challenges.[55] The concept of Environmental Social Sciences and Humanities (ESSH) has entered circulation to argue for the significant potential that these knowledge fields can provide.[56] In the past, much of the research within the human and social sciences was engaged to build the kinds of incentive and value structures that characterize the

Urban Planning 6(2021):2, 314–324. Nina Wormbs & Maria Wolrath Söderberg, "Thinking Structures of Climate Delay: Internal Deliberations among Swedes with Sustainable Ambitions," *Environment, Development and Sustainability* (online July 31, 2023), https://doi.org/10.1007/s10668-023-03618-x (retrieved August 17, 2023).

[55] J. J. Williams, "The New Humanities," *The Chronicle of Higher Education* 65(2019), December 13.

[56] Noel Castree, William M. Adams, John Barry, et al., "Changing the Intellectual Climate," *Nature Climate Change* 4(2014):9, 763–768.

environmentally destructive states that we now have. There is, hence, an emerging potential for reform and new contributions. Religion, ideas, values, social and psychological drivers, history, media, communication, thought, and affect – such dimensions will likely play larger roles. Design, architecture, and social planning will also widen their scope.

There will be continued politicization of climate and environment, perhaps even increased polarization. Geopolitical conflict in the twenty-first century will likely be hybrid and complex, involve natural resources, and put pressures on limited Earth Systems properties (atmosphere, oceans, forests, glaciers, soils, etc.). These are life-sustaining governable objects of the Earth that are required for the well-being of all but are affected to various degrees by states, firms, social groups, and other types of actors. Their abuse and devastation is not (yet) considered criminal behavior, but regardless of their status they are driving potential conflict, further necessitating transformation in order to lower the risk of conflict. Polarization on the national level will thus be accompanied by deepened polarization at the global and regional level.

The fact that politics are hybridizing will also play out in a new kind of "politics with other means." These were Clausewitz' famous words for "war." Distinctly environmental war, or climate, resource, or energy wars, or combinations of these, cannot be excluded; they will be part of hybrid conflicts. A similar politics with other means will take place on the national, local, and community level, where certain societal units, prominently including cities, will keep defining livable and sustainable futures. This may include lifestyle choices, values, registers of affect and attitude, ideas of family life, or evolving and emerging versions of modernity. Ways of life are, inevitably, also ways of (lived and applied) environmental governance. This may become politics that have so far had less articulation, and with most of it yet to come.

The trajectory of Stockholm since the mid-twentieth century remains a template for how such politics with other means could look in a world undergoing multiple crises and facing tremendous challenges. It at least provides a series of experiences, and of institutional and political mechanisms – and bold moves – that we think can be useful as new ideas emerge in an increasingly complex world, at least as unforeseeable as it has ever been.

Selected Bibliography

All references for this volume are presented in full bibliographic detail in the footnotes, including both printed and archival primary sources, interviews, web pages, and other nonprinted information, and secondary published sources (books and articles). In this selected bibliography, we have compiled a selection of published material from the references that we find particularly illustrative of the main thrust of the volume. The selection focuses on relatively recent work, with a preference for books, on the role of Stockholm and Sweden, on the UN 1972 Stockholm Conference, on the rise of global environmental governance and its relation to Earth System science, and on the significance of cities as knowledge-dense hubs of policy innovation. In addition, this list includes some of the central theoretical and conceptual work that has been important for us.

Andersson, Jenny, "Choosing Futures: Alva Myrdal and the Construction of Swedish Futures Studies 1967–1972," *International Review of Social History* 51(2006):2, 277–295.

Andersson, Jenny & Mary Hilson, "Images of Sweden and the Nordic Countries," *Scandinavian Journal of History* 34(2009):3, 219–228.

Anker, Peder, *The Power of the Periphery: How Norway Became an Environmental Pioneer for the World* (Cambridge: Cambridge University Press, 2020).

Aronowsky, Leah V., *The Planet as Self-regulating System: Configuring the Biosphere as an Object of Knowledge, 1940–1990*, PhD diss. (Harvard University, Graduate School of Arts & Sciences, 2018).

Bäckstrand, Karin & Annica Kronsell, eds., *Rethinking the Green State: Environmental Governance towards Climate and Sustainability Transitions* (Abingdon, Oxon: Routledge, 2015).

Benson, Etienne, *Surroundings: A History of Environments and Environmentalisms* (Chicago, IL & London: The University of Chicago Press, 2020).

Bergquist, Ann-Kristin & Kristina Söderholm, "Green Innovation Systems in Swedish Industry, 1960–1989," *Business History Review* 85(2011):4, 677–698.

Bernhard, Carl Gustaf, *The Beijer Institute: 1977–1989* (Stockholm: The Royal Swedish Academy of Sciences, 1991).
Biermann, Frank, *Earth System Governance: World Politics in the Anthropocene* (Cambridge, MA: MIT Press, 2014).
Biermann, Frank & Eva Lövbrand, eds., *Anthropocene Encounters: New Directions in Green Political Thinking* (Cambridge: Cambridge University Press, 2019).
Biermann, Frank & Rakhyun E. Kim, eds., *Architectures of Earth System Governance: Institutional Complexity and Structural Transformation* (Cambridge: Cambridge University Press, 2020).
Bohn, Maria, "Concentrating on CO_2: The Scandinavian and Arctic Measurements," In: J. R. Fleming & Vladimir Jankovich, eds. *Osiris 26: Klima* (Chicago, IL: The University of Chicago Press, 2011), 165–179.
Bolin, Bert, ed., *The Atmosphere and the Sea in Motion: Scientific Contributions to the Rossby Memorial Volume* (New York: Rockefeller Institute Press & Oxford University Press, 1959).
Bolin, Bert, *A History of the Science and Politics of Climate Change: The Role of the Intergovernmental Panel on Climate Change* (Cambridge: Cambridge University Press, 2007).
Borowy, Iris, *Defining Sustainable Development for Our Common Future: A History of the World Commission on Environment and Development (Brundtland Commission)* (London & New York: Routledge, 2013).
Caldwell, Lynton Keith, *International Environmental Policy: Emergence and Dimensions* (Durham, NC & London: Duke University Press, 1984).
Christensen, Miyase, Annika Nilsson & Nina Wormbs, eds., *Media and Arctic Climate Politics: Breaking the Ice* (New York: Palgrave MacMillan, 2013).
Chu, Pey-Yi, *The Life of Permafrost: A History of Frozen Earth in Russian and Soviet Science* (Toronto, ON: University of Toronto Press, 2021).
Conca, Ken, *An Unfinished Foundation: The United Nations and Global Environmental Governance* (Oxford: Oxford University Press, 2015).
Connolly, William E., *Facing the Planetary: Entangled Humanism and the Politics of Swarming* (Raleigh, NC: Duke University Press, 2017).
Croeser, Eve, *Eco-Socialism and Climate Justice: A Neo-Gramscian Analysis* (Abingdon, Oxon: Routledge, 2021).
Dasgupta, Partha, *The Economics of Biodiversity: The Dasgupta Review* (London: HM Treasury, 2021).
Dryzek, John, *The Politics of the Earth: Environmental Discourses* (Oxford: Oxford University Press, 1997).
Duit, Andreas, ed., *State and Environment: The Comparative Study of Environmental Governance* (Cambridge, MA: The MIT Press, 2014)
Eckersley, Robyn, *The Green State: Rethinking Democracy and Sovereignty* (Cambridge, MA: The MIT Press, 2004).
Ekberg, Kristoffer & Martin Hultman, "A Question of Utter Importance: The Early History of Climate Change and Energy Policy in Sweden, 1974–1983," *Environment and History* 29(2021):3, 399–421.
Ekberg, Kristoffer & Viktor Pressfeldt, "A Road to Denial: Climate Change and Neoliberal Thought in Sweden, 1988–2000," *Contemporary European History* 31(2022):4, 627–644.

Ekström, Anders & Staffan Bergwik, eds., *Times of History, Times of Nature: Temporalization and the Limits of Modern Knowledge* (New York: Berghahn Books, 2022).

Engfeldt, Lars-Göran, *From Stockholm to Johannesburg and Beyond: The Evolution of the International System for Sustainable Development Governance and Its Implications* (Stockholm: Ministry of Foreign Affairs, 2009).

Engh, Sunniva, "The Conscience of the World?: Swedish and Norwegian Provision of Development Aid," *Itinerario* 33(2009):2, 65–82.

Ernstson, Henrik & Sverker Sörlin, eds., *Grounding Urban Natures: Histories and Futures of Urban Ecologies* (Cambridge, MA: MIT Press, 2019).

Ernstson, Henrik, Sverker Sörlin & Thomas Elmqvist, "Social Movements and Ecosystem Services: The Role of Social Network Structure in Protecting and Managing Urban Green Areas in Stockholm," *Ecology and Society* 13(2008):2, 703–722.

Fleming, James Roger, *Inventing Atmospheric Science: Bjerknes, Rossby, Wexler, and the Foundations of Modern Meteorology* (Cambridge, MA: The MIT Press, 2016).

Gieryn, Thomas F., *Truth-Spots: How Places Make People Believe* (Chicago, IL & London: The University of Chicago Press, 2018).

Goossen, Benjamin W., "A Benchmark for the Environment: Big Science and 'Artificial' Geophysics in the Global 1950s," *Journal of Global History* 15(2020):1, 149–168.

Griffith Spears, Ellen, *Baptized in PCBs: Race, Pollution, and Justice in an All-American Town* (Chapel Hill, NC: University of North Carolina Press, 2014).

Hajer, Maarten A., *The Politics of Environmental Discourse: Ecological Modernization and the Policy Process* (Oxford: Oxford University Press, 1995).

Harper, Kristine, *Weather by the Numbers: The Genesis of Modern Meteorology* (Cambridge, MA: MIT Press, 2008).

Höhler, Sabine, *Spaceship Earth in the Environmental Age, 1960–1990* (London: Pickering & Chatto, 2015).

Ivanova, Maria, *The Untold Story of the World's Leading Environmental Institution: UNEP at Fifty* (Cambridge, MA: The MIT Press, 2021).

Jänicke, Martin, "Trend-setters in Environmental Policy: The Character and Role of Pioneer Countries," *European Environment* 15(2005):2, 129–142.

Jasanoff, Sheila & Marybeth Long Martello, *Earthly Politics: Local and Global in Environmental Governance* (Cambridge, MA: MIT Press, 2004).

Javaudin, Enora, "Environmental Problem-Solvers?: Scientists and the Stockholm Conference," In: Wolfram Kaiser & Jan-Henrik Meyer, eds. *International Organizations and Environmental Protection: Conservation and Globalization in the Twentieth Century* (New York: Berghahn Books, 2017), 74–102.

Jørgensen, Dolly, Finn Arne Jørgensen & Sara B. Pritchard, eds., *New Natures: Joining Environmental History with Science and Technology Studies* (Pittsburgh, PA: University of Pittsburgh Press, 2013).

Kaijser, Arne, "Combatting 'Acid Rain': Protecting the Common European Sky," In: Anna-Katharina Wöbse & Patrick Kupper, eds. *Greening Europe: Environmental Protection in the Long Twentieth Century – A Handbook* (Berlin & Boston: De Gruyter Oldenbourg, 2021), 363–388.

Kaiser, Wolfgang & Jan-Henrik Meyer, eds., *International Organizations and Environmental Protection: Conservation and Globalization in the Twentieth Century* (New York & Oxford: Berghahn Books, 2017).

Kjellén, Bo, *A New Diplomacy for Sustainable Development: The Challenge of Global Change* (New York: Routledge, 2007).

Kronsell, Annika, "Sweden: Setting a Good Example," In: Mikael Skou Andersen & Duncan Liefferink, eds., *European Environmental Policy: The Pioneers* (Manchester: Manchester University Press, 1997).

Lachmund, Jens, *Greening Berlin: The Co-Production of Science, Politics, and Urban Nature* (Cambridge, MA: The MIT Press, 2013).

Larsson Heidenblad, David, *The Environmental Turn in Postwar Sweden: A New History of Knowledge*, transl. Arabella Childs (Lund: Lund University Press, 2021).

Linnér, Björn-Ola, *The Return of Malthus: Environmentalism and Post-War Population–Resource Crises* (Cambridge: White Horse Press, 2003).

Luccarelli, Mark & P. G. Røe, eds., *Green Oslo: Visions, Planning, Discourse* (London: Elgar, 2012).

Lundgren, Lars J., *Acid Rain on the Agenda: A Picture of a Chain of Events in Sweden, 1966–1968* (Lund: Lund University Press, 1998).

Lundqvist, Lennart J., *The Hare and the Tortoise: Clean Air Policies in the United States and Sweden* (Ann Arbor: University of Michigan Press, 1980).

Lundqvist, Lennart J., "A Green Fist in a Velvet Glove: The Ecological State and Sustainable Development," *Environmental Values* 10(2001):4, 455–472.

Lundqvist, Lennart J., *Sweden and Ecological Governance: Straddling the Fence* (Manchester: Manchester University Press, 2004).

Macekura, Stephen, *Of Limits and Growth: The Rise of Global Sustainable Development in the Twentieth Century* (Cambridge: Cambridge University Press, 2015).

Mårald, Erland & Christer Nordlund, "Modern Nature for a Modern Nation: An Intellectual History of Environmental Dissonances in the Swedish Welfare State," *Environment & History* 26 (2020):4, 495–520.

Marklund, Carl, "Double Loyalties?: Small-State Solidarity and the Debates on New International Economic Order in Sweden during the Long 1970s," *Scandinavian Journal of History* 45(2020):3, 384–406.

McCormick, John, *Reclaiming Paradise: The Global Environmental Movement* (Bloomington & Indianapolis: Indiana University Press, 1989).

McNeill, John R. & Peter Engelke, *The Great Acceleration: An Environmental History of the Anthropocene since 1945* (Cambridge, MA: The Belknap Press of the Harvard University Press, 2016).

Merchant, Carolyn, *The Anthropocene and the Humanities: From Climate Change to a New Age of Sustainability* (New Haven, CT & London: Yale University Press, 2020).

Oppenheimer, Michael, Naomi Oreskes, Dale Jamieson, Keynyn Brysse, Jessica O'Reilly, Matthew Shindell & Milena Wazeck, *Discerning Experts: The Practices of Scientific Assessment for Environmental Policy* (Chicago, IL: The University of Chicago Press, 2019).

Owens, Susan, *Knowledge, Policy, and Expertise: The UK Royal Commission on Environmental Pollution 1970–2011* (Oxford: Oxford University Press, 2015).

Paglia, Eric, *The Northward Course of the Anthropocene: Transformation, Temporality and Telecoupling in a Time of Environmental Crisis*, PhD diss. (Stockholm: KTH Royal Institute of Technology, 2016).

Paglia, Eric, "The Swedish Initiative and the 1972 Stockholm Conference: The Decisive Role of Science Diplomacy in the Emergence of Global Environmental Governance," *Humanities and Social Sciences Communications* 8(2021), 2.

Paglia, Eric & Sverker Sörlin, "Greening Our Common Fate: Stockholm as a Node of Global Environmental Memory," In: Glenda Sluga, Kate Darian-Smith & Madeleine Herren, eds. *Sites of International Memory* (Philadelphia, PA: University of Pennsylvania Press, 2023), 237–265.

Radkau, Joachim, *Nature and Power: A Global History of the Environment* (German orig. 2002) (Cambridge: Polity, 2008).

Robin, Libby, Sverker Sörlin & Paul Warde, eds., *The Future of Nature. Documents of Global Change* (New Haven, CT: Yale University Press, 2013).

Rockström, Johan, Will Steffen, & Kevin Noone, et al., "A Safe Operating Space for Humanity," *Nature* 461 (2009), 472–475.

Ruth, Arne, "The Second New Nation: The Mythology of Modern Sweden," *Daedalus* 113(1984):2, 53–96.

Rydgren, Jens & Sara van der Meiden, "The Radical Right and the End of Swedish Exceptionalism," *European Political Science* 18(2019):3, 439–455.

Sarasini, Steven, "Constituting Leadership via Policy: Sweden as a Pioneer of Climate Change Mitigation," *Mitigation and Adaptation Strategies for Global Change* 14(2009):7, 635–653.

Schleper, Simone, *Planning for the Planet: Environmental Expertise and the International Union for Conservation of Nature and Natural Resources, 1960–1980* (New York & Oxford: Berghahn Books, 2019).

Schmelzer, Matthias, *The Hegemony of Growth: The OECD and the Making of the Economic Growth Paradigm* (Cambridge: Cambridge University Press, 2016).

Schmelzer, Matthias, "Born in the Corridors of the OECD: The Forgotten Origins of the Club of Rome, Transnational Networks, and the 1970s in Global History," *Journal of Global History* 12(2017):1, 26–48.

Selcer, Perrin, *The Postwar Origins of the Global Environment* (New York: Columbia University Press, 2018).

Söderholm, Kristina & Ann-Kristin Bergquist, "Firm Collaboration and Environmental Adaptation: The Case of the Swedish Pulp- and Paper Industry 1900–1990," *Scandinavian Economic History Review* 60(2012):2, 183–211.

Söderqvist, Thomas, *The Ecologists: From Merry Naturalists to Saviours of the Nation: A Sociologically Informed Narrative Survey of the Ecologization of Sweden 1895–1975* (Stockholm: Almqvist & Wiksell, 1986).

Sörlin, Sverker, "Ice Diplomacy and Climate Change: Hans Ahlmann and the Quest for a Nordic Region beyond Borders," In: S. Sörlin, ed., *Science, Geopolitics and Culture in the Polar Region: Norden beyond Borders* (Farnham: Ashgate, 2013), 23–54.

Sörlin, Sverker, "The Environment as Seen through the Life of a Journal: *Ambio* 1972–2022," *Ambio: A Journal of Environment and Society* 50(2021):1, 10–30.

Sörlin, Sverker, "Environmental Times: Historical and Scientific Temporalities from Annales to Anthropocene, ca 1920–2020," In: Anders Ekström &

Staffan Bergwik, eds. *Times of History, Times of Nature: Temporalization and the Limits of Modern Knowledge* (New York: Berghahn, 2022), 64–101.
Sörlin, Sverker ed., *Resource Extraction and Arctic Communities: The New Extractivist Paradigm* (Cambridge: Cambridge University Press, 2023).
Sörlin, Sverker, Paul Warde, Isobel Akerman, Thomas Harbøll Schrøder, Jasmin Höglund Hellberg, Sabine Höhler, Erik Isberg, Eric Paglia & Gloria Samosír, "The Great Dispersal: The Fall and Rise of Global Environmental Governance," (in review).
Speth, James Gustav & Peter Haas, eds., *Global Environmental Governance: Foundations of Contemporary Environmental Studies* (Washington, DC: Island Press, 2006).
Steffen, Will, Katherine Richardson, Johan Rockström, et al., "Planetary Boundaries: Guiding Human Development on a Changing Planet," *Science* 347(2015):6223, 736–747.
Steffen, Will, et al., "The Trajectory of the Anthropocene: The Great Acceleration," *The Anthropocene Review* 2 (2015):1, 81–98.
Steffen, Will, Johan Rockström, Katherine Richardson, et al., "Trajectories of the Earth System in the Anthropocene," *Proceedings of the National Academy of Sciences of the United States of America* 115 (2018):33, 8252–8259.
Stone, Peter, *Did We Save the Earth at Stockholm?: The People and Politics in the Conference on the Human Environment* (London: Earth Island, 1973).
Strong, Maurice, *Where on Earth Are We Going?* (New York & London: Texere, 2000).
Thomas, Julia Adeney, "Using Japan to Think Globally: The Natural Subject of History and Its Hopes," In: Ian Jared Miller, Julia Adeney Thomas & Brett L. Walker, eds. *Japan at Nature's Edge: The Environmental Context of a Global Power* (Honolulu: University of Hawai'i Press, 2013), 293–310.
Thomas, Julia Adeney, ed., *Altered Earth: Getting the Anthropocene Right* (Cambridge & New York: Cambridge University Press, 2022).
Thomas, Julia Adeney, Mark Williams & Jan Zalasiewicz, *The Anthropocene: A Multidisciplinary Approach* (Cambridge: Polity, 2020).
Thunberg, Greta, *No One Is Too Small to Make a Difference* (London: Penguin Books, 2019).
van Munster, Rens & Casper Sylvest, eds., *The Politics of Globality Since 1945: Assembling the Planet* (London & New York: Routledge, 2016).
Walker, Richard, *The Country in the City: The Greening of the San Francisco Bay Area* (Seattle, WA: University of Washington Press, 2008).
Ward, Barbara & René Dubos, *Only One Earth: The Care and Maintenance of a Small Planet* (New York: W.W. Norton, 1972).
Warde, Paul & Sverker Sörlin. "Expertise for the Future: The Emergence of Environmental Prediction c. 1920–1970," In: Jenny Andersson & Eglė Rindzevičiūtė, eds. *The Struggle for the Long Term in Transnational Science and Politics: Forging the Future* (Abingdon: Routledge, 2015), 38–62.
Warde, Paul, Libby Robin & Sverker Sörlin, *The Environment: A History of the Idea* (Baltimore, MD: Johns Hopkins University Press, 2018).

Index

Page numbers in *italic* refers to figures.

350 advertisement, 217, *218*, 220
350-ppm planetary boundary, 217, 256

Abisko, Sweden, 30
acid rain, 123–124, 126–129, 142, 148–149, 176
adaptive management, 246
Advisory Group on Greenhouse Gases (AGGG), 191–192, 194, 197, 198, 200–201
Aeronautical Meteorology, 95–96
Aga Khan IV, 188
Agenda 2030 decade, 16, 20, 289
aggregative expertise, 242
Ahlmann, Hans W., 86, 105, 106, 108, 109, 110–111, 112
air pollution, 29, 101, 176
Air Research and Development Command, 116
Alfredson, Hans, 70, 71
Alfvén, Hannes, 85, 115, 168
Allen, Myles, 256
Ambio journal, 39, 90, *164*, 163–167, 170, 178, 179, 209, 228
Amsterdam Declaration on Earth System Science, 209, 227
Andersson, Magdalena, *294*, 292–295
Anér, Kerstin, 74
Ångström, Anders, 119
Anker, Peder, 42
Annan, Kofi, 29
Antarctic Treaty, 86
Anthropocene, 5, 38, 42, 43, 164, 224, 226–227, *228*
Anthropocene Review journal, 223
Arctic
  as hot zone in Cold War, 86, 110
  warming of, 110
Arrhenius Laboratory, University of Stockholm, 118
Arrhenius, Svante, 93, *94*, 101, 243
artificial intelligence, 285
arts, contribution to widening the social space of the environment, 68–70, 72, 125
Askö workshops, 215, 247, 248
Åström, Sverker, *134*
  call for scientific report to support UN initiative, 138
  diplomatic strategy at UN, 134–135, 136–138
  on national sovereignty, 30
  preparations for Stockholm Conference, 143–45
  speech to UNGA, 35, 122, 123, 140, 142
  speeches to ECOSOC, 138, 139–140, 142
atmospheric chemistry, 128, 148
Atomic Energy Commission, 96

Bali Road Map, 220
Baltic Sea Declaration, 166

Baltic Sea Joint Comprehensive Environmental Programme, 166
Baltic Sea Prime Ministers Conference, Ronneby 1990, 165
Banduri, Tariq, 222
Bastida, Xiye, 288
Baumol, William, 248
Beijer Foundation, 214
Beijer Institute, 39, 154, 165, 167–169, 180, 185, 186–189, 190, 191, 195, 197, 212
Beijer Institute of Ecological Economics, 167, 210, 213–216, 247, 248
Beijer, Kjell, 167–168, 305
Bengtsson, Ingemund, 155
Bennet, Carl, 311
Bergen School, 105, 113
Bergeron, Tor, 105, 109
Berkner, Lloyd V., 84, 86
Bernhard, Carl Gustaf, 168, 186
BESK (*Binär Elektronisk Sekvenskalkylator*) computer, 97–100, 114, 117
Biermann, Frank, 16
Big World, Small Planet: Abundance within Planetary Boundaries (Rockström & Klum), 265
Bildt, Carl, 60
Billner, Börje, 135, 138, 145
biogeochemistry, 271
biosphere, 87
Bjerknes, Jacob, 109
Bjerknes, Vilhelm, 91, 105, 109
Blavatsky, Anna Helena, 272
Blix, Hans, 28, 155
Board of Computing Machinery, 107
Bohn, Viveka, 74
Bolin, Bert, 99, *219*
  acid rain case study, 148
  and AGGG, 192
  approach to climate change, 220
  atmospheric chemistry, 102
  as authority on biogeochemistry, 206–207
  on carbon tax, 64
  career, 89, 121–122
  on computer modelling, 170
  conference reports, 182
  death, 249
  early years and education, 119–120
  formation of climate governance institutions, 183
  and IGBP, 207–210
  and IMI, 88–89, 90, 119
  and IPCC, 35, 200, 202, 203–204
  legacy, 94
  as meta-expert, 243
  organization of conferences, 169, 179
  and politicization of climate change, 188
  reputation, 105, 119–120, 121, 169
  as research student, 98–99, 100
  as science diplomat, 121, 202, 220
  SCOPE reports, 207
  and SEI, 211, 214
  and Villach-Bellagio conference 1987, 192
Bolin, Richard, 119
Borgström, Georg, 166, 238, 243
Brand, Stewart, 299
Brazil, 138
Bretherton Diagram, 204
Broadgate, Wendy, 268
Brodin, Gunnar, *213*
Brohult, Sven, 129
Bruce, Jim, 192, 196
Brundtland Commission, 184, 189, 190, 196, 279
Brundtland Report, 156, 184–185, 196, 279
Brundtland, Gro Harlem, 184, 196
Buch, Kurt, 102
Budyko, Mikhail, 150
Bush, Vannevar, 111
Byers, Horace, 105

Caldwell, Lynton, 6
Callendar, Guy Stewart, 101
carbon dioxide
  effect on global warming, 97, 101–104, 142
  tracking of atmospheric concentrations, 89
carbon taxes, 31, 64
Carl Gustaf XVI, *King*, *142*, 174–175, 187, 294
Carson, Rachel, 243
Center Party, Sweden, 32
Chadwick, Michael, 187
Chakrabarty, Dipesh, 22
Challenges of a Changing Earth Conference, Amsterdam 2001, 209, 227, 252
Chalmers University of Technology, Gothenburg, Sweden, 281
Charney, Jule, 97, 105, 120, 121
Childs, Marquis, 5, 30, 61
Christian Democrats, Sweden, 60

Christian, David, 230
CIA, Office of Scientific Intelligence (OSI), 113
cities
competition between, 323
global cities, 9–11
global green capitals, 9
green cities, 82, 324
role in policy, 323
why they matter, 22–26
Clark, William, 192, 215, 250, 251, 252
climate change
2 degree temperature target, 198–199, 257
anthropogenic climate change, 97, 104
carbon dioxide and global warming, 97, 104, 142
and climate crisis discourse, 220
drivers of, 170
as IGY topic of concern, 88
intertwined science and politics of, 122
mitigation and adaptation policies, 180
polar climate change, 110
policy implications of, 182
politicization of, 179–184, 190, 191
promotion as issue of global political relevance, 151
transnational climate change, 94–97
climate science
advancing through scientific activism, 191–201
cutting-edge research at Stockholm University, 120
institutional foundations, 89–90
institutionalization, 119, 121–122, 124
Club of Rome, 175
Earth3 report, 282
Limits to Growth report, 147, 235, 273, 278, 279, 282
Cold War, Arctic as hot zone, 86, 110
Committee for Research and Factual Issues (*Kommittén for forskning och andra sakfrågor*, CRF), 145, 147, 152
Committee on Atmospheric Sciences, 120
Commoner, Barry, 159, 160, 162, 166, 243
conceptualizing work, 38, 309–310
connecting with others, 38–39, 306–307, 317–318
connection, concept of, 37–41
contributary expertise, 38
convening power, 39, 107, 192, 197, 252–253, 254, 307–308, 317–318
Convention on Long-Range Transboundary Air Pollution, 29, 176
con-words, 37–41, 306–310
COP15, 256
Copenhagen Accord, 199, 257
Cornell, Sarah, 267
Croll, James, 223
Crosby, Alfred W., 228–229
Crumley, Carole, 254
Crutzen, Paul, 63, 90, 163, 164, 194, 199, 224, 257–258, 268
cryosphere, 87

Dag Hammarskjöld Foundation, 39
Dahl, Birgitta, 74, 202, 210, 212
Dahlem Conference, 2005, 224, 226, 230
Dalai Lama, 288
Daléus, Lennart, 32
Daniel Guggenheim Fund for the Promotion of Aeronautics, 95–96
Danielsson, Tage, 70, 71
Dasgupta, Partha, 210, 214, 215, 247, 248
Davos Man, 265, 287
Davos, Switzerland, 265
Dawkins, Richard, 271
Declaration on the Human Environment (Stockholm Declaration), 28, 35, 125, 145, 154–155
Degefa, Worknesh, 192
Denmark, 87, 114, 313
Developing Policies for Responding to Climate Change report, 198
Development Dialogue journal 32
Diamandis, Peter, 270
DIVERSITAS, 209
Döös, Bo, 183
Doughnut Economics, 263, 276, 282
Dryzek, John, 41
Dubos, René, 27, 156, 158

Earth System science. *See also* Planetary Boundaries
enduring links to Stockholm, 5, 164
establishing international hub in Stockholm, 206–210
institutionalization, 107, 204–205
integrative approach, 224
meetings, 224
policy relevance of global change research, 205
rise of, 40, 68

Earth System science (cont.)
understanding of human-earth relationship, 229, 231
Earth System Science Partnership (ESSP), 209, 210
Earth System trends, 226
Earthwatch, 154
Easterbrook, Gregg, 62
ecocide, 157, 293
ecological economics, 247
Ecological Economics journal, 214
ecological modernization, 277
Ecological Research Committee, 153
ecological resilience, 215, 216
eco-modernism, 57–58, 273
Edberg, Rolf, 125, 168
Egnér, Hans, 127
Ehrlich, Paul, 159, 166, 214, 215, 243, 248
Ekman, Bo, 220, 237
Engelke, Peter, 226
Engfeldt, Lars-Göran, 35, 141, *142*, 145, 146, 174
Engström, Arne, 145
ENIAC computer, 99
environment
concept of human environment, 1–4, 136, 160, 162, 296
and development, 160, 178
history of concepts of, 124–125, 228, 272
integrative aspect, 37–38
as issue of social change, 66
neoliberalization of, 277–279
as problematic and challenging, 10–11
US armed forces understanding of the, 109
environmental expertise, 71–72
environmental governance. *See also* global environmental governance (GEG)
adaptive management, 246
human dimension of its history, 3–4
place of 'governance' in, 276
planetary framing of challenges facing humanity, 288
environmental issues, public awareness in 1960s, 124–125
environmental journalism, 70, 74
environmental movement, co-existence of Swedish state with, 59
environmental politics, globalisation of, 141
environmental problems, transboundary nature of, 123, 126, 139, 149, 166, 172, 304
environmental protection
concerns of developing countries over, 147
measures derived from economics, 276
environmentalism
critiques of, 66
early contributions in Sweden, 54
feminism and, 74, 131, 133
of the rich, 321–324
and science in Sweden, 32–35
Eriksson, Erik, 127, 128, 150, 153, 207, 208
Erlander, Tage, 109, *134*, 317
European Air Chemistry Network (EACN), 127–128
European Union, climate policy, 199

Fagerberg, Sven, 66
Falkenmark, Malin, 5, 74, 166
Faucheux, Sylvie, 251
Febvre, Lucien, 272
Fehrm, Martin, 71
feminism and environmentalism, 74, 131, 133
Ferguson, Howard, 192, 196
Finland, 87
Flohn, Hermann, 150
Folke, Carl, 233, 234, 236, 237, 248, 249, 288
Forsse, Anders, 153
Fourth Industrial Revolution (4IR), 284
France, 138
Francis, *Pope*, 262
Fridays for Future movement, 290
Friends of the Earth, 202
friluftsliv (out of doors), 51–52
Future Earth research program, 39, 210
Future Earth Secretariat Stockholm, 208, 210
future studies, 65

Gaffney, Owen, 223, 232, 269–270, 285
Gaia, 271–273
Gandhi, Indira, 160, *161*
Gapminder Foundation, 5
Garcia, Rolando V., 120
GARP. *See* Global Atmospheric Research Program (GARP)
gender diversity, and environmental progress, 72–75
geoengineering, 257–258, 260–261
geophysical research, Swedish reputation for, 87–88
George, Henry, 133
Gerholm, Tor Ragnar, 66

German Marshall Fund, 195
Gillberg, Björn, 32
Gleick, Peter, 198
Global Atmospheric Research Program (GARP), 120, 121
 1967 conference, 119, 121–122
 1974 conference, 169–171
Global Change magazine, 209
global change movement, 215
global cities, 9–11
global environmental diplomacy, 162
global environmental governance (GEG)
 conceptual developments, 38
 contibutions of Stockholm, 315–318
 contributions of Stockholm to, 179, 297–298
 convening of meetings, 39
 fundamental principle, 238
 in the future, 324–326
 history of advances in, 15–18
 international networks, 38
 key arenas, 19, 21
 legacy of Stockholm Conference, 2
 literature on history of, 41–44
 periods in evolution of, 35–37
 primary method of, 300
 progress towards sustainable environment, 299–300, 301
 role of key 'architects', 37–41
 role of Stockholm in evolution of, 300–306
 studying historically, 18–22
 Sweden's role in, 31
global environmental monitoring, 151–154, 162
global green capitals, 9
Global Network of Environmental Monitoring (GNEM), 153–154
global warming, 97, 101–104, 110, 142
Goldsmith, Edward, 235
Goodman, Gordon
 and Beijer Institute, 169, 184–187, 189, 212
 and Brundtland Commission, 189, 190, 191
 death, 249
 at Rättvik Conference, 188
 and SCOPE report 1, 153
 and SEI, 198, 201, 212
 and Villach-Bellagio conference 1987, 192, 194, 197
Gordon, Line, 254
Gore, Al, 103, 122, 217, 220, 288
Gothenburg Protocol, 29
Gothenburg University, 281
Granat, Lennart, 148
Great Acceleration
 coining of term, 226
 emergence of idea, 224
 Keeling curve and, 104
 natural resource extractivism during, 273
 tipping points and, 238, 285
Great Acceleration graphs, 226, 268
green cities, 82
Green *Folkhem*, 60–63, 244–245
Green Party, Sweden, 32, 60, 63, 75, 211
Greenhouse effect, 63, 101
greenhouse gas emissions, 180
greenhouse gases, research into, 103–104
Greenpeace Nordic, 32, 319
Gren, Ing-Marie, 248
Guhu, Ramachandra, 6, 42
Gustafson, Torsten, 109
Guterres, António, 294

Hajer, Maarten, 57, 62, 277–279
Halbwachs, Maurice, 21
Hambraeus, Birgitta, 74
Hamilton, Clive, 258–259, 260–261
Hammarskjöld, Dag, 309, 315
Hansen, Jim, 217, 221, 222
Hansson, Per Albin, 61
Haraway, Donna, 56
Hardin, Garrett, 243
Hare, Ken, 192
Hayashi, Y., 188
Heidenstam, Verner von, 49
Heimat movement, 60
Helsinki Convention 1974, 166
Heyerdahl, Thor, 158
Higgins, Polly, 157
history
 and memory, 21
 timescales of human history, 230
Hofsten, Angelica von, 71
Holdren, John, 166
Hollander, Jack, 165, 188
Holling, C.S., 215, 246, 248, 253, 254
Holocene stability, 231
homeostasis, 272
Hultqvist, Bengt, 85, 115
human environment
 dimensions of, 296
 history of concept, 1–4, 136, 228

human environment (cont.)
history of disasters, decline and decay, 11–15
humanities, contribution to widening the social space of the environment, 72
hydropower dams, 69

India, 138
Ingelstam, Lars, 147, 148
Integrated History and Future of People on Earth (IHOPE network), 224
Intergovernmental Panel on Climate Change (IPCC), 5, 23, 35, 121, 170, 200
first assessment report, 200, 201–204
second assessment report, 203
Sundsvall meeting 1990, 201–204
International Biological Program (IBP), 153, 205
International Council of Scientific Unions (ICSU), 39, 120, 121, 170, 171, 183, 206, 253
International Federation of the Institutes of Advanced Studies (IFIAS), 177–179
International Geophysical Year (IGY) programme 1957/58, 84–87, 102, 104, 205, 207
International Geosphere-Biosphere Programme (IGBP), 5, 39, 207–210, 224
International Human Dimensions Programme on Global Environmental Change, 209
International Institute for Applied Systems Analysis (IIASA), 9, 30, 169, 193
International Institute for Energy, Resources and the Human Environment, 167
International Institute of Environmental Affairs, 158
International Meteorological Institute (IMI), 39, 88, 89, 90, 97, 119, 180
International Polar Year (IPY) program, 84
International Population Institute, 158
International Society for Ecological Economics, 214
International Study Conference on the Physical Basis of Climate and Climate Modelling, 170
International Union of Geodesy and Geophysics, 120

Jaeger, Jill, 183, 192, 193–194, *193*, 198, 199, 201, 212, 251
Jagger, Bianca, 217
Jansson, Ann-Mari, 214, 215, 248
Jansson, Bengt-Owe, 214, 248
Jenkins, Dale, 153
Johnson, Keith, 156
Journal of Meteorology, 97
Journal of the Atmospheric Sciences, 97
Joyce, J. Wallace, 85

Karolinska Institute, 85, 117
Kassas, Mohamed, 158
Keeling, Charles David, 103
Keeling curve, 89, 103–104
Kellogg, William, 150
Kenya Academy of Science, 189
Kenya Ministry of Energy, 189
Kenyatta, Uhuru, 294
King, Alexander, 176
Kiruna Geophysical Observatory (KGO), 84–86, 87, 115, 116
Kjellén, Bo, *219*
Klum, Mattias, 265
knowledge
environmental expertise, 71–72
environmental knowledge stimulated by media controversies, 65–66
history of, 64
role in Sweden's environmental turn, 64–68
situated nature of, 23, 24
knowledge clusters, 77
Koch, Karin, 132
Kristoferson, Lars, 167, 168, 212, 214
Kropotkin, Peter, 133
KTH Environmental Humanities Laboratory, 282, 311
KTH Royal Institute of Technology, 117
Kurzweil, Ray, 270
Kyoto Protocol, 77, 203, 220

Lachmund, Jens, 25
Lagerlöf, Selma, 45–48
Latour, Bruno, 23
Lee, James, 152
Lenton, Tim, 237, 271
Levin, Simon, 247, 248
Liberal Party, Sweden, 63
Lie, Trygve, 309
Limits to Growth report, 234, 273, 278, 282

Lindahl-Kiessling, Kerstin, 214
Lindh, Anna, 28, 29, 74
Lindhagen, Anna, 73
Linköping University, 282
Linnaeus, 58
Liverman, Diana, 221, 222
Lönnroth, Ami, 74
Lönnroth, Måns, 199
Lovelock, James, 217, 271, 272
Lovins, Amory, 166
Lubchenco, Jane, 251
Lund University, 282
Lundholm, Bengt, 123, 129, 152–154, 162–163, 186
Lynas, Mark, 217, 237, 258–260, 261–262

Mabogunje, Akin, 251
MacNeill, Jim, 178, 190, 191, 199
Mäler, Karl-Göran, 64, 210, 214, 215, 247, 248
Malone, Thomas, 120, 206, 207
Manabe, Syukuro, 150
Margulis, Lynn, 271, 272
Massachusetts Institute of Technology (MIT), 149
Mathews, Jessica, 199
McCormick, John, 6, 42, 123
McKibben, Bill, 217, 243
McNeill, John R., 164, 225, 226, 268
Mead, Margaret, 137, 158
memory and history, 21
meta-expertise, 242–246
Meteorological Institute of Stockholm University (MISU), 97, 113, 114, 120
Mistra foundation, 175, 232, 240, 249, 250, 311
Moberg, Åsa, 74
Moderate Party, Sweden, 60
Moi, Daniel arap, 174
Montreal Protocol, 12, 31, 63, 196
Munro, Robert, 178
Museum of National History, Stockholm, 117
Myrdal, Alva, 73, 133, 135, 306, 316, 317
Myrdal, Gunnar, 158

Nairobi, Kenya, 161
NASA (National Aeronautics and Space Administration, United States), 204
Nash, Roderick, 13
national parks, 46, 280
National Science Foundation (NSF), 84, 85
Natural Science Research Council Ecological Research Committee, 152
nature
  connections with welfare, 51–53, 54
  political arithmetic's portrayal of, 58
  race and biology, 54–56
nature conservation
  in German Nazi ideology, 55
  as Swedish national issue of highest order, 48–50
neoliberalism, 62, 276
Neumann, John von, 96, 97, 98, 100, 270
Nilsson, Sam, 176
nitrogen and eutrophication, 101
Nobel Foundation, 76, 288, 310–311
Nobel Institute, 107
Nobel Laboratory of Physical Chemistry, 117
Nobel prizes, 76, 168
Noone, Kevin, 233, 236, 241
Nordic Conference on Soil and Water Pollution, 1966, 125
Nordic Defense Union, 114
Nordic Environmental Protection Convention, 125
Nordic exceptionalism, 312
Nordic model, 312, 313
Nordström, Ludvig, 70
Norinder, Harald, 106, 108, 109
Norway, 87, 114, 313
Norwegian-British-Swedish Antarctic Expedition (NBSX), 111
nuclear weapons testing
  need for scientific intelligence, 111, 116
  need for weather predictions, 96
Nykvist, Björn, 241

O'Riordan, Tim, 251
Observation Hill, Stockholm, 106
Odén, Birgitta, 71, 74
Odén, Svante, 71, 124, *127*, 126–129, 148
Odum, Howard T., 167
OECD
  on acid rain, 129
  International Conference on Environment and Economics, 1984, 278
Office of Naval Research (ONR), 112
Only One Earth (Stockholm Conference report), 27–28, 156
Oppenheimer, J. Robert, 96
Oppenheimer, Michael, 192

Örn, Torsten, 142
Oslo, Norway, 82, 117
Ostrom, Elinor, 254
Our Future in the Anthropocene Biosphere report, 288–289
Our Planet, Our Future summit, 2021, 288–289
Owens, Susan, 44, 240
ozone depletion, 62–63, 90, 181

Paglia, Eric, 297
Palme, Olof, 34, 58, 62, *134*, 143, 157, 158, 293, 295, 303, 317
Palmstierna, Hans, 122, 126, *130*, 129–131, 133, 138–139, 140, 145, 163, 295
papal encyclical on the environment, 262
Paris Agreement, 257
Peccei, Aurelio, 158, 176
Pekeris, Chaim, 102
Persson, Åsa, 232, 233, 236, 237, 241
Persson, Göran (Swedish government administrator of environmental issues), 123
Persson, Göran (Swedish Prime Minister, 1995–2006), 61, 62, 217, 244
Petri, Lennart, *142*
Phillips, Norman, 99
Pisharoty, Pisharoth Rama, 150
Planet Under Pressure Conference, London 2012, 210, 263
Planetary Boundaries 2.0, 262–264, 265, 268
Planetary Boundaries concept, 222–225, 231–238, 239, 240–242, 245–246, 273, 289
Planetary Boundaries debate
  lack of level playing field, 274
  politics of boundary concept, 274
  roots of, 272–273
Planetary Boundaries diagram, 241–242, *241*, 282
Planetary Boundaries framework, 5, 38, 222, 223, 239, 241, 245, 255, 256, 262–264, 268, 276
Planetary Boundaries project, 254–257
Planetary Boundaries research, 267
planetary metaphysics, 267–271
Plass, Gilbert, 97
Plunder, Famine, Poisoning (Palmstierna), 122, 126, 129–131, 140
polar climate change, 110
populist politics, 298
Princeton Computer Project, 96, 98, 99, 100, 113, 114
public intellecturals, contribution to widening the social space of the environment, 70–71

Ramel, Stig, 168
Randers, Jörgen, 282
Raskin, Paul, 188
Rättvik Conference on Environmental Research and Management Priorities for the 1980s, 177–179, 188
Raworth, Kate, 263, 276, 282, *283*
realist sustainability, 56–60
regime shifts, 234, 238
Reichelderfer, Francis W., 95
resilience, 215, 216, 246–248
Resilience 2008 Conference, 253–254
Resilience Alliance, 246, 247, 248, 253
Revelle, Roger, 97, 103
Richet, Charles, 234
Right Livelihood Award, 311
Rio Earth Summit, 175, 203, 212
Rio+20 summit, 210
Ripert, Jean, 203
Rockefeller Brothers Fund (RBF), 190–191, 195, 201
Rockefeller Foundations, 310
Rockström, Johan, 222, 242, 266
  Big World, Small Planet, 265
  Breaking Boundaries, 285
  as meta-scientist, 243
  on Planetary Boundaries, 223, 232, 242, 245
  and SEI, 221, 232, 249
  and SRC, 240, 249
  and Tällberg Forum, 233
  and Villa Brevik meeting, 237
Rodhe, Henning, 63, 148
Roosevelt, Theodore, 56
Rosling, Hans, 5, 301
Rossby, Åke, 99
Rossby, Carl-Gustaf, 35, 92
  career, 94–97, 98, 100, 101–103, 127
  death, 88, 104
  legacy, 91, 94
  reputation, 91, 96, 243
  as research director, 105
  return to Stockholm, 105–106, 108–114
  as visionary, 104, 106–108

Rosswall, Thomas, 208, 214, 253
Rotschild, Emma, *213*
Royal Carl XVI Gustav professorship, 175, 240
Royal Colloquium, 175
Royal Institute of Technology, KTH, Stockholm, 98
Royal Swedish Academy of Engineering Sciences, 149, 165
Royal Swedish Academy of Sciences
  Ambio journal, 90, 163
  as authority on nature protection, 46
  and Beijer Institute, 168, 187, 319
  conferences organized by, 63, 149, 166, 177
  Future Earth Secretariat Stockholm, 39, 210, 263, 269, 319
Rydbeck, Olof, *142*

Saltsjöbaden symposium 1982, 214
Saltsjöbaden Workshops, 29
Samfundet för hembygdsvård (SfH) (Society for Heimat), 60
Sami people, 33, 55, 59, 84, 295
San Francisco Bay Area, environmental history, 24, 82
Sápmi, 84
Sassen, Saskia, 9
Schellnhuber, Hans Joachim, 239, 257, 258, 262
Schneider, Stephen, 150, 166, 215
Schragger, Mike, 232, 233
Schumacher, E. F., 167
science
  and environmentalism in Sweden, 32–35
  history of, 64
  situated nature of, 23, 24
scientific activism, advancing climate science, 191–201
Scientific Committee on Problems of the Environment (SCOPE), 168
  Commission on Monitoring, 153
  report on global environmental monitoring, 151–154, 186
  report on greenhouse effect, climate change, and ecosystems, 183
  report series, 152, 208
Second World Climate Conference, Geneva 1990, 196, 201
Seeds of a Good Anthropocene project, 263
de Seynes, Phillipe, 133
Sievert, Rolf, 85, 117
Silent Spring (Carson), 122, 124
Singularity University, 270
Social Democratic Party, Sweden, 34, 57–58, 60, 211, 295
Social nature protection, 51–52
social-ecological systems, 246, 247
Söder, Karin, *213*
solar radiation management, 257–258
Soller, Barbro, 70–71
Solow, Robert, 188
Sörlin, Sverker, 233, 234, 236, 237, 254, 297
Soviet Academy of Sciences, 165
Soviet-Swedish Symposium on the Pollution of the Baltic, 165
Soviet-Swedish Working Group on Problems of the Human Environment, 165
Speth, James Gustave, 178, 190
Stafford-Smith, Mark, 275
Steffen, Will, 164, 221, 222, 223–224, 225, 236, 237, 241, 268, 269, 274
Stemme, Erik, 98
Stockholm accelerator, 107
Stockholm Conference. *See* United Nations Conference on the Human Environment 1972 (Stockholm Conference)
Stockholm Conference on the Acidification of the Environment, 1982, 176
Stockholm Convention on Persistent Organic Pollutants, 29
Stockholm Declaration. *See* Declaration on the Human Environment (Stockholm Declaration)
Stockholm Environment Institute (SEI), 5, 39, 107, 167, 187, 188, 198, 201, 210–213, 232, 248, 319
Stockholm exhibition, 1930, 52, 53
Stockholm Högskola. *See* Stockholm University
Stockholm International Institute of Environmentally Sound Technologies Assessment (SIIESTA), 212
Stockholm International Water Institute (SIWI), 318
Stockholm Resilience Centre (SRC), 5, 39, 107, 167, 216, 222, 232, 233, 248, 252, 255, 265, 267, 281, 282, 319

Stockholm University
  as central node in evolution of climate science, 89, 94
  Department of Meteorology, 109, 148, 163, 183, 194
  location, 117
Stockholm, Sweden
  as a centre of knowledge production, 76–81
  contributions to evolution of GEG, 3–4, 297–298
  Elm Conflict, 26, 320
  as global hub for Earth System science, 206–210
  as global hub for GEG, 8–11, 179, 244, 315–318
  as Green Capital of Europe, 82, 324
  as place of environmental memory, 16–18
  research and higher education institutions, 78–81
  rise as global environmental hub, 117–118
  role in evolution of GEG, 2, 300–306
  significance for GEG, 4–7
  in summer of 1972, *18*
  as ultimate smart and creative city, 81–83
  Vallhallavägen, 80
Stockholm+10 Conference, Nairobi, 173–174
Stockholm+25, 28
Stockholm+40, 28
Stockholm+50, 28, 292–295, 298
Stråth, Bo, 61
Strindberg, August, 48
Strong, Maurice, *144*
  and Beijer Institute, 188
  keynote address at Rättvik Conference, 178
  and Stockholm Conference, 143, 153, 156, 158
  and Stockholm Declaration, 155
  and Stockholm+25, 28
  and UNEP, 161, 162
  UNEP Gold Medal award, 174
Study of Man's Impact on Climate (SMIC), 149–151, 170
Sundberg, Kaj, *142*
Sundsvall, Sweden, 202
Suomi, Verner, 120
sustainability science, 250–252
Sustainability Science Workshop, Friibergh Manor 2020, 251–252
sustainable development, 62, 156, 184
Suzuki, David, 243
Svalbard, Norway, 117
Svedin, Uno, 237, 251, 254
Svensson, Göte, 183, 211
Svensson, Sören, 153, 186
Sverdrup, Harald Ulrik, 112
Sweden. *See also* Stockholm, Sweden
  carbon tax, 31, 64
  deferring from state to society, 318–321
  as democratic template, 30
  devolution, 321–324
  environment and climate policy, 4, 28, 33, 277–279, 296
  environmental politics, 64, 66–68
  environmental reawakening, 211
  environmental turn, 64, 66–68, 71, 124, 125, 131
  and evolution of GEG, 30
  foreign policy, 30, 188, 303, 304, 321
  gender diversity and envrionmental progress, 72–75
  Green Folkhem, 60–63, 244–245
  hollowing out of green state, 289–291, 322
  industrialization and emigration, 48
  internationalism, 30, 33, 76, 316–317
  military strength, 87, 309
  neutrality, 65, 75, 87, 114, 309, 312
  public awareness of environmental issues in 1960s, 124–125
  research funding, 311–312
  role in GEG, 31
  science and environmentalism, 32–35
  science and scholarship policy, 3
  science diplomacy at UN, 131–137
  scientific excellence, 76, 113
  security policy collaboration with the US, 106, 114
  self-conception as a progressive pioneering country, 75
  solidarity, 303–304, 312
  state of nature in early twentieth century, 47–48
  Torekov Compromise, 79–80
  as world conscience, 302
Sweden–America Foundation, 95
Swedish Air Force, 98, 100, 106, 107, 114
Swedish Board of Computing Machinery, 99
Swedish environmental exceptionalism, 312–315

Swedish Environmental Research Institute, 125
Swedish exceptionalism, 53–54
Swedish *Folkhem*, 244
Swedish identity, 50
Swedish Institute for Space Physics, 115
Swedish International Development Cooperation Agency (Sida), 302, 316
Swedish Meteorological and Hydrological Institute (SMHI), 93, 106, 119
Swedish model, 61, 75
Swedish National Defence Radio Establishment (FRA), 99
Swedish National Environmental Protection Agency, 125, 165
Swedish nationalism, role of nature in, 49–56
Swedish Research Council, 30, 251
Swedish Society for Nature Conservation (SSNC), 46, 51, 60, 64, 318, 319
Switzerland, seminar in Founex, 156–157

Talbot, Lee, 152
Tällberg Forum gatherings, 220
  2006, 232
  2007, 232, 244
  2008, 221–222, 233, 237–238, 239, 244
Tällberg Foundation, 220, 232
Tällberg, Sweden, 220, *221*
Tamm, Elisabeth, 71
Targets and Indicators of Climate Change report, 198
Taube, Evert, 71
Tellus journal, 39, 89, 97, 104, 107, 128, 271
Tepper, Morris, 120
Thomas, Julia Adeney, 43
Thompson, Philip D., 113
Thorsson, Inga, 73, *132*, 131–137, 163, 166, 316
Thunberg, Greta, 5, 72, 104, 243, 245, 284, 286, 285–287, 290, 319
tipping points, 139, 221, 233, 237, 238
Tiselius, Arne, 175
Tolba, Mostafa, 174, 180, *181*, 183, 192, 196, 211
transformation, 288–289
Transition Town movement, 324

Ulriksdal Palace, Stockholm, 175
Umeå University, 282
UN Climate Action Summit, New York 2019, 287
UN Committee on the Peaceful Uses of Outer Space, 163
UN Economic and Social Council (ECOSOC), 133, 135, 140
UN Economic Commission for Europe, 133
UN Intergovernmental Panel on Climate Change (IPCC), 89
UN Secretariat Office for Science and Technology, 146
UN Sustainable Development Goals (SDGs), 20, 210, 276, 288, 289
UN World Population Conference 1974, 166
United Kingdom, 138
United Nations Conference on the Human Environment 1972 (Stockholm Conference)
  Action Plan, 151, 154, 170, 174
  big picture brainstorming event, 158
  boycott by most Eastern Bloc countries, 155
  Declaration on the Human Environment, 28
  delegates and attendendees, 155
  demonstrations near venues by activists, 160
  Distinguished Lecture Series for Conference delegates, 158
  Environment Forum, 157, 158–160
  human environment concept, 1
  Indira Gandhi address to plenary session, 160, *161*
  institutional legacy, 161
  intellectual basis, 135
  legacy, 28, 292, 298, 308
  national environmental reports and case studies, 146–147
  objectives, 136, 137, 145, 146
  opening speech, 157
  overarching principle, 145
  People's Forum, 157–158
  place in evolution of GEG, 2, 8
  plenary, 158
  pre-conference regional seminars, 156
  preparations for, 122, 129, 143–147, 156
  proposal for, 34, 122
  report (*Only One Earth*), 27–28, 156
  Sweden's case study on acid rain, 148–149
  transport options, 160

United Nations Environment Programme (UNEP), 23, 35, 39, 63, 161–162, 170, 171, 174, 183
United Nations Framework Convention on Climate Change (UNFCCC), 198, 199, 200, 203
United Nations General Assembly
adoption of Stockholm Declaration, 35, 155
resolution 2398 *Problems of the human environment*, 35, 122, 141
resolution 45/212 *Protection of global climate for present and future generations of mankind*, 203
resolution to establish UNEP, 161
Swedish science diplomacy, 135–137
United Nations Symposium on Population, Resources and Environment, Stockholm 1973, 166
United States, military funding for MISU, 113, 114
Uppsala University, 94, 120, 282
Uppsala, Sweden, 76, 78, 106, 109
US Air Force, 113, 115, 116
US Air Force Cambridge Research Center Space Group, 116
US Air Research and Development Command, 112
US Navy, 112
US Weather Bureau, 93, 95

van Allen, Abbie, 84
van Allen, James A., 84
Vellinga, Pier, 198, 199
Vernadsky, Vladimir, 272
Victoria, *Crown Princess*, 74, 175, 294
Vienna Convention for the Protection of the Ozone Layer, 181
Vienna, Austria, 9
Villa Brevik meeting, Lidingö 2008, 237
Villach Conference, 1985, 179–184, 197
Villach, Austria, 9
Villach-Bellagio conference 1987, 191–201
Vindelälven River, Sweden, 69
Vogt, William, 243
Volvo Environment Prize, 311

W. Alton Jones Foundation, 195, 201
Wägner, Elin, 68–69, 71, 131, 132, 133
Wahman, Tom, 190, 196
Walker, Brian, 247
Walker, Richard, 24
Wall, Anders, 212
Wallén, Axel, 110
Wallén, C. C., 192, 199
Wallenberg Foundation, 214
Ward, Barbara, 27, 156, 158
warfare, weather forecasting and, 109
Warrick, Richard, 183
Water Stress Indicator, 5
welfare, connections with nature, 51–53, 54
White, Gilbert, 188
Wijkman, Anders, 283
Wikström, Jan-Erik, 212
Wilson, Carroll, 150, 158, 169
women, role in environmental progress, 72–75
The Wonderful Adventures of Nils (Lagerlöf), 45–46
Woodwell, George, 192
World Climate Conference, Geneva 1979, 171
World Climate Programme, 39, 121, 171, 180, 197
World Climate Research Programme, 121, 207, 209
World Commission on Environment and Development (WCED) (Brundtland Commission). *See* Brundtland Commission
World Conference on the Changing Atmosphere, Toronto 1988, 196
World Economic Forums (WEF), Davos, 265, 284, 287
World Meteorological Organization (WMO), 23, 35, 121, 170, 171, 183
Wormbs, Nina, 254
World Resources Institute, 179, 190
WWF Sweden, 174, 175, 319

Yates, Dan, 111

Zuckerman, Solly, 124, 158

Studies in Environment and History *(continuation of page ii)*

Peder Anker *The Power of the Periphery: How Norway Became an Environmental Pioneer for the World*
David Moon *The American Steppes: The Unexpected Russian Roots of Great Plains Agriculture, 1870s–1930s*
James L. A. Webb, Jr. *The Guts of the Matter: A Global Environmental History of Human Waste and Infectious Intestinal Disease*
Maya K. Peterson *Pipe Dreams: Water and Empire in Central Asia's Aral Sea Basin*
Thomas M. Wickman *Snowshoe Country: An Environmental and Cultural History of Winter in the Early American Northeast*
Debjani Bhattacharyya *Empire and Ecology in the Bengal Delta: The Making of Calcutta*
Chris Courtney *The Nature of Disaster in China: The 1931 Yangzi River Flood*
Dagomar Degroot *The Frigid Golden Age: Climate Change, the Little Ice Age, and the Dutch Republic, 1560–1720*
Edmund Russell *Greyhound Nation: A Coevolutionary History of England, 1200–1900*
Timothy J. LeCain *The Matter of History: How Things Create the Past*
Ling Zhang *The River, the Plain, and the State: An Environmental Drama in Northern Song China, 1048–1128*
Abraham H. Gibson *Feral Animals in the American South: An Evolutionary History*
Andy Bruno *The Nature of Soviet Power: An Arctic Environmental History*
David A. Bello *Across Forest, Steppe, and Mountain: Environment, Identity, and Empire in Qing China's Borderlands*
Erik Loomis *Empire of Timber: Labor Unions and the Pacific Northwest Forests*
Peter Thorsheim *Waste into Weapons: Recycling in Britain during the Second World War*
Kieko Matteson *Forests in Revolutionary France: Conservation, Community, and Conflict, 1669–1848*
Micah S. Muscolino *The Ecology of War in China: Henan Province, the Yellow River, and Beyond, 1938–1950*
George Colpitts *Pemmican Empire: Food, Trade, and the Last Bison Hunts in the North American Plains, 1780–1882*
John L. Brooke *Climate Change and the Course of Global History: A Rough Journey*
Paul Josephson et al. *An Environmental History of Russia*
Emmanuel Kreike *Environmental Infrastructure in African History: Examining the Myth of Natural Resource Management*
Gregory T. Cushman *Guano and the Opening of the Pacific World: A Global Ecological History*
Sam White *The Climate of Rebellion in the Early Modern Ottoman Empire*
Edmund Russell *Evolutionary History: Uniting History and Biology to Understand Life on Earth*
Alan Mikhail *Nature and Empire in Ottoman Egypt: An Environmental History*
Richard W. Judd *The Untilled Garden: Natural History and the Spirit of Conservation in America, 1740–1840*
James L. A. Webb, Jr. *Humanity's Burden: A Global History of Malaria*
Myrna I. Santiago *The Ecology of Oil: Environment, Labor, and the Mexican Revolution, 1900–1938*

Frank Uekoetter *The Green and the Brown: A History of Conservation in Nazi Germany*
Matthew D. Evenden *Fish versus Power: An Environmental History of the Fraser River*
Alfred W. Crosby *Ecological Imperialism: The Biological Expansion of Europe, 900–1900, second edition*
Nancy J. Jacobs *Environment, Power, and Injustice: A South African History*
Edmund Russell *War and Nature: Fighting Humans and Insects with Chemicals from World War I to Silent Spring*
Adam Rome *The Bulldozer in the Countryside: Suburban Sprawl and the Rise of American Environmentalism*
Judith Shapiro *Mao's War against Nature: Politics and the Environment in Revolutionary China*
Andrew Isenberg *The Destruction of the Bison: An Environmental History*
Thomas Dunlap *Nature and the English Diaspora*
Robert B. Marks *Tigers, Rice, Silk, and Silt: Environment and Economy in Late Imperial South China*
Mark Elvin and Tsui'jung Liu *Sediments of Time: Environment and Society in Chinese History*
Richard H. Grove *Green Imperialism: Colonial Expansion, Tropical Island Edens and the Origins of Environmentalism, 1600–1860*
Thorkild Kjærgaard *The Danish Revolution, 1500–1800: An Ecohistorical Interpretation*
Donald Worster *Nature's Economy: A History of Ecological Ideas, second edition*
Elinor G. K. Melville *A Plague of Sheep: Environmental Consequences of the Conquest of Mexico*
J. R. McNeill *The Mountains of the Mediterranean World: An Environmental History*
Theodore Steinberg *Nature Incorporated: Industrialization and the Waters of New England*
Timothy Silver *A New Face on the Countryside: Indians, Colonists, and Slaves in the South Atlantic Forests, 1500–1800*
Michael Williams *Americans and Their Forests: A Historical Geography*
Donald Worster *The Ends of the Earth: Perspectives on Modern Environmental History*
Robert Harms *Games against Nature: An Eco-Cultural History of the Nunu of Equatorial Africa*
Warren Dean *Brazil and the Struggle for Rubber: A Study in Environmental History*
Samuel P. Hays *Beauty, Health, and Permanence: Environmental Politics in the United States, 1955–1985*
Arthur F. McEvoy *The Fisherman's Problem: Ecology and Law in the California Fisheries, 1850–1980*
Kenneth F. Kiple *The Caribbean Slave: A Biological History*

www.ingramcontent.com/pod-product-compliance
Ingram Content Group UK Ltd.
Pitfield, Milton Keynes, MK11 3LW, UK
UKHW030354050125
452959UK00006B/14

9 781009 177801